The EAAP Series is published under the direction of Siem Korver and Jean Boyazoglu

Organizing committee

J.H.M. Metz (chairman), Institute of Agricultural Engineering, Wageningen, Netherlands
J.G. de Wilt (secretary), Directorate for Agricultural Research, Wageningen, Netherlands
A.C. Smits (finance), Institute of Agricultural Engineering, Wageningen, Netherlands
P.L. Bergström, Research Institute for Animal Production 'Schoonoord', Zeist, Netherlands
Tj. de Boer, Commodity Board for Livestock and Meat, Rijswijk, Netherlands
A.E. van de Braak, Denkavit Nederland B.V., Voorthuizen, Netherlands
J. den Hartog, Commodity Board for Feedstuffs, The Hague, Netherlands
I.P. Jorna, D.M.V. Campina, Veghel, Netherlands
G.J.M. van Kempen, TNO Institute for Animal Nutrition and Physiology, Wageningen, Netherlands
D. Oostendorp, Extension Service for Cattle Production, Lelystad, Netherlands
M.W.A. Verstegen, Department of Animal Nutrition, Wageningen Agricultural University, Wageningen, Netherlands

Sponsors

Protein Technologies International, St. Louis, M0, USA (main sponsor)
Ministry of Agriculture, Nature management and Fisheries, The Hague, Netherlands
Directorate for Agricultural Research, Wageningen, Netherlands
Alpuro B.V., Uddel, Netherlands
Denkavit Nederland B.V., Voorthuizen, Netherlands
Commodity Board for Livestock and Meat, Rijswijk, Netherlands
Gardenbroek's Veevoederfabriek 'Navobi' B.V., Elspeet, Netherlands
N.V. Nukamel S.A., Olen, Belgium
Real Veal Inc., Ixonia, Wi, USA
C.V. 'Sloten', Deventer, Netherlands

New trends in veal calf production

New trends in veal calf production

Proceedings of the International Symposium on Veal Calf Production, Wageningen, Netherlands, 14 - 16 March 1990
(EAAP Publication No. 52, 1991)

J.H.M. Metz and C.M. Groenestein (Editors)

Pudoc Wageningen 1991

CIP-data Koninklijke Bibliotheek, Den Haag

ISBN 90-220-1016-3
NUGI 835

Printed in the Netherlands.

Contents

5. Welfare legislation

6. Marketing

7. Meat quality and growth stimulants

8. Nutrition: protein sources

9. Nutrition: antinutritional factors

10. Nutrition: supplementation

Preface

Approximately 160 people from eleven countries attended the first ever International Veal Symposium in March 1990. This symposium was triggered by the fact that veal calf production in Europe and North America is strongly influenced by developments related to animal welfare, milk quotas, nutrition, automation, waste management and marketing. Therefore, it seemed timely for experts from various countries and disciplines to discuss the potential and prospects for this branch of industry. Three days of presentations resulted in a tremendous amount of material and discussions. People from different backgrounds (research institutes, the industry and policy-makers) participated in the exchange of facts, ideas and expectations. Hopefully, this symposium has contributed to further improvement and to new developments in veal calf production and related activities.

It is a pleasure to acknowledge the help of the organizing committee in preparing this symposium. Thanks to their intensive efforts the symposium was held about six months after the organization started. In this context, the contribution of A.C. Smits and J.G. de Wilt as local organizers, and that of A.E. van de Braak and G.J.M. van Kempen in the programme organization deserve special mention. T. Tomkins played a significant role in the initiative to hold the symposium. Further, the organization was invaluably helped by the financial support of a number of organizations.

Due to the short preparation time, it was impossible to have a proceedings published at the time of the meeting. Although draft manuscripts were ready, much work still needed to be done. Members of the organizing committee and also G.H. Tolman are acknowledged for their help with reviewing the papers. Karin C.M. Groenestein did a tremendous job by taking responsibility for the final stages of reviewing, revising and editing the whole series of papers. Trees E.M. Hettinga (AHoF, Friesland) handled the preparation of the computer version. Finally, with one exception all papers could be included in this proceedings.

The Institute of Agricultural Engineering at Wageningen acted as the host institute. Director A.A. Jongebreur is acknowledged for the institute's facilities, services and whole-hearted support.

J.H.M. Metz

1. Veal production over the world

Dutch policy on veal production

R. Woudstra
Director of Animal Husbandry and Dairying. Ministry of Agriculture, Nature Management and Fisheries, The Hague, NL

Introduction

Cooperation is increasing throughout the world, as can be seen from the developments in the EC towards the completion of the internal market in 1992, recent developments in Eastern Europe, and the GATT negotiations. Research and policy should be oriented broadly and internationally. The symposium 'New trends in veal calf production' could contribute to future cooperation, to the benefit of all its participants.

The long-term agricultural policy of the Netherlands was laid down last year in the Memorandum on the Future Structure of Agriculture. The objective of this policy is a safe, sustainable and competitive agriculture. The keywords are quality, welfare, health and the environment.

The veal calf sector in the Netherlands

In the Netherlands, the development of the veal calf sector is closely linked to developments in the dairy sector. In the past 30 years, the dairy sector has become the largest agricultural sector, with almost 1.9 million cows and a total production of 11 million tonnes milk per year. Veal production became an important economic activity in the 1950s, based on the revaluation of the by-products of the dairy industry: milk powder and a surplus of calves. But when the milk quota system was introduced in 1984, the position of the veal calf sector became more difficult. The number of dairy cows fell, resulting in fewer new-born calves available to the veal industry. At the same time, feed costs rose because of the reduction in the volume of milk powder.

Eighty-five percent of total veal production from the Netherlands is exported, largely to Italy, France and West Germany. Some 90 percent of production comes from farms that have financial contracts with the calf milk industry. The sector is therefore characterized by the limited independence of the producers.

Animal welfare policy and research

Public concern about farm animal welfare has increased in recent decades, partly because of intensification of production methods. In response, various institutes started research on animal welfare, in order to develop criteria to measure animal welfare. Later, research focused on the development of alternative housing systems.

The Dutch industry recognized the importance of research on animal welfare quite early on. In 1978, the Farm Animal Welfare Fund was instituted by the Commodity Board for Feedstuffs in order to stimulate the development of alternative housing systems. The Fund contributed about 10 million Dutch guilders in 10 years, which was about 10 per cent of the total costs of research on animal welfare in the Netherlands over that period. The resources of the Fund are allocated by the Farm Animal Welfare Committee. Producers, suppliers of milk replacer, the animal welfare organization and government officials were represented in this Committee.

Various research institutes have worked on a group housing system for veal calves. Group housing is now used on about 13 per cent of the veal farms and it is functioning satisfactorily. The further improvement and commercialization of these systems by the industry will be supported by additional research. As regards individual housing, research has been done on the minimum ethologically acceptable crate sizes.

Dutch policy on farm animal welfare and the harmonization of this policy within of the EC are largely based on the results of this research. A balanced policy on animal welfare should take into account the interests of animals as well as of the producers. Therefore at national level, the government has consulted the industry and the Dutch Society for the Protection of Animals about policy proposals for legislation, and extension. The EC Council's proposal for minimum standards for the housing of veal calves, which focuses on group housing was important in these discussions. Welfare measures that might affect the competitive position of the sector should be regulated within the framework of the EC. Furthermore, present-day housing systems cannot be abolished before there are alternatives acceptable, in terms of animal welfare and farming practice. Finally, the destruction of capital investment should be avoided. This can be done by allowing for sufficiently long terms of depreciation so that investments can be slotted into normal farm management.

Now that the principles and perspectives of the group housing system are known, it is primarily the industry's own responsibility to develop these systems further and to introduce them on farms. In the coming years, research must focus on improving the health of calf group-housing systems.

Feeding

The diminishing supply of skimmed milk powder has led to a rise in feed costs. This may necessitate changes in the feed, for example replacing of animal protein by vegetable protein. From the point of view of welfare, the supply of some roughage is desirable.

In the near future, research must focus on appropriate feeding systems, adequate rations and improved feeding equipment.

Growth promoters

An aspect which has regularly caused negative publicity is growth promoters. The Dutch policy on growth promoters complies with the policy on registering feed additives at European level. According to this, growth promoters should

- be effective;
- have no adverse effect on animal health;
- leave no residues which are or may be harmful to public health in the end products;
- have no harmful impact on the environment.

In the future, the regulations about residues in the end-product will become stricter. Product safety and the consumer's attitude will be taken into account. Furthermore, the safeguarding of animal health and animal welfare will be emphasized.

In addition, the EC policy is aimed at approving products for which no waiting period is required. Products requiring a waiting period increase the risk of unintentional or intentional mistakes on the part of the farmer. This will make the approval of new growth promoters more difficult.

It is very important for Dutch farmers to comply with the EC regulations, not only to safeguard public health, animal health, and the environment, but also to ensure the undisturbed export of animal products. The industry should bear responsibility for the proper use of registered growth promoters and develop an adequate quality assurance system.

Quality

To improve the quality of the veal, it is vital to measure quality aspects in the various links in the production chain. It is essential to obtain relevant information and to pass it on to the preceding and subsequent phases of production. Objective quality parameters must be developed: several projects to do this are currently in progress. When the results are available, an integrated system of quality control will be introduced throughout the production chain. The aim of this system will be to enable the different links to coordinate their management with the help of information from the other links in the production chain.

Environment

The Netherlands is one of the first countries to be confronted with problems of environmental pollution resulting from a high concentration of animal production. The environmental problems consist of a surplus of manure and of NH_3 emission. In the veal calf sector there is a surplus of 1.5 million tonnes of calf slurry, out of a total production of 2.5 million tonnes. Most of this surplus is treated centrally in calf-slurry-processing units. This is a very expensive process. The veal calf sector contributes little to NH_3 emission.

Prospects

The veal industry faces important choices both at sector level and at farm level. Animal welfare and health, the environmental problems and requirements concerning product quality will have an important influence on the prospects for this sector. The necessary investments will put heavy pressure on the sector's profitability. In the long term, however, there are sufficient prospects for a veal production sector, although perhaps on a smaller scale, that takes market trends and consumer opinion into account. In the Netherlands investments in regulations to deal with environmental pollution will be relatively high. This means that the cost price of veal must be reduced if Dutch veal is to maintain a competitive position in the EC. There are considerable differences in farm results, suggesting there is scope for improvements that will reduce costs. The farmer's own responsibility to implement such improvements will have to be emphasized.

Group housing and a limited supply of roughage may be part of a production system that is more in line with the needs of the animals and the demands of society.

The choices indicated above should be based on information. An important part of this information is generated through research. The exchange of research results will lead to new ideas and new initiatives. This symposium will undoubtedly contribute to the adaptations and changes needed in the veal calf sector.

Veal production in the European Community

Tj. de Boer
Commodity Board for Livestock and Meat, P.O. Box 5805, 2280 ZH Rijswijk, NL

Summary

Veal production in the European Comunity (EC) depends much on the dairy production. For years, the production of veal increased with the increased number of dairy cows and with the availability of skimmed milk powder as a result of EC policy. Since 1984, a milk quota system has caused an annual decline in dairy and veal production. In 1989, the total veal production was about 790 000 tons (10% less than in 1988). West Germany and France (the major producing country, with 39% of total EC veal production) showed the sharpest decline. In Italy and the Netherlands (with 25% and 19% of the total production), the decline was only small. A further reduction of veal production is forecast. Also the decline in consumption of the last few years (9% less in 1989) will continue. Most of the veal is being consumed in France and Italy (44% and 32% of total EC consumption, respectively). The EC has a self-sufficiency rate for veal of 115. This causes an export of veal. Within the EC, the Netherlands is the major exporter (over 90% of their production); Italy is the major importer. For several reasons, the profitability of veal production is being reduced for all parts of the total production column.
Descriptors: veal; production; consumption; European Community.

Introduction

In the EC, veal production is closely related to dairy production. For years, production of veal had increased with the increased number of dairy cows and with the EC policy on skimmed milk powder. After several fruitless attempts to introduce measures to limit milk production, a milk quota system was introduced in March 1984. This quota system affected veal production to a large extent.

The quota system has also affected the whole dairy industry. The availability of skimmed milk powder (e.g. from intervention stocks) particularly stimulated veal production for years. Veal production is namely based on the feeding of only, or almost only, milk or milk replacers. This results in a white or pink coloured meat. The meat is very tender as long as young animals are slaughtered at weights between 100 kg and 200 kg.

Milk quota system

With the introduction of the milk quota system in the EC, the number of dairy cattle has decreased significantly and this decline will continue until the year 2000, as experts predict. By 1991 there might be a reduction of 20% relative to 1984 (almost 26 million dairy cattle in 1984, about 20 million in 1991). This reduction results in about 75% less calves available from the dairy herds for beef and veal production. Besides, the number of beef cattle has only slightly increased. Increased milk yield of individual cows will also cause an additional decrease of dairy cattle in coming years. The smaller number of dairy cows will particularly affect the veal production, where it is almost entirely dairy calves that enter this sector of meat production. In 1989, 23% of all the calves born entered veal production (still 25.2% in 1988). In 1989, 10.9% of total beef and veal production comprised veal, compared to 11.6% in 1988. Moreover the dairy farmer is more interested in crossbreeding some of his cattle with beef bulls. The crossbred calves are destined for beef production to a large extent. In other words, this tendency will also reduce the number of calves intended for veal production. Of course, the EC reduced the options for intervention for beef. Nevertheless a small increase of beef production in the EC is expected. For the last two years, the beef market has proved to function fairly well without the underpinning of the intervention system.

All the effects work in only one direction: a decrease of the veal production. In the last 3 years, the number of slaughters decreased considerably: in 1989, again by 7.5% compared to 1988.

Veal consumption

The trend in veal consumption in the EC is of major interest for the whole of veal production. In the EC most of the veal is being consumed in France and Italy (299 000 and 218 000 tons of carcass weight in 1989, respectively, which is 44% and 32% of total consumption in the EC). Total consumption of veal was about 690 000 tons in 1989. A sharp decline in consumption in 1988 was repeated in 1989 (estimated at 8.6% less). This means that over a two year period the market fell by 18%. A survey of veal consumption is given in Table 1.
The decline resulted from supply and demand factors:

- the substantial decline of the production resulted in marked rises in prices and so reduced consumption, particularly in France
- the image of veal has been affected by consumer concerns about the use of illegal hormones and intensive production techniques.

Trends in the major consuming countries were as follows.

- In France, there was another sharp fall in consumption (–8%) as a result of lower production and a switch to competing meats
- In Italy, a marked fall occurred as a result of consumer concern about illegal use of hormones
- The veal market in West Germany collapsed (–25%); the decline over a two-year period has been almost –40%.

Forecasts indicate another market decline in 1990 (–4.1%). In 1988, the consumption per head was in France 5.8 kg, in Italy about 4 kg and in West Germany and the Netherlands less than 2 kg. The decreased consumption is also reflected in the development of the consumption per head in the main veal-producing countries (Figure 1).

Production of veal

What are the veal production figures like, as a result of the two main influences:

- the milk quota system
- the reduced consumption of veal?

In 1989, the total production of veal was about 790 000 tons. Only 5 member countries of the EC were responsibly for nearly 95% of the total production. France was the largest producer, accounting for 39%. The other main producers were Italy (25%), the Netherlands (19%), West Germany (7%) and Belgium (5%).

Relative to 1988, there was a sharp decline in net production in all major countries; from 24% in West Germany, with a sharp crisis in the veal sector, and 12% in France to only 2% in Italy. Another reduction of about 3% is forecast for 1990 (Table 2).

From Table 2, it will be clear that only 5 member countries of the EC are of real importance to consider on behalf of veal production. It might have been noticed before that less attention is paid to other countries. Just a few remarks on those other areas of the EC.

Table 1. Veal consumption in the EC (in 1000 tons of carcass weight).

	Consumption (in 1000 tonnes)				Relative change (%)	
	1980	1985	1988	1989	1989/88	1990/89F
France	367	357	326	299	– 8.3	– 5.4
Italy	213	231	227	218	– 4.0	– 2.3
Netherlands	17	24	22	24	+ 9.1	– 9.1
West Germany	117	106	91	68	–25.3	– 8.8
Belgium/Lux.	31	32	28	29	+ 3.6	+ 3.4
Greece	71	39	24	21	–12.5	-
Denmark	2	2	2	2	-	-
United Kingdom	12	8	6	5	–16.7	-
Ireland	1	-	-	-		
EC-10	760	799	726	666	– 8.3	– 4.2
Spain		13	16	15	– 6.3	-
Portugal		10	10	7	–30.0	-
EC-12		822	752	688	– 8.6	– 4.1

F: Forecast
Source: GIRA based on EUROSTAT, PVV.

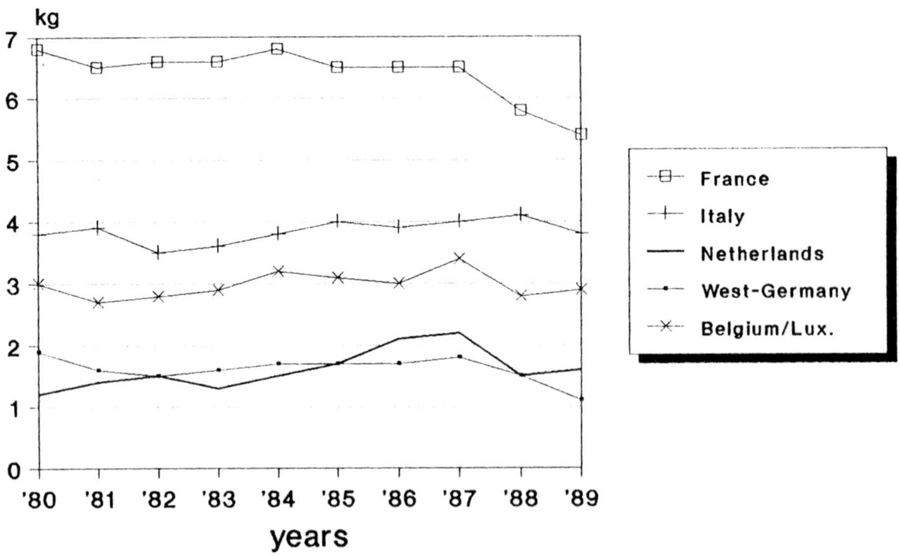

Figure 1. Veal consumption in the EC (kg per head per year, 1980–1989). Source: EUROSTAT.

Table 2. Veal production in the EC (in 1000 tonnes of carcass weight).

	Veal production (in 1000 tonnes)				Relative change (%)	
	1980	1985	1988	1989	1989/88	1990/89F
France	405	387	351	309	-12.0	- 5.5
Italy	146	175	202	198	- 2.0	- 5.1
Netherlands	141	183	164	148	- 9.8	+ 6.8
West Germany	78	83	75	57	-24.0	-10.5
Belgium/Lux.	33	43	44	42	- 4.5	+ 2.4
Greece	18	6	6	6	-	-
Denmark	2	2	2	2	-	-
United Kingdom	14	10	11	11	-	- 9.1
Ireland	1	-	-	-		
EC-10	838	889	855	773	- 9.6	- 3.0
Spain		12	16	15	- 6.3	+25.0
Portugal		9	4	4	-	
EC-12		910	875	792	- 9.5	- 2.8

F: Forecast
Source: GIRA-OFIVAL based on EUROSTAT, PVV.

1. Spain is somewhat difficult to judge. From the consumption per head, the production of veal seems to be rather important, but several statistics include figures of heavier beef animals. In fact, production of veal comprises only a small proportion of white veal. Most of the calves are slaughtered as young animals of 300–350 kg carcass weight ('terneros', 'terneras' and 'anojos', producing a 'red' veal). Recently production statistics have been modified to conform to EC requirements. These young animals, previously classified as veal, are now classified as beef. The production of white veal is concentrated around Madrid (Segovia - Avila), the production of red veal in Catalonia (Huesca - Lérida).
2. In Portugal, production and consumption of veal are of minor interest. There has never even been any encouragement.
3. Veal production has never really existed in Greece. A certain amount is imported, mainly from the Netherlands. Since September 1989, channels for import improved and imports might increase. In particular, the catering industry is demanding veal.
4. In Denmark and in Ireland, there is no real veal production and consumption.
5. The United Kingdom produces very little veal. Most of the veal consumed is imported, largely from the Netherlands. Moreover Britain is not an important veal-consuming country. Significant is the number of exported calves (about 300 000 in 1989).

As the self-sufficiency rate of the EC is 116 (Table 3), there is export of veal to other countries (mainly the Middle East). Within the EC, the Netherlands is the major exporter (over 90% of their production); Italy is the major importer.

Table 3. Self-sufficiency rates for veal in the EC.

	1980	1985	1988	1989
France	110	108	108	103
Italy	69	76	89	77
Netherlands	725	647	745	617
West Germany	67	78	82	84
Belgium/Lux.	110	134	157	145
EC-10	110	111	118	116
EC-12		111	116	115

Source: EUROSTAT

Table 4. Production of milk repacer in the EC (in 1000 tonnes).

	Production (in 1000 tonnes) 1980	1985	1987	1988	Relative change (%) 1988/87
France	877	794	841	777	– 7.4
Italy	325	270	258	252	– 2.3
Netherlands	512	607	652	651	– 0.2
West Germany	378	273	192	181	– 5.7
Belgium/Lux	60	59	35	39	+11.4
Others	91	91	61	60	– 1.6
EC-10	2 243	2 094	2 039	1 960	– 4.9
EC-12		2 173	2 089	2 003	– 4.1

Source: EUROSTAT.

Structure of veal production

Production of veal is situated near some dairy regions, where also the production of skimmed milk powder and milk replacers has been concentrated. The main regions are west and south-west France, the north of Italy, in West Germany near the Dutch border and the Netherlands.

As another consequence of the different developments in the EC and as a consequence of the structure of veal production, trade in live calves has changed in the last few years. There was an increase in import demand in veal-producing countries. Italy is the main importer of young calves. On the one hand, the United Kingdom is an important supplier of young calves, On the other hand, there has been a notable increase in exports from Eastern Europe (especially Poland). These imports may increase with the political change. However in the longer term in Eastern Europe, local production of meat may increase, with less export of live animals.

The role of the milk-replacer industry is very important. This industry dominates veal production and the most dominant manufacturing companies operate EC-wide. They already operate more or less in the future EC of 1992; the EC with one common market and without borders. Some companies originated in the Netherlands have a leading position.

To a large extent, veal production is based on a contract between producers and the milk-replacer-manufacturing industry. In France and the Netherlands, 90% of the veal producers has such a contract, in Belgium 80% and in West Germany 70%. In Italy too, contracts are very common, but often the relationship between producer and milk-replacer industry is even closer, as the manufacturer itself is the owner of the farm and the veal-calf fattener gets a salary. This salary will partly depend on the quality of his work, since the fattener can receive a bonus, for example based on growth, mortality and carcass quality of the veal calves. Such a payment system has been included in many contracts too.

The structure of veal production is affecting the size of the calf-fattening farms. In Italy, some farms are very large, mostly the ones owned by the milk-replacer manufacturers. About 75% of the 1300 farms have a capacity of 100 fattening positions or more. These farms account for over 96% of the total Italian veal production; only 200 farms are responsible for 65% of the total production. In 1989, in the Netherlands 2340 farms had an average capacity of 281 calves. In France, 4300 farms had a capacity of 50 calves or more. They had an average capacity of 150 calves and accounted for 78% of total French veal production.

Milk replacers

Production of milk replacers is decreasing. Table 4 shows the production until 1988. In the EC, production of milk replacers has decreased for years. In 1988, the decline was 4.1% relative to 1987. Only in the Netherlands did production increase until 1987, a demonstration of the position of this Dutch industry. The Netherlands is the major exporter of milk replacers; 31% of the total production in 1987 and even 39% in 1988. Nevertheless a decline in production was also inevitable there.

In 1988, milk production in Europe decreased by 2.6%, whereas there was an increase in all other areas of the world. More and more attention will have to be paid to profitable feeding-stuffs and even alternative feeding systems, since use of dairy products is discouraged in EC veal production. There is pressure on the whole milk-replacer industry now. The structure of veal production causes a direct response to this influence by veal producers.

Profitability

Veal production has been less profitable during the last few years, from 1986 on. Profitability of veal production is being reduced by the following factors:

- increasing cost of skimmed milk powder, the main feed ingredient, as intervention stocks have been eliminated and production has declined
- reduced availability of calves, resulting in increased calf prices in 1988 and 1989
- ban on the use of growth hormones in all EC countries, thereby reducing feed conversion efficiency.

The situation is likely to remain fairly difficult in 1990 and, as mentioned before, another decline in net slaughter by 3% is forecast by experts.

In future, there will be additional costs as a consequence of EC legislation on animal welfare and housing systems. At present, individual pens for veal calves are common. In France and Italy, hardly any veal calf is housed in groups, in the Netherlands only 10% but in West Germany about 25% are already housed in groups. In other words, present proposals of the EC will affect veal production in Europe to a large extent, because even in the last few years higher veal prices could not increase profitability.

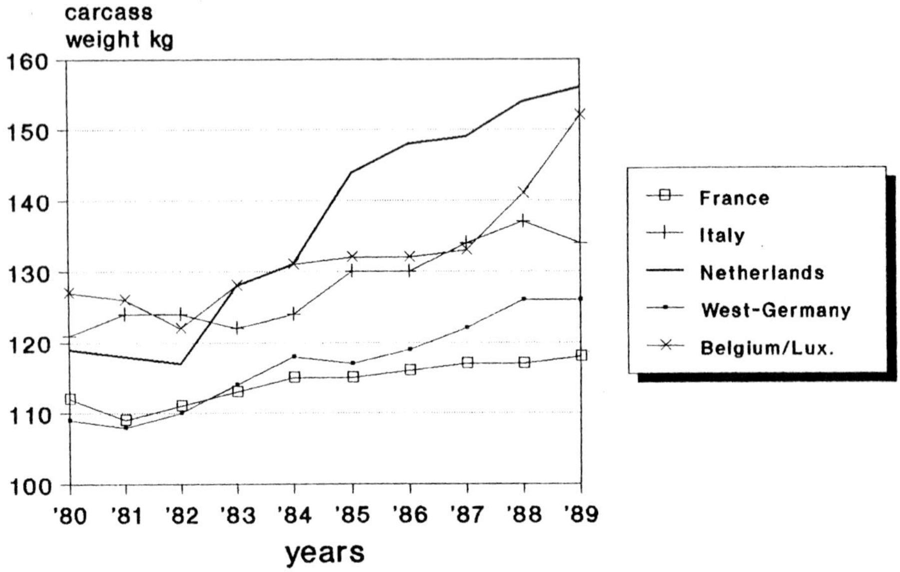

Figure 2. Average dressed carcass weight of veal in the EC (1980–1989). Source: EUROSTAT, PVV.

The further sharp fall in production resulted in another marked rise in the EC price of veal calves (8%) in 1989. Variations between countries were considerable.

- In France, prices increased sharply by 13.5% as demand held up rather better than in other countries and production again declined sharply
- There was a sharp price rise by 17.5% in the Netherlands as the market recovered, particularly in the first half of the year when production was especially low
- In West Germany and Italy, price rises were marked, 12.5% and 14%, respectively, as the fall in production was the more important factor, despite weak demand.

The average EC market price of 1989 was 245 ECU (about 646 NLG) per 100 kg liveweight.

The EC market can be expected to move further upwards in 1990. However, the price rise may be more modest (5.5%) because of the forecast smaller decline in production. Such price rises put veal increasingly into the 'luxury' category, with relative prices turning consumers away from the product.

Slaughterhouses

In the total production system, slaughterhouses are not having an easy time either. With the decline in number of slaughters, profitability decreased sharply. The competition between the slaughterhouses is strongly responsible for the high veal prices. Especially in the Netherlands, a few slaughterhouses specialize in the slaughter of veal calves. Only 7, most of them in the fattening areas, are involved in almost a million slaughters. Four slaughterhouses did more than 100 000 slaughters a year in 1989 and 4 are closely associated to a milk-replacer manufacturer too.

The decrease in the number of slaughters is not completely reflected in decreased production of veal, because the average carcass weight has increased for years. However carcass weight stabilized last year at an average of 130 kg. There are great differences between countries. The Netherlands produced the heaviest veal (average carcass weight in 1989 156 kg) and weight is still increasing (Figure 2). In Belgium, carcass weight is rapidly increasing to the same level as in the Netherlands. In France, however, weights remain low, with an average of only 118 kg in 1989 and hardly any increase.

Conclusions

From developments in European Community, some conclusions about veal production may be derived

1. Veal production in the EC is closely related to the dairy production and to the EC policy on dairy production.
2. France, Italy and the Netherlands are the main veal-producing countries; Italy and France consume the most
3. The consumption of veal is decreasing in most EC countries. This decline is a result of common concerns about the production of veal and of the high prices compared to other kinds of meat.
4. Since 1984, a milk quota system has caused an annualy decline in dairy and veal production. A further decline has been forecast. The decrease in number of calves available for veal producion will be most significant. Even imports of calves from Eastern Europe may decrease in the longer term.
5. The milk-replacer industry dominatest production of veal throughout the EC. Availability of skimmed milk powder and production of milk-replacers is decreasing. This is causing pressure on the profitability of the industry.
6. Most veal production is based on a contract between producers and milk-replacer manufacturer. Since 1986, the veal producer has had low profits. Forecasts for the next few years are no better. Proposed EC legislation on animal welfare and housing systems will cause even more pressure on profits.
7. With the decline in number of slaughters profitability of slaughterhouses has also decreased. As the average carcass weight of veal has increased for years, the decrease in number of slaughters is not completely reflected in decreased production of veal.

Veal production in North America

R. Groenevelt
Real Veal, Inc., P.O. Box 347, Ixonia, Wisconsin 53036, US

Summary

Veal production in the United States had a slow start in the sixties, but expanded during the seventies. At the moment, there are about 1300 veal-calf enterprises in the US. This paper describes the history and the market situation of US veal production, developments from 1970–1990 and an outlook to the future.

Introduction

This paper will concentrate mostly on veal production in the US. Canadian veal production, mainly concentrated in the Provinces of Quebec and Ontario, is similar to that of the US. Canada has veal calf facilities for about 40 000 calves, of which 15% are in loose-housing systems. An estimated 100 000 formula-fed calves were marketed in 1989. Exports were minimal.

Unlike the US, the majority (90%) of veal producers in Canada are labour-lease growers, paid a fixed weekly income for their labour, supplying the facilities and the utilities. No substantial growth is expected for the Canadian veal industry.

This present description of US veal production will focus primarily on the area east of the Rocky Mountains, where 95% of today's formula-fed veal is produced.

Production history

Commercial production of formula-fed veal calves in the US began more than a decade after its introduction in Holland. The huge surplus of non-fat dry milk made it an inexpensive and ideal ingredient for milk replacer. Whey, a logical carbohydrate source, was plentiful and mostly dumped in the fields. Furthermore, baby dairy bull calves were hardly utilized. They were either killed and buried on the farm, or if there was a demand, slaughtered and boned for veal. In other words, production costs of 'European-Style' veal seemed very attractive in the sixties.

The start was very slow, however, and it was hard to convince meat-packers and prospective veal producers that there could be a great future in raising and marketing formula-fed veal calves. The main reason that the veal business got off the ground was the strategy to focus on the large New York City Jewish population and the promotion of 'fancy-kosher' veal.

By the end of the sixties, the number of veal producers started to increase remarkably and by 1970, close to 20 000 pens in environmentally controlled barns had been erected for fancy veal production. The largest expansion of veal pens occurred during the seventies, resulting in a total number of 350 000 in 1980.

Loose housing became a hot issue in the early 1980s, promoted by some to try to pacify the animal rights groups and exploited by others to enter the market with mechanical feeders and the like. A new generation of growers converting barns, garages etc., at low cost and putting in automatic feeding equipment, found out that calf feeding could not be done while sipping a beer and watching television. At the same time, many misinformed existing growers eliminated their veal pens to raise veal calves more economically, finding out that the opposite was the case. This surge of loose housing religion caused the market to drop because of a fast-increasing oversupply of often veal calves of inconsistent quality. Many of those new growers put their cars back in the garage and the others, if not broke, went back to raising their animals in individual pens. Today, not even 1% of US veal growers raise their calves in loose housing. Certainly noteworthy is that during the last two decades the average and better grower has always been able to make healthy profits year after year.

Market analysis

There are about 1300 veal-calf-raising enterprises in the US. The veal-pen count as of January 1990 was 425 000, of which less then 300 000 pens seemed to be filled. Over 50 000 of the 425 000 pens might be considered non-refillable (Table 1). Over the last few years, the number of enterprises has decreased, while the size of the average operator has increased. In the very near future, many older veal barns will become obsolete because of disrepair or unsuitability for housing heavier calves.

The typical and average independent US veal producer holds 300 calves and raises 2 cycles annually. The profile of a veal grower differs substantially from area to area and from farm to farm. Contrary to most Canadian and European growers, the majority of US veal producers (70%) are financially independent from meat-packers, feed companies and other investors. In 1988, labor-lease operations housed in excess of 35% of the US formula-fed veal population. Low profits in early 1989 caused elimination of mostly the poorer labour-lease growers. A large number of veal farms in the State of Wisconsin are at present inactive because of this. Except on a small or local scale, total integration has had limited success in the USA and accounts for less than 9%. Almost all Midwestern veal producers contract each group of calves for a guaranteed minimum market price, while most growers in the East do without such contracts. By number of veal pens, the following are the major production States: Wisconsin (158 000), Indiana (63 000), Pennsylvania (56 000) and Ohio (43 000).

There are about 200 veal-calf enterprises equipped for liquid feeding. Producers using liquid feed in Indiana, Ohio and Michigan are on the decline. In Northeast Wisconsin, where more than 50% of the liquid users are, veal producers have been profitable. Normally, starting and finishing the calves on powdered feeds and using 80% liquid feed, those growers living close to a quality-cheese plant seem to obtain decent results.

Table 1. Distribution of veal farms in the US.

	Upper midwest		Mid central		North east		South	
Veal farms	496		252		358		92	
Number of pens	177 800		106 400		85 200		28 900	
Active veal farms	391	(79%)	208	(83%)	297	(83%)	71	(77%)
Number of pens	103 900		91 700		74 800		21 400	
Veal farms used for dairy beef	34	(7%)	23	(9%)	19	(5%)	9	(9%)
Number of pens	8 600		6 350		3 350		2 400	
Inactive veal farms	71	(14%)	21	(8%)	42	(12%)	12	(11%)
Inactive veal pens	65 900		8 350		7 050		5 100	
Average number of pens per farm	265		440		250		300	

Veal consumption in the US has been decreasing, but consumption of formula-fed veal has been growing slowly but steadily. In the US, 'veal' is a collective name for many kinds of meats obtained from 'calf' slaughter (Table 2).

Formula-fed finished veal calves increased 40 pounds (about 20 kg) in liveweight between 1987 and 1990; an average of 350 pounds (about 160 kg) in 1987; and close to 400 pounds (about 180 kg) now. Because of the decreasing number of available and suitable bob calves and the increasing purchase cost, it will be necessary and more economical to raise even heavier veal calves in the future.

Developments 1970-1990

Since 1970, the average price of non-fat dry milk, whey products and other ingredients of veal-milk have increased in price over fourfold. In the same period, US prices of veal feed only doubled. Prices of baby calves increased sixfold. Successfully lowering content of protein in finisher feeds and eliminating non-fat dry milk from most formulas, together with raising a lighter bob calf to a heavier weight with minimum sacrifices in feed conversion, death losses and health care kept veal production profitable during the last two decades. Table 3 shows how US formula-fed veal production has changed between 1970 and 1990.

Table 2. Number of calves slaughtered by class under Federal inspection (1000 head).

Year	Bob veal ≤ 150 lbs	150–400 lbs formula fed	150–400 lbs non-formula	Other ≥ 400 lbs	Total
1987	1 207.8	1 002.7	171.4	297.5	2 679.4
1988	1 065.9	1 003.3	155.9	185.1	2 410.2
1989	873.2	932.6	108.9	196.9	2 111.6

Table 3. Formula-fed veal production in 1970 and 1990 in the US.

	1970	1990
Calf's starting weight (kg)	50	40
Marketing liveweight (kg)	125	175
Growth cycle (week)	12	17
Feed consumption (kg)	120	240
Feed conversion	1.6	1.8
Average daily gain (kg)	0.89	1.14
Feed cost ($)	66.00	264.00
Initial calf cost ($)	25.00	150.00
Live market price/kg ($/kg)	1.01	2.97
Gross profit ($/week)	2.94	6.24

Future outlook

Demand for veal is expected to stay strong at least 10 months out of the year over the next couple of years. From the point of view of health, veal is getting and expecting to get a lot of positive publicity in the US and Canada.

Successful application of non-dairy proteins in veal feeds and raising an even heavier high-quality veal calf at the lowest possible production cost, with ample profits for the veal producer, make the meat processor and the feed manufacturer the key issues for the 1990s.

2. Animal welfare

Needs and welfare of housed calves

D.M. Broom
Department of Clinical Veterinary Medicine, University of Cambridge, Madingley Road, Cambridge CB3 OES, GB

Summary

The concept of welfare is introduced and basic biological functions are described in relation to the needs of calves. These form mechanisms that have evolved in the species and persist in domestic cattle today. Indicators of poor welfare in calves are reviewed. Current systems for rearing calves in crates result in poor welfare but welfare is much better in a well managed group-housing system. Public attitudes are starting to affect the veal industry considerably and a solution to the problem involving a change to group-housing systems with a diet including adequate roughage and iron is necessary if the industry is to survive.
Descriptors: welfare; needs; behaviour; physiology; calf rearing.

Welfare

Animals encounter a variety of difficulties during their lives and they use physiological and behavioral methods to try to cope with these. The extent to which coping methods, such as adrenal activity, are used can be measured, as can the consequences of failure to cope, for example reduced ability to grow or reproduce. The welfare of an individual is its state as regards its attempts to cope with its environment (Broom, 1986). Welfare can vary from very poor to very good, and can be measured. Measurements relevant to the assessment of welfare include those of injury level, disease incidence, pain responses, consequences of lack of control, especially frustration, and the effects of fear.

Situations or human actions that lead to poor calf welfare are direct abuse; neglect which is deliberate, accidental or due to lack of knowledge; provision of inadequate conditions during housing, transport or marketing; disease; failure to provide for emergencies and improper procedures during and before slaughter. Direct abuse includes hitting or frightening the animal and carrying out surgical operations. Neglect might involve failure to feed or clean out adequately, to treat it diseased, or to care for if injured, disturbed or likely to be harmed in some way. Emergencies which could arise on any animal unit are fire, power failure or extreme weather conditions. Welfare can be very poor in these emergency situations and provision should be made for early warning, back-up power supplies, preventing the freezing of water supplies, providing ventilation or showers in very hot conditions, fire prevention, firefighting, and rapid evacuation of all animals in case of fire. The welfare of most diseased animals is poor, sometimes very poor, so disease prevention, without general exacerbation of disease problems by too widespread a use of antibiotics, is desirable. Procedures that minimize pathogen contact with the calves and the treatment of diseased individuals is the best strategy here. Conditions and procedures when animals are handled, transported, marketed and slaughtered can often lead to very poor welfare but the topic that will be discussed in most detail in this paper is the housing and associated management of calves.

The factors affecting welfare outlined above fall into three somewhat overlapping categories in terms of human responsibility. The design of the housing and management system is largely the responsibility of the owner of the unit and the site manager. These people have a substantial affect on the welfare of large numbers of individual animals. The men and women who care directly for the animals in that they feed them, clean up after them, check them to see if they have problems, handle them for treatment or movement to another place, and load them onto vehicles, have a very important effect on animal welfare. Contributions are also made by specialists such as veterinary surgeons, vehicle drivers and abattoir staff. None of these alone can insure that welfare is good. Each has a moral obligation towards the animals.

Needs

A need is a deficiency in an animal that can be remedied by obtaining a particular resource or responding to a particular environmental or bodily stimulus. Needs are dictated by the basic biology of the animal, for they relate to its various functional systems. Farm animals, including calves, have an array of needs similar to those of their wild ancestors. The mechanisms which evolved within the species are still present in our domestic animals although some responses, such as those associated with flight from man, are reduced. Various functions are described below with explanation of the need and what the animal must do in order to satisfy them.

1. In order to obtain adequate nutrients and to reduce disease incidence, the young calf must ingest and digest, first colostrum and then milk. The biological mechanisms that have evolved in order to achieve this are searching behaviour which should result in finding a teat and then the licking and sucking of a teatlike object. The behaviour of the calf is not just organized so that all such behaviour is terminated if the gut is filled with milk but the licking and sucking behaviour itself is of great importance to the calf. The calf needs both to rectify any nutrient or energy deficiency and to suck on a teat or teatlike object.
2. The functions of recuperation and avoiding danger during the time of day when feeding is less efficient in range conditions are served by resting and sleeping. Adequate rest and sleep require that the calf adopts a sleeping posture on a suitable surface for a period during which there is not too much disturbance.
3. The avoidance of danger also requires that animals explore their surroundings so as to find out about sources of danger, and to hide or show escape responses if necessary. Exploration is also of value in the efficient exploitation of resources. The need to explore exists for all animals whilst those to hide or to escape exist if the calf perceives that there is some danger. Man is often treated as a dangerous predator by young calves and many events in the life of a young calf elicit attempts to hide or escape.
4. In order that bone and muscle can develop normally, and that injury can be avoided, the calf needs to take sufficient exercise and to move or adopt postures that do not cause discomfort. If bones are not loaded by muscle action and body movement then osteopenia will develop. A small amount of exercise at frequent intervals may be sufficient to insure normal bone development. Extreme inactivity causes abnormal muscle development and may also cause joint problems. The biological mechanism to prevent or remedy these problems is for animals to want to take regular exercise and to avoid discomfort such as that which results from difficulty in making certain movements.
5. Important aspects of gut development in calves are the changes in the rumen etc., which result in efficient utilization of grass and other fibrous plants that are the staple food of cattle. Hence the calf needs to eat roughage and to ruminate. It has a very strong preference to eat fibrous material and to attempt to ruminate, even if no roughage has been ingested. Any inadequacies of nutrients, including iron, also make calves want to eat a variety of materials that might supply the missing nutrient.
6. Disease and parasitism levels are minimized by keeping the body clean. Grooming also helps in thermoregulation and other functions. Hence animals need to respond to stimuli which indicate that grooming is necessary, principally by licking themselves. These stimuli emanate from the whole body and if an animal initiates a grooming sequence, the whole body has to be groomed if it is to be satisfied with what it has done. Hence inability to groom the hind part of the body causes problems.

7. Cattle are social animals, and ability to deal with social situations is important and demanding to them. In order to develop social ability, each calf needs to approach and interact with other calves.

Whilst a general study of the biology of calves tells us that each of these needs is of importance to the animals, measurements of behaviour and physiology give more information about how important they are. In some cases, frequent and vigorous behaviour or attempted behaviour, such as sucking or grooming, indicates the strength of the need. It may be impossible for calves in certain farm conditions to show some kinds of behaviour because of the restrictions placed upon them by those conditions. Hence it is necessary for as wide as possible a range of signs of poor welfare, for example abnormalities of behaviour or physiology, to be used to recognize the importance of the need.

Indicators of poor welfare in relation to rearing conditions

High disease incidence

Calves are particularly susceptible to respiratory and gastro-intestinal disease. Van Putten (1987) quotes data from Postema showing that 67% of batches of veal calves in the Netherlands required treatment for respiratory disease by 10 weeks and 20% had enteric disorders after one week in a veal-production unit. Similarly, van der Mei (1987) found that 25% of veal calves had respiratory disease and 14–18% had severe diarrhoea before 10 weeks of age, the higher level referring to group-housed calves. Webster (1984) reported less disease in group-housed calves and it seems likely that disease might be lower in well managed group housing than in crates but higher in less well managed grouphousing. It is clear, however, that in both crate housing and group housing, these high incidences of disease in veal calves are a particularly serious welfare problem. Calves purchased and brought into a unit are five times more likely to require treatment for disease (Webster, 1984), so this is an important cause of the problem.

With the exception of a few suckler calves, veal calves are separated from their mothers at a very early age. This separation causes some problems for the calf and inevitably makes it rather more difficult for the farmer to keep it healthy. Calves left with their mothers stand earlier and are more active than those which are separated (le Neindre & Signoret, 1987). The reduced vigour and lack of interest in their environment shown by some calves after separation can result in failure to obtain adequate immunoglobulin and nutrients. Even if some time is spent with the mother, a variety of factors affect the likelihood that sufficient colostrum will be ingested and immunoglobulin absorbed from it (Broom, 1983). Much greater disease problems arise when young calves are transported, sometimes via a market, to another farm so that they encounter new pathogen challenges and often a new diet.

Stereotypies

Tongue-rolling or tongue-playing and other 'purposeless oral behaviour' (Webster et al., 1985) is abnormal behaviour that is not shown by calves that can cope well with their environment. These are examples of stereotypies, repeated relatively invariate sequences of movements that have no obvious purpose (Fraser & Broom, 1990), and just as in man, zoo animals or sows, stereotypies are indicators of poor welfare. Tongue-rolling is much more frequent amongst crate-reared calves than amongst those reared in groups (Andreae et al., 1980; Webster et al., 1985; Wierenga, 1987). There is individual variation in its occurrence but there is always much variation in the methods used for trying to cope with inadequate rearing conditions (Broom, 1986; 1988). A reduction in the occurrence of stereotyped licking, manipulation etc.

occurs when straw is provided (van Putten & Elshof, 1978; Webster, 1984) and higher levels are observed at low light level of 2 lux than at 20, 100 or 130 lux (Dannemann et al., 1984).

Table 1. Oral behaviour of calves at two weeks of age (%).

	Suckler calves	Crate-housed	Group-housed
Purposeless oral activity	1	14	3
Grooming	6	15	6
Ruminating	8	2	5

Table 2. Calf behaviour when fed from bucket or teat (after Hammell et al., 1988).

	Bucket	Teat
Duration of drinking (min)	18	44
Time sucking dummy teat (min)	13	1
Drinking interrupted by sucking dummy teat	yes	no

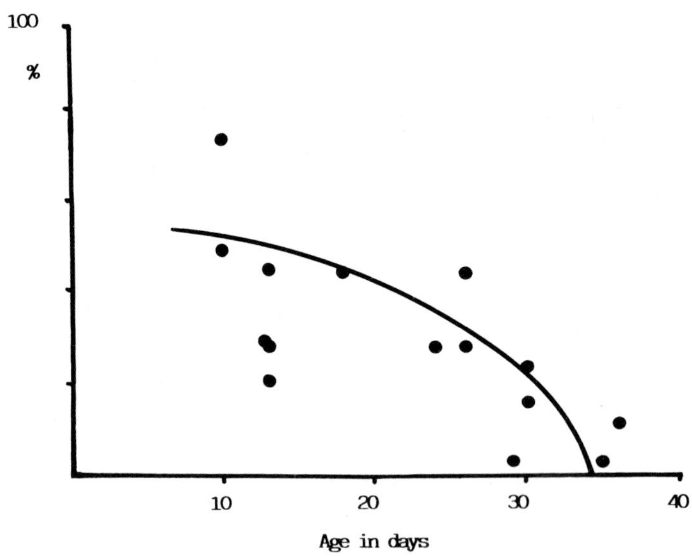

Figure 1. Percentage of time sucking or licking a novel non-nutritive teat or another calf during 10 min after drinking milk from a bucket (Broom & van Praag, in prep.).

Sucking and licking

Some sucking and licking is stereotyped but other behaviour of this kind, although too variable in form to be called a stereotypy, has a high enough frequency or duration to be abnormal. Data from Webster et al. (1985) is shown in Table 1. Similarly, Ketelaar-de Lauwere & Smits (1989) reported such oral activities for 20–30% of the time in crates ranging in size from 70 cm × 150 cm to 170 cm × 250 cm. It would seem from this result that absence of suitable stimuli for normal sucking, lack of social contact or inadequate diet are likely to be more important than lack of space in causing this behaviour.

Non-nutritive sucking is likely to arise in the first place if the calf's need to suck is not satisfied during feeding. Its occurrence depends upon what is available for the animal to suck. Calves may want very much to suck but have no object present so they may increase licking behaviour or show other signs of poor welfare. If in groups, calves that wish to suck may suck the penis, navel or ears of other calves and in doing so may cause problems to themselves by drinking urine, or to other calves by causing soreness. Individually housed calves that can reach one another may suck another calf and 'kissing' pairs are frequent in occurrence. The sucking of the pen or of other calves occurs most just after feeding and is much more frequent if the calf is fed from a bucket than if it is fed from a teat.

When a non-nutritive teat is provided in a calf's pen, the calves spend much time sucking it, especially after feeding (Waterhouse, 1978; Broom, 1982). Such sucking is reduced if milk is supplied more frequently (Metz, 1984). It is more prolonged and may interrupt drinking when milk is drunk from a bucket (Hammell et al., 1988; Table 2). In an experiment in which a novel non-nutritive teat was presented to individually housed calves at different ages, calves more than 30 days of age did not suck it within two minutes of finishing drinking from a bucket (Broom & van Praag, in prep.). Figure 1 shows the decline in interest in the novel non-nutritive teat with age. The desire of calves to suck a teat declines with age but animals which associate feeding with teat-sucking will continue sucking to a later age.

Excessive self-licking occurs frequently in crate-housed calves and results in ingestion of hair and formation of hair balls or bezoars. It could be that inability to groom the hind part of the body increases the desire to groom those parts of the body that can be reached. Self-licking was reduced by feeding 75g of straw per day and numbers of hair balls dropped from means of fifteen to less than one (van Putten & Elshof, 1978). In the same study, sham rumination occurred in crate-housed milk-fed calves for two hours per day. Provision of straw allowed normal rumination in that study and in that by Webster et al. (1985).

Adrenal and other physiological indicators

In a study of 8-week-old Friesian calves, Dantzer et al. (1983) assessed previous adrenal activity by means of challenge with adrenocorticotrophic hormone (ACTH). Tethered calves showed a significantly greater cortisol response to ACTH challenge than did group-housed calves, suggesting that the tethered calves had used their emergency adrenal cortex response more often during the rearing period. A similar result was obtained by Friend et al. (1985a), who also found significantly higher basal cortisol levels, thyroid hormone levels and neutrophil/lymphocyte ratios in tethered or individually housed calves than in group-housed calves (Table 3).

Response to disturbing stimuli

Calves are much more severely affected by contact with man if they have had little previous human contact. Frequent gentle contact is desirable for calves. Excessive fearfulness occurs when calves are kept in poorly lit buildings (Webster & Saville, 1981). The responses of calves to handling and transport are substantially greater if they have been kept in crates than if they

Table 3. Physiological 6-week-old calves after different housing. Data from Friend et al. (1985a).

Physiological indicator	Tether-stall 0.56 m × 1.2 m	Individual pen 1.2 m × 1.5 m	Group-housed
Tri-iodothyronine (ng dl^{-1})	106	78	58
Thyroxine (μg ml^{-1})	6.0	5.0	4.3
Basal cortisol (ng ml^{-1})	5.4	4.2	3.4
Cortisol after ACTH challenge (ng ml^{-1})	103	98	75
Neutrophil/lymphocyte ratio	0.69	0.37	0.36

have been kept in groups (Trunkfield et al., this book). So it would appear that crate-reared calves are less well able to cope with difficulties such as those presented by those procedures.

Difficulties of posture and locomotion

The postures adopted by most resting calves with no space restriction cannot be adopted by a calf in a narrow crate (de Wilt, 1985). The most preferred head position is difficult to adopt and hind-leg extension cannot occur. This study, and a subsequent detailed investigation by Ketelaar-de Lauwere & Smits (1989) has shown that calves try very hard to adopt postures that are comfortable and that result in good thermoregulation, particularly in hot conditions. Hind-leg stretching occurs for 2–8% of the time and the common lying posture in crates with one leg extended along the body does not occur if more space is available. Behaviours like leg stretching that are infrequent may still be of great importance to an individual. An example of another behaviour that is infrequent but important is defaecation. The minimum width that allowed comfortable lying for calves was found to be 90 cm. A further problem reported in small crates was that calves had difficulties standing up and lying down. Calves choose to walk at intervals if they are able to and crate-housed calves show much locomotor behaviour when released from the crate (Warnick et al., 1977; Friend et al., 1985b). After six months in a crate, many calves have severe locomotory problems (Trunkfield et al., this book).

Social and maternal inadequacy

Calves reared in isolation performed badly in competition with group-reared calves when mixed after 6–8 weeks (Warnick et al., 1977), 12 weeks (Waterhouse, 1979), and 6 months (Broom & Leaver, 1978). Some elements of social interaction are still shown after isolation but signals such as ear movements are abnormal and ability to compete effectively within a mixed herd is clearly impaired (Broom, 1982); the effects of isolation for 6 months were still apparent one year later. Some signs of maternal inadequacy were evident in calves reared in isolation for six months (Broom & Leaver, 1977) although le Neindre (pers. comm.) was not able to see impairment of maternal behaviour after shorter periods of isolation. Although crate-reared veal calves do not encounter other calves except for the period before slaughter, it is clear that their normal social development is impaired substantially by isolation. Waterhouse (1979) found that effects on social ability were much greater if the calves were in visual isolation than if they could see and touch other calves through a hurdle partition (also Broom, 1982). Calves in 'enriched' pens with independently moving objects in them were more normal in their social behaviour.

Group-housed calves show little competitive behaviour that might be considered harmful. There can be uneven growth rates but these problems can be solved by better provision of food. Social facilitation aids weaker calves in group-housing as long as there are enough feeding places. Where milk was provided from teats to calves 0–5 weeks old, good results were obtained by putting five milk-supplying teats in a row 20 cm apart (Barton, 1983; Barton & Broom, 1985). Group-housed calves also learn readily to drink from responder-controlled feeders (Maatje, this book).

Injuries and anaemia

These topics are covered by Webster and others in this volume. Anaemia results in poor welfare. Severe ulceration does so too but there are doubts about the effects of small ulcers, which are frequent after feeding on milk and straw (Welchman & Baust, 1987). There may be links between behavioral responses to difficult conditions and abomasal ulceration (Wiepkema, 1985). Claw problems can be bad on inadequate or poorly drained floors and tail-tip necrosis can be very high on slats.

Conclusions

It is clear that the crate used for veal calf production does not meet the needs of the calves and results in a wide range of indicators of poor welfare in calves.

What change is necessary in order that the welfare of calves will be good? If the crate were larger, for example 90–120 cm wide, some of the problems would be solved. The calf would be able to turn round, stretch and adopt more comfortable lying postures. However it would still be unable to exercise normally by walking, interact socially, chew solid food, ruminate, carry out investigatory behaviour or respond appropriately if frightened. Hence calves housed individually in larger solid-sided pens still show clear signs of poor welfare. If the solid walls of individual crates are replaced by open sides, the environment of the calf is considerably improved because some social interaction can occur and the calf can move in a little more space. However social behaviour is still very much limited by the crate and the other deficiences listed above are not removed. The change that results in the removal of these deficiences is to keep the calves in a group in a larger pen.

A major problem in the veal industry is the demand for 'white' meat. The indications from a wide range of studies are that the diet composition which is necessary to produce 'white' veal will always cause some severe welfare problems. In order to avoid clinical anaemia in all the calves and to provide roughage that the calves can manipulate, chew and ruminate on, the meat must become a little pinker on average. Such an improved diet for veal calves should be cheaper, should result in better feed conversion and should lead to lower disease levels. With improved diet, including adequate roughage and iron, and a group-housing system with measures taken to minimize intersucking, veal can be produced humanely. Disease levels in group housing can be high but this does not occur in good group-housing systems, which have good care by stock persons, a good diet and usually materials such as straw for calves to manipulate and chew. Fully slatted floors are less likely to work well and disease is more often a problem in such systems, especially where adequate iron and roughage are not provided in the food. Individual crates that have open sides for social contact and that are large enough for calf comfort (more than 90 cm wide) are certainly better for calves than normal crates but are not as good as full group housing and it may well be cheaper to use a group-housing system.

The veal industry has to take note of the fact that veal consumption is dropping rapidly, even in countries like Italy where there is a strong tradition for veal dishes. The public perception of the poor welfare of the veal calf is a major factor in the accelerating decline in veal consumption. There is no reason, however, why this decline should not be stopped and reversed if changes in diet and housing result in improved welfare and the public is informed of this. There is a tendency for the public to associate white veal with poor welfare so a new improved image of veal could include some change in colour together with explanation from the industry that new veal comes from better fed calves whose welfare is better because they live in groups and because their other biological needs are also met.

References

Andreae,V., J. Unshelm & D. Smidt, 1980. Handhabung van Kälbern in Gruppenhaltung. Arbeiten zur artgemäßen Tierhaltung. 254:89–96. K.T.B.L. Darmstadt.

Barton, M.A., 1983. The effects of management and behavioral factors on intake of acidified milk and concentrates by group-reared calves. Animimal Production 36:512.

Barton, M.A. & D.M. Broom, 1985. Social factors affecting the performance of teat-fed calves. Animimal Production 40:525.

Broom, D.M., 1982. Husbandry methods leading to inadequate social and maternal behaviour in cattle . In: Disturbed behaviour in farm animals. Ed. W. Bessei. Hohenheimer Arbeiten, 121:42–50. Eugen Ulmer, Stuttgart

Broom, D.M., 1983. Stereotypies as animal welfare indicators. In: Indicators relevant to farm animal welfare. Ed. D. Smidt. Current Topics in Veterinary Medicine & Animal Science 23:81–87. Martinus Nijoff, The Hague.

Broom, D.M., 1986. Indicators of poor welfare. British Veterinary Journal 142:524–526.

Broom, D.M., 1988. Welfare considerations in cattle practice. Proceedings of the British Cattle Veterinary Association for 1986-7:153–164. B.C.V.A., London.

Broom, D.M. & J.D. Leaver, 1977. Mother-young interactions in dairy cattle. British Veterinary Journal: 133, 192.

Broom, D.M. & J.D. Leaver, 1978. The effects of group-rearing or partial isolation on later behaviour of calves. Animal Behaviour 26:1255–1263.

Dannemann, K., D. Buchenauer & H. Fliegner, 1984. The behaviour of calves under four levels of lighting. Applied Animal Behaviour Science 13:243–258.

Dantzer, R., P. Mormede, R.M. Bluth & J. Soissons, 1983. The effect of different housing conditions on behavioural and adrenocortical reactions in veal calves. Reproduction Nutrition & Development 23:67–74.

Fraser, A.F. & D.M. Broom, 1990. Farm animal behaviour and welfare. Tindall, Bailiere, London.

Friend, T.H., G.R. Dellmeier & E.E. Gbur, 1985a. Comparison of four methods of calf confinement. 1. Physiology. Journal of Animal Science 60:1095–1101.

Friend, T.H., G.R. Dellmeier & E.E. Gbur, 1985b. Comparison of four methods of calf confinement. 2. Behaviour. Journal of Animal Science 60:1102–1109.

Hammell, K.L., J.H.M Metz & P. Mekking, 1988. Sucking behaviour of dairy calves fed milk and libitum by bucket or teat. Applied Animal Behaviour Science 20:275–285.

Ketelaar-de Lauwere, C.C. & A.C. Smits, 1989. Onderzoek naar de uit ethologisch oogpunt minimaal gewenste boxmaten voor vleeskalveren met een gewicht van 175 tot 300 kg. IMAG Rapport 110. IMAG, Wageningen.

le Neindre, P. & J.P. Signoret, 1987. Ethological results of welfare research in veal calves. In: Welfare aspects of housing systems for veal calves and fattening bulls. ed. M.C. Schlichting & D. Smidt: 77–88. C.E.C., Luxemburg. EUR 1077 EN.

Metz, J.H.M., 1984. Regulation of sucking behaviour of calves. In: Proceedings of the International Congress on Applied Ethology of Farm Animals. Kiel. ed. J. Unsholm, G. van Putten & K. Zeeb. p. 70–73. K.T.B.L., Darmstadt.

van der Mei, J., 1987. Health aspects of welfare research in veal calves. In: Welfare aspects of housing systems for veal calves and fattening bulls. ed. M.C. Schlichting & D. Smidt, 99. C.E.C., Luxemburg. EUR 1007 EN.

van Putten, G., 1987. Housing and management concepts in today's veal production. In: Welfare aspects of housing systems for veal calves and fattening bulls. ed. M.C. Schlichting & D. Smidt. p. 45–59. C.E.C., Luxemburg. EUR 1077 EN.
van Putten, G. & W.J. Elshof, 1978. Zusatzfuttering von Stroh an Mastkalber. Aktuelle Arbeiten zur artgemäßen Tierhaltung 233:210–219. K.T.B.L., Darmstadt.
Warnick, V.D., C.V. Arave & C.H. Mickelsen, 1977. Effects of group, individual and isolated rearing of calves on weight gain and behaviour. Journal of Dairy Science 60:947–953.
Waterhouse, A., 1978. The effects of pen conditions on the development of calf behaviour. Applied Animal Ethology 4:285–286.
Waterhouse, A., 1979. The effects of rearing conditions on the behaviour and growth of dairy calves. Ph.D. thesis, University of Reading.
Webster, A.J.F. & C. Saville, 1981. Rearing of veal calves. In: Alternatives to intensive husbandry systems. p. 86–94. U.F.A.W., London.
Webster, A.J.F., C. Saville, B.M. Church, A. Gnanasakthy & R. Moss, 1985. Some effects of different rearing systems on the development of calf behaviour. British Veterinary Journal 141:249–264.
Webster, A.J.F., 1984. Calf husbandry, health and welfare. Collins, London.
Welchman, D. de B. & G.N. Baust, 1987. A survey of abomasal ulceration in veal calves. Veterinary Record 121:586–590.
Wiepkema, P.R., 1985. Over gedragsstoringen bij dienen in de veehouderij. Tijdschrift voor Diergeneeskunde 110:12–20.
Wierenga, H.K., 1987. Behavioural problems in fattening bulls. In: Welfare aspects of housing systems for veal calves and fattening bulls. ed. M.C. Schlichting & D. Smidt, p. 105–122. C.E.C., Luxemburg. EUR 1007 EN.

Effects of breed and early social environment on calf behaviour

P. le Neindre
Institut National de la Recherche Agronomique, Theix, 63122 Cey rat, FR

Summary

At pasture on the first day after birth, the calf is with its mother away from the other members of the herd. Some days after birth its dam rejoins the group of cows and only later the calf is integrated into its peer groups. Most of the calf's social interactions are directed towards its dam during the first months of life. The effects of isolation on postnatal and subsequent social and maternal behaviours were studied. Mothered calves are much more active than isolated ones. Subsequent maternal behaviour and the social behaviour of Friesian animals are not influenced by early management. However in a beef breed, isolated animals are dominated by mothered ones. The effects of isolation can therefore vary between breeds of cattle.
Descriptors: social behaviour; maternal behaviour; welfare.

Introduction

The welfare problems of veal calves can be better analysed and understood by observation of calves in an undisturbed environment. The activities of calves suckled by their mothers on pasture and the effects of isolation during the early stages on subsequent social behaviour are reported.

Ontogeny of social behaviour

Thirty-nine primiparous cows were observed with their calves at pasture. They were divided into four herds in paddocks of five to eight hectares. Each herd was observed for five whole days. The nearest neighbour was recorded every 30 minutes for each animal. Social interactions were observed continuously.

Just after birth, the calf and its dam were together and spent most of their time isolated from the other animals of the herd. After the first week the dam spent most of the day in the herd (Figure 1a) but the calf still remained alone more than 10% of the total time (with no other animal within 40 metres). It was then alone very seldom and most of the time it had a calf as a nearer neighbour (Figure 1b). During the first months of life, the calf interacted mainly with its dam. It suckled only her and was licked only by her.

Social environment of the calf in an undisturbed environment is then quite complex. From the first days to the postweaning period, there is a slow change in the social life of the calf.

Effects of early isolation on subsequent social behaviour

Veal calves in intensive farms are far from these natural conditions. In particular, they have no contact with their dams and sometimes none with their peers. Two sets of experiments were designed: the first to analyse the effects of isolation on calf behaviour during the first few hours after birth; the second to study the long term effect of isolation.

Very different types of cattle exist from the dairy breeds to the beef breeds. Dairy calves are usually isolated from birth whereas beef calves remain with their mothers until the age of 8 months. We therefore tested wether the effects of isolation might be different for the two types of calves. We studied animals from two breeds: Friesian, a dairy breed; Salers, a beef breed.

Postnatal activity

Fifty-six calves (Friesians n = 29; Salers n = 31) were observed for 4 hours after birth. Thirty-one were isolated from their mothers after birth and the others were mothered. Their activities were filmed on video. The interval between birth and first standing up and the total time spent standing up were recorded.

Mothered animals stood up much earlier than isolated ones. Only 40% of the Salers calves stood up when isolated compared with 75% of the Friesians (Table 1).

Mothered Salers calves drunk more colostrum when suckling their mothers than when feeding from a bottle (8.9% vs. 6.3% of their bodyweight). Like Metz & Metz (1984) but in contrast with the studies of Selman et al. (1971) and Fallon (1978), we found that mothering had no effect on the calf's absorption of colostrum from the intestine.

Salers calves drank less colostrum than Friesians (39 vs. 82 % drunk more than 2 kg) and for the same amount of immunoglobulin drunk, Salers had a lower blood level of immunoglobulin (12.5 mg/kg vs. 18.2).

Social behaviour

Fourteen artificially reared heifers and 14 mothered ones were observed at pasture when five months old. The number of interactions directed towards the other calves was higher in the artificially reared animals than in the others (agonistic interactions: 16.6 vs. 1.0, non-agonistic interactions and licks: 6.6 vs. 0.3).

Thirty-six 3-year-old cows were observed at pasture. Twenty had been isolated from birth until the age of 3 months (Friesians n = 12, Salers n = 8) after which they were grouped. The other animals were mothered by nurse cows (Friesians n = 8, Salers n = 8). There was no significant difference between the social behaviour of isolated and mothered dairy cows. However mothered beef cows were significantly more active and dominated those that had been isolated (Table 2).

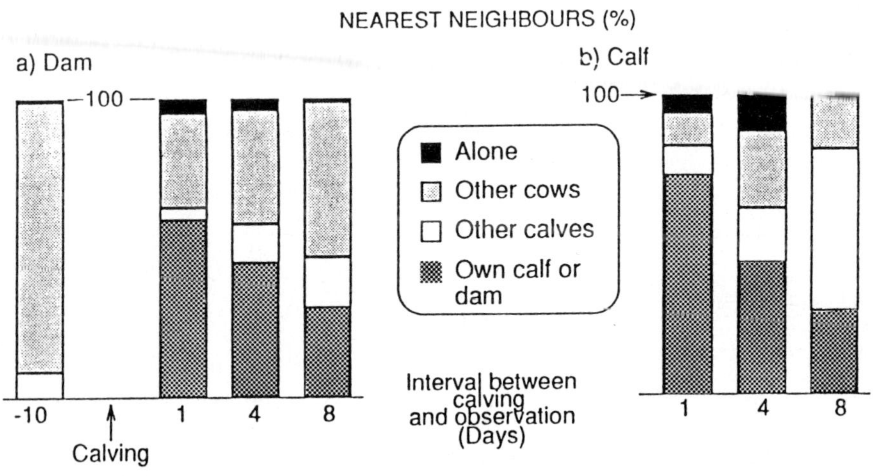

Figure 1. Change in nearest neighbours with the interval between calving and observation. Expressed as relative frequencies of scans (%, one scan every thirty minutes).

Table 1. Influence of breed and environment on the interval between birth and standing up (in min = 240 if the calf didn't stand up) (the number of calves to stand up/total number). [a,b,c,] significant difference P < 0.05. From Le Neindre & Signoret (1987).

	Salers	Friesians
Isolated	204 ± 58^{a}	117 ± 86^{b}
	(6/15)	(12/16)
Mothered	64 ± 60^{c}	45 ± 53^{c}
	(15/16)	(9/9)

Table 2. Influence of breed and rearing conditions on hierarchialp osition (the higher the data, the higher the hierarchial position). From Le Neindre (1989b).

Herds	FN	FM	SN		SM
SN/SM			4.7	**	12.0
FN/FM	8.7	6.5			
SM/FM		6.1	**		12.7
SN/FN	3.0	**	9.9		

Mann-Whitney test: * $P < 0.05$
** $P < 0.01$

F Friesian S Salers
N Non-mothered; M Mothered

Table 3. Effects of breed and rearing condition on cow-newborn calf relationships (4-h period after birth; Mann-Whitney test). From Le Neindre (1989a).

	Breed		Rearing condition		Total	
	Friesian	Salers	non-mothered	mothered	mean	s.d.
Number of data	6	7	7	6	13	13
Number of attempts before first suckling	2.2***	8.0	3.7*	7.2	5.3	5.0
Birth-first suckling interval (min)	69.5	75.3	73.7	71.3	72.6	24.0
Latency between claving and first licking (min)	6.6	5.0	7.7*	3.7	6.0	4.2
Total licking time (min)	44.3	41.9	51.4**	33.2	43.0	17.7
Number of kicks at the calf	0.5	1.7	0.9	1.5	1.2	1.9

* $P < 0.05$ ** $P < 0.01$ *** $P < 0.001$

Maternal behaviour

The maternal behaviour of the cows described above was observed. Thirteen were observed during the four hours following birth and all animals were observed for 5 whole days during the first three months following calving.

Early social isolation of female calves had no important effects on their subsequent maternal behaviour. The detrimental effects (in particular aggressive behaviour towards newborns) of not being mothered at birth or of being isolated after birth both of which were reported by Harlow & Harlow (1969) in monkeys, were not observed. On the contrary, isolated cows licked their calves longer than mothered cows (Table 3). When the calves were older, the only significant difference was that observed mothered dams suckled more often than those from isolated dams (4.69 suckling bouts per day vs. 4.05) .

Conclusion

Calves in a suckling herd have a very complex social environment and interact with all the members of the herd. It seems correct to keep this social environment as much as possible. Traditional French production of veal calves is a good way of maintaining it. In this husbandry system, calves suckle their mothers twice a day (three times a day when they are young) and they are not usually isolated from the other calves.

The effects of early isolation on calves are not as great as they are in some species of monkeys and not so great as is often said. However beef calves are very disturbed by isolation even if the effects are not as important on dairy calves. Calves from the dairy herds must be separated from their mothers but how their welfare is better if they have some contacts with other calves is unclear. We do not know if the mere presence of human beings is important for the calf and can compensate the absence of congeners. However it is clear that some contacts with human beings during the early ages ease the subsequent handling ability of the animals.

References

Fallon, R.J., 1978. The effect of immunoglobulin levels on calf performance and methods of artificially feeding colostrum to the new-born calf. Ann. Rech. Vet. 9:347–352.

le Neindre, P., 1989a. Influence of rearing conditions and breed on social relationship of mother and young. App. Anim. Behav. Sci. 23:117–127.

le Neindre, P., 1989b. Influence of rearing conditions and breed on social behaviour and activity of cattle in novel environments. App. Anim. Behav. Sci. 23:129–140.

le Neindre, P. & J.P. Signoret, 1987. Ethological results of welfare research in veal calves. In M.C. Schlichting & D. Smidt (ed.): The welfare of calves. EUR 10777 p. 77–88.

Metz, J. & J.H.M. Metz, 1984. Die Bedeutung der Mutter in der Umwelt der neugeborenen Kalbes. Kuratorium für Technik und Banwesen in der Landwirtschaft 307:188–198.

Selman, I.E., G. Mc Ewan, & E.W. Fisher, 1971. Studies on dairy calves allowed to suckle their dams fixed times post partum. Res. Vet. Sci. 12:1–6.

Ethological and physiological reactions of veal calves in group-housing systems

Christiane Müller & M.C. Schlichting
Institute of Animal Husbandry and Animal Behaviour FAL, Trenthorst, D-2061 Westerau, DE

Summary

The effects of group size, space allotment and supplementary roughage feeding was studied in 80 veal calves over a fattening period of 20 weeks. Behavioral activity, resting behaviour and preputial sucking was recorded every five weeks. The ethological results showed that locomotion was more frequent in the uncrowded groups and that uncrowded calves had less problems lying down. After 6 weeks and at the end of the study, adrenal reaction to ACTH (1–24) stimulation (Friend, 1980) showed only small differences between groups with group size and space allotment. Supplementary feeding on roughage, however, exerted a larger effect on the adrenal reaction.
Descriptors: veal calves; group size; space allotment; roughage supplement; preputial sucking; resting behaviour; adrenal reaction.

Introduction

Traditionally, veal production means keeping calves separately in crates with a slatted floor during an unbalanced intensive fattening period (Andreae et al., 1980; de Wilt, 1985). Meeting the high requirements of the young calf is a big problem for most farmers. The tendency towards group housing of veal calves is obviously welcome for welfare reasons. It is necessary, however, to improve some aspects of the different group-housing systems. With this incentive, we studied different combinations of group size and space allotment.

Material and methods

In two consecutive experiments, a total of 80 veal calves were housed in groups of five or ten for a fattening period of 20 weeks. From the age of two weeks, all calves were kept on concrete slatted floors and fed on a milk-replacer from buckets with nipple twice daily. After each feeding the calves were restrained for 15 minutes in the feeding stanchion (Graf et al., 1989). In the first experiment, each calf was provided with 200 g straw cobs per day and in the second experiment with 200 g straw per day. For each group size, space allotment was 1.0 and 1.5 m^2 per animal. In addition, one group of 10 calves was kept on deep litter with 1.5 m^2 per animal as a reference system.

Results and discussion

Locomotion depended on age or body weight of the animals and, to some degree, on available space and floor type (Figure 1). Several activities (such as bucking, running, and jumping) depended upon space allotment and were only possible in groups with 1.5 m^2 per animal. Most locomotion was observed in the reference group with deep litter. In the other groups, the concrete slatted floor was relatively non-slippery at the beginning of the fattening period. Later, the slatted floor became increasingly slippery through accumulation of manure.

In all groups, daily weight gain was not influenced by group size or space allotment. Contrary to common opinion, increased locomotion in the reference group did not reduce weight gain either. Although the straw intake from the deep litter was not recorded, its nutritive value can be ignored. Since weight gain was similar in calves fed with normal straw and calves fed with straw cobs, it is questionable whether straw cobs have a higher nutritive value than regular straw.

At the beginning of the studies, resting behaviour was similar in all groups. With increasing body weight, however, calves on slatted floor displayed increasing difficulty during lying down. After reaching a body weight of about 150 kg, calves on slatted floor showed a decreased frequency of lying bouts. Similar changes were observed in older heifers kept tether (Müller et al., 1989). These changes in resting behaviour were somewhat more pronounced in the crowded groups.

The main problem in keeping veal calves in groups is the occurrence of mutual sucking. Depending on the feeding system, young calves compensate their need for sucking by orienting the behaviour towards animate and inanimate objects (Sambraus, 1985). Preputial sucking is especially frequent, possibly because of the similarity between the mammary gland and the preputial region (Metz & Mekking, 1987). With increasing age, the conditioned sucking reflex becomes less marked (Figure 2).

In our studies, calves kept on straw displayed significantly less preputial sucking over the whole fattening period. For calves on slatted floor, preputial sucking was highest in crowded groups with 10 animals and tends to be less frequent in the uncrowded groups with 5 animals.

In addition to behavioral observations, the possible effect of stress from the housing conditions on the adrenocortical reaction was tested after ACTH(1–24) stimulation (Ladewig & Smidt, 1989). In the 6th and 22nd week, all animals were equipped with intravenous catheters and, on the following day, given 1.98 iu of ACTH per kg $BW^{0.75}$. Blood samples for

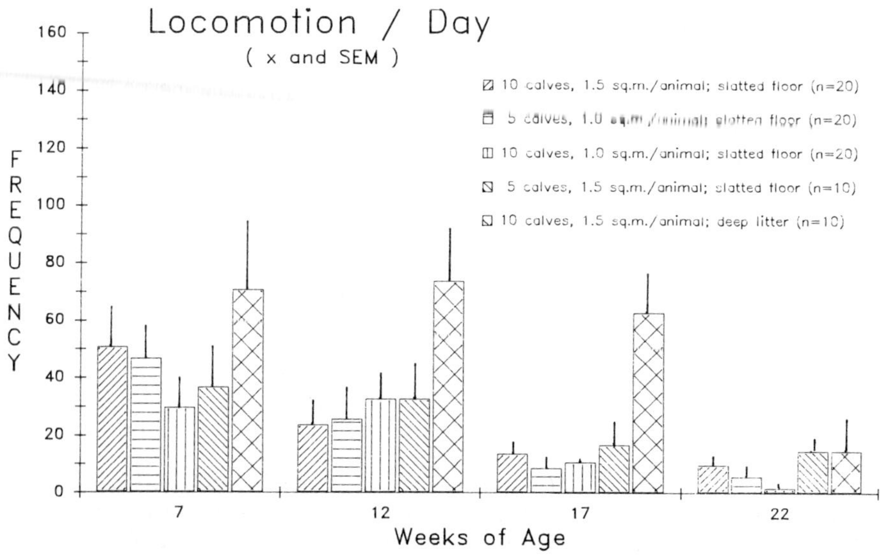

Figure 1. Frequency of locomotion (bucking, running and jumping) in 24 h of observation during the 7th, 12th, 17th and 22nd week of age of the first experiment.

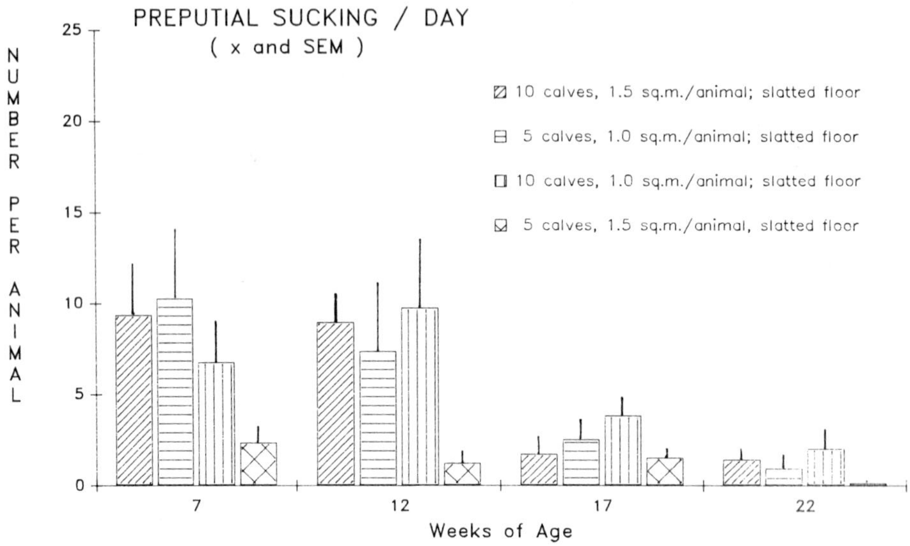

Figure 2. Frequency of preputial sucking in 24 h of observation period in the 7th, 12th, 17th and 22nd week of age in both experiments.

cortisol determination were collected at irregular intervals from 1 h before till 5 h after stimulation.

In the first test, calves kept on concrete slatted floor at a high stocking rate showed a tends to be higher cortisol response. In the second test, all groups except the reference group were at the same lower level. The difference in plasma cortisol between the group of ten calves with 1.0 m^2 per animal and the reference group was significant ($P \leq 0.01$).

This result indicates that supplementary roughage feeding is the decisive factor. Only small differences were found between groups with different group size and space allotment.

The addition of negative influences from the housing and feeding system caused stress reaction in behaviour and physiology. A similar interpretation has been reported in other studies (Dantzer et al., 1983; Philipps et al., 1982).

Conclusion

Our data indicate that the greatest difficulty in veal calf production is the unnatural feeding of young cattle. There were no conspicious differences in the measured parameters with group size and space allotment. Only animals of the reference group on deep litter, in which roughage was available ad libitum revealed a different behavioral and physiological reaction. For the high weight at which calves are slaughtered, feeding adult cattle only on milk replacer liquid food without additional roughage constitutes a welfare problem.

As long as the high demand for veal for export exists, slaughter weights are likely to increase further. The main task for producers will be a general change in feeding of veal calves until they are young bulls (e.g. Barley beef). In this process, it seems to be useful to separate the fattening periods to accommodate the changing needs of the calves.

References

Andreae, U., J. Unshelm & D. Smidt, 1980. Handhabung von Kälbern in Gruppenhaltung. KTBL-Schrift 254:89–96.

Dantzer, R., P. Mormède, R.M. Bluthé & J. Soissons, 1983. The effect of different housing conditions on behavioural and adrenocortical reactions in veal calves. Reprod. Nutr. Dévelop. 23:501–508.

de Wilt, J.G., 1985. Behaviour and welfare of veal calves in relation to husbandry systems. Thesis, Landbouwuniversiteit. IMAG, Wageningen, NL.

Friend, T.H., 1980. Stress: What is it and how can it be quantified? Int. J. Stud. Anim. Prob. 1:366–374.

Graf, B., N. Verhagen & H.H. Sambraus, 1989. Reduzierung des Ersatzsaugens bei künstlich aufgezogenen Kälbern durch Fixierung nach dem Tränken oder Verlängerung der Saugzeit. Züchtungskunde 61:384–400.

Ladewig, J. & D. Smidt, 1989. Behavior, episodic secretion of cortisol, and adrenocortical reactivity in bulls subjected to tethering. Horm. Behav. 23:344–360.

Metz, J.H.M. & P. Mekking, 1987. Reizqualitäten als Auslöser für Saugen bei Kälbern. KTBL-Schrift 319:228–236.

Müller, C., J. Ladewig, H.-H. Thielscher & D. Smidt, 1989. Behaviour and heart rate of heifers housed in tether stanchions without straw. Physiol. Behav. 46:751–754.

Phillips, W.A., R.P. Wettemann & F.P. Horn, 1982. Influence of preshipment management on the adrenal response of beef calves to ACTH before and after transit. J. Anim. Sci. 54:697–703.

Sambraus, H.H., 1985. Mouth-based anomalous syndromes. In: Fraser, A. Ethology of farm animals. Elsevier Sci. Publ., Amsterdam. Chapter 31. p. 391–422.

Effects of housing on responses of veal calves to handling and transport

H.R. Trunkfield,* D.M. Broom,* K. Maatje,** H.K. Wierenga,** E. Lambooy **
& J. Kooijman**
* *Dept. of Clinical Veterinary Medicine, University of Cambridge, Madingley Road, Cambridge, GB*
** *Research Institute of Animal Production, 'Schoonoord', P.O. Box 501, 3700 AM Zeist, NL*

Summary

The responses to handling and transport of male Dutch Friesian veal calves housed in different systems and reared with different feeding regimes have been investigated. Calves reared to six months of age in isolation in crates and bucket-fed were compared with groups of 15 or 30 calves fed by automatic teat-feeder. Behavioral observations and physiological welfare indicators show that the calves reared individually in crates were more stressed in response to handling and transport than the calves reared in groups. Other studies have been carried out with calves reared in groups of five on different feeding regimes. Basal cortisol levels were lower on a milk-and-straw diet than on milk only or milk and hay, but after handling and transport, plasma cortisol levels increased on each regime.
Descriptors: veal calves; housing; handling; transport; feeding regime; welfare.

Introduction

The life experience of an animal in its housing environment will influence its subsequent response to a stressful procedure. We have investigated the effects of different housing and feeding systems currently used for rearing veal calves on the subsequent responses of the calves to transport. Transport is a complex mixture of adverse stimuli including experience of a novel noisy environment, handling by people during loading and unloading, mixing of unfamiliar calves often with crowding, deprivation of food and water, exposure to fluctuations in temperature and humidity, and vehicle movement. Physiological responses of calves to handling and transport have been found (e.g. Kent & Ewbank, 1983; Fell & Shutt, 1986). Behavioral responses have also been reported in cattle (Kenny & Tarrant, 1987) and calves (Friend et al., 1981). Calf welfare during transit may vary from good to poor with conditions and methods of transport, and factors such as age, breed and sex of the calf and duration of journey (Trunkfield & Broom, 1990).

Study 1

Numerous investigations have shown that calves reared in isolation in crates are less active, less socially experienced and behave in more ways that are considered to be abnormal than group-reared calves because of the restrictions imposed by their environment. These findings suggest that calves reared in crates are more chronically stressed than calves reared in groups. Because of these differences, two trials were carried out to investigate whether there was a difference in response to the acute stress of transport between group-reared and isolation-reared calves.

Study 2

Veal calves have been shown to behave in various ways that are considered to be abnormal including stereotypies, such as tongue-rolling, which are a sign of poor welfare (Fraser & Broom, 1990). One of the functions of stereotypies is thought to be to reduce stress. Supplying roughage to veal calves appears to reduce both the development and occurrence of abnormal behaviour (Kooijman et al., this volume). A study was therefore carried out to determine whether calves reared on different levels of roughage responded differently to subsequent acute stress during transport.

Methods

Study 1

Both trials involved 6-month-old calves undergoing production monitoring (Maatje & Verhoeff, this volume). Twenty calves were housed individually in crates, 170 cm by 70 cm and were bucket-fed on a ration of 9 kg of milk replacer followed by 100 g straw pellets twice daily. Four groups of 15 calves (Trial 1) and one group of 30 (Group 3) plus two of 15 (Groups 1 and 2) in Trial 2, at a space allowance of about 1.5 square metres per animal were fed by automatic teat feeder on a ration of 18 kg of milk replacer and were provided with free access to straw pellets. Behavioral observations were made during one week before transport to determine any differences in behaviour in situ between the two housing systems. Observations of responses to loading were made on the day of transport. The duration of journey was one hour followed by about half an hour spent in lairage. The crate-reared calves were kept separately in two groups in the lorry and in one group in the lairage. Group-reared calves were all mixed on the lorry and during lairage.

Study 2

Calves were reared to eight months in groups of five on different feeding regimes: milk only (M), milk and hay ad libitum (MH), and milk and straw pellets ad libitum (MS). They were compared in their responses to handling and transport. The journey duration was one hour. All groups were mixed on the lorry and during 45 minutes spent in lairage.

In both studies, simultaneous saliva and blood samples were taken on the day before transport at the estimated time of loading, and after transport at slaughter. Saliva was analysed for cortisol by an ELISA technique (Cooper et al., 1989). Saliva contains the free fraction of plasma cortisol. Plasma was analysed for total cortisol by radioimmunoassay. Arginine vasopressin (AVP) also appears to be involved in the stress response in some species, showing an eventual depression in levels in response to adverse stimuli (Matthews, personal communication). AVP plasma levels were also determined by radioimmunoassay. The enzyme lactate dehydrogenase (LDH) consists of five isomers which can be analysed by electrophoresis. Isomer 5 appears to leak into the blood from skeletal muscle as a result of damage such as bruising, in response to stress hormones and during strenuous exercise (Jones, personal communication).

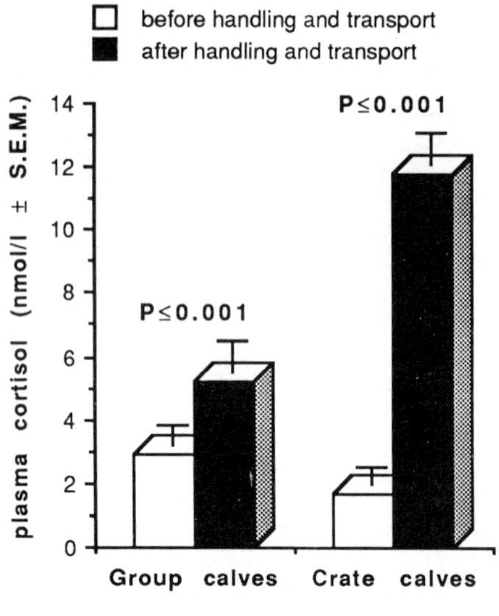

Figure 1. Mean cortisol levels before and after handling and transport (Study 1, Trial 1). Increase in crate-reared vs. increase in group-reared, $P \leq 0.001$.

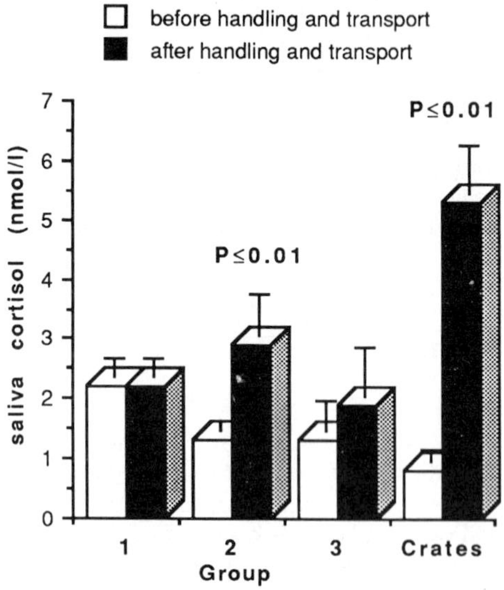

Figure 2. Mean saliva cortisol levels before and after handling and transport (Study 1, Trial 2). Increase in crate-reared vs. group-reared, $P \leq 0.001$.

Results

Study 1

Calves reared in crates lay down for a greater percentage of the time than did calves reared in group pens. Visual observations during loading showed that crate-reared calves had difficulty in boarding a walk-on ramp but were loaded much more quickly than group-reared calves. Group-reared calves had higher basal cortisol levels than calves reared in crates perhaps due to sampling stress. An increase in both plasma (Figure 1) and saliva (Figure 2) cortisol levels following handling and transport was found to be significantly greater in crate-reared calves than in group-reared calves. Arginine vasopressin showed a larger depression in crate-reared calves. Crate-reared calves only showed a significant increase in rectal temperature after handling and transport.

Study 2

The feeding regime resulting in the lowest saliva cortisol levels before transport was milk plus straw pellets, $P = 0.06$ (Figure 3). The increase in cortisol with handling and transport was similar for all calves on all three feeding regimes. LDH-5 levels were similar after transport, however the increase from basal levels was significant only for the calves fed on milk and hay (Figure 4) and approached significance for those calves fed on milk and straw pellets.

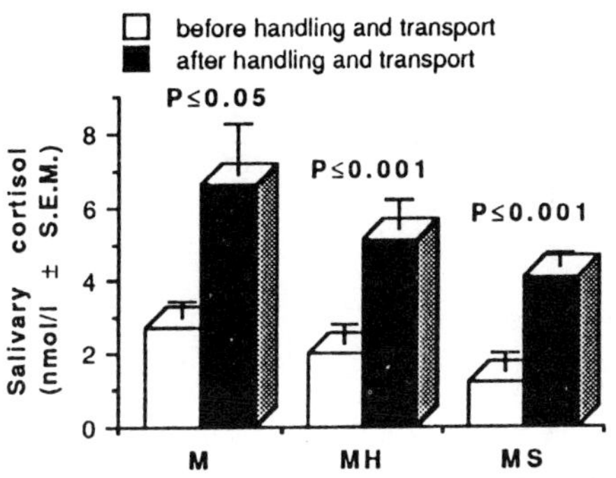

Figure 3. Mean saliva cortisol levels before and after handling and transport (Study 2). Basal levels lower in MS treatment, P = 0.06.

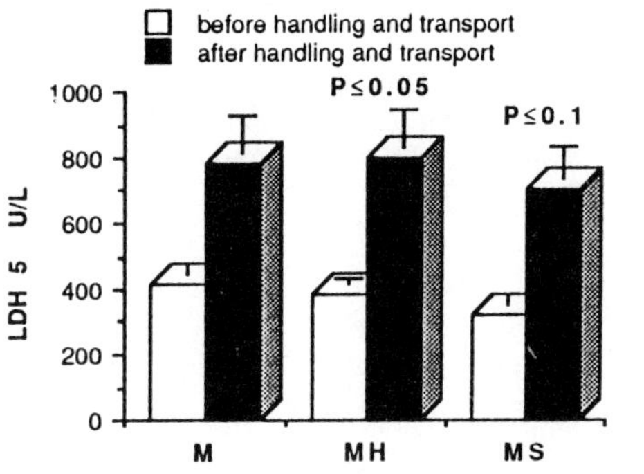

Figure 4. Mean plasma LDH-5 levels before and after handling and transport (Study 2).

Conclusions

The previous physical and social environment in which calves have been reared has been shown to affect their responses to transport. Calves reared in crates cope less well with handling and transport than those reared in groups, thus providing further evidence that such housing systems should be discontinued. The results of the second study show that differences in feeding regime, which have been shown to produce differences in behaviour in the rearing environment appear, to result in different physiological responses to transport stress.

References

Cooper, T.R., H.R. Trunkfield, A.J. Zanella & W.D. Booth, 1989. An enzymelinked immunosorbent assay for cortisol in the saliva of man and domestic farm animals. J. Endocrinol. 123:R13–16.

Fell, L.R. & D.A. Shutt, 1986. Adrenocortical responses of calves to transport stress as measured by salivary cortisol. Can. J. Anim. Sci. 66:637–641.

Fraser, A.F. & D.M. Broom, 1990. Farm Animal Behaviour and Welfare. London: Bailliere, Tindall.

Friend, T.H., M.R. Irwin, A.J. Sharp, B.H. Ashby, G.B. Thompson & W.A. Bailey, 1981. Behaviour and weight loss of feeder calves in a railcar modified for feeding and watering in transit. Int. J. Stud. Anim. Prob. 2:129–137.

Kenny, F.J. & P.V. Tarrant, 1987. The reaction of young bulls to short haul road transport. Appl. Anim. Behav. Sci. 17:209–227.

Kent, J.E., 1977. The effect of road transportation on young cattle. MSc. Thesis, University of Liverpool.

Kent, J.E. & R. Ewbank, 1983. The effect of road transportation on the blood constituents and behaviour of calves. I. Six months old. Br. Vet. J. 139:228–235.

Trunkfield, H.R. & D.M. Broom, 1990. The welfare of calves during handling and transport. Appl. Anim. Behav. Sci., in press.

Behaviour of veal calves in individual stalls and group pens

J.L. Albright *, D.K. Stouffer * & N.J. Kenyon **

* *Department of Animal Sciences, Purdue University, West Lafayette, Indiana, US*

** *School of Agricultural and Forest Sciences, University College of North Wales, Bangor, GB*

Summary

A field study was conducted in Northern Indiana during the winter of 1988-89. Eighty Holstein male calves; 40 in stalls (0.9 m^2) and 40 in 8 pens (1.3 m^2/calf) were observed at 20-minute intervals at 5, 10, and 17 weeks old. Stalls and pens with slatted floors were made of wood. Each calf was categorized for lying and standing and 1 of 40 postural and behavioral activities later grouped for analysis, showing % time spent in each.

We observed that sternal recumbency increased with age from 47 to 57.6% for stall calves and from 45.5 to 55% for penned calves. Other non-erect resting postures decreased from 21.6 to 12.2% for stall calves and 26.6 to 12.3% for penned. Time spent standing decreased with age from 30.5 to 29.5% for stall calves and increased from 27 to 32% for grouped penned calves. Behaviours compiled into categories show that time spent performing grooming, investigating, and idling activities decrease with age from 13.3 to 3.1% for stall calves and 12.5 to 2.3% for penned. These behaviours were partly replaced by head and body movements, increasing from 2.4 to 13.4% in stall calves and from 2.8 to 14.9% in penned calves. Oral activities were similar; stall and pen calves both showed 5% tongue playing at 17 weeks. Other behaviours noted were alike or low in frequency in all calves. For each calf an assessment of ease of movement was made the morning of trucking to the abattoir. All calves moved freely; stall calves attempting to leave the alley to get back in line whereas pen calves showed more sniffing and investigation behaviour of the alley floor.

Descriptors: behaviour; veal calves; individual housing; group housing; handling; movement; welfare.

Introduction

Most North American veal and dairy producers believe that individual penning is, on average, better for calf health and well-being than group housing systems. (Agriculture Canada, 1988; Albright, 1983; Baker et al., 1981; Curtis et al., 1988; Smith et al., 1990). There remains a need for research and development of housing systems that consider the well-being of veal calves, so that the problems associated with restricted movement and lack of outlet for natural behaviour can be saved (Agriculture Canada, 1988).

According to Webster (1989), the conventional system of confining veal calves in individual wooden crates and feeding them a liquid milk-replacer diet twice daily from buckets has received much criticism on welfare grounds. He claims the most obvious insult to the welfare of the calf is to confine it to a stall where, in the latter stages of growth, it cannot turn around, groom itself properly, adopt a normal sleeping position, or even stand up and lie down without difficulty. British Friesian veal calves in boxes (stalls) over the age of 10 weeks were unable to adopt normal sleeping positions with their heads tucked into their sides; and to overcome this, crates would need to be at least 91 cm wide (Webster et al., 1986).

In their studies van Putten & Elshof (1982) concluded that veal calves with a body weight of 200 kg need stalls at least 70 cm wide and 170 cm long. Difficulties arose with smaller stall sizes (55 cm), particularly in the last month of the growth period, when lying was restricted, and getting up and lying down required great effort, as did grooming. According to de Wilt

(1985, 1986), calves in individual (1.2 m^2) crates spent more time lying than those in the group (7.3 m^2) pens (74% against 68%) due to the greater variation of stimuli in the group pens and perhaps also because lying calves were occasionally mounted and trodden on by penmates. In the group pens calves spent 2% of lying time lying on their side with both hindlegs stretched; these postures only rarely occurred in the crates by the end of the 20 week fattening period.

In Canada new and renovated facilities for calves must comply with the current minimum recommended individual pen size: 65 cm × 165 cm (1.1 m^2) for calves weighing up to 202 kg and 1.4 m^2 for group pens. This recommendation may change as a result of further research (Agriculture Canada, 1988). Without giving dimensions, the American Veterinary Medical Association (1985) supports confinement rearing, provided proper husbandry practices are followed. Recently, they (Boyce, 1989) stated calves should be able to stretch, stand, and lie down completely, and that they may or may not be allowed to turn around, depending on management, housing, and health considerations. (The statement calls for adequate lighting and ventilation in veal calf barns, and for regular monitoring of haemoglobin values to prevent calves from becoming anaemic.) The American Veal Association revised guidelines call for individual stall sizes for calves weighing up to 182 kg to be 61 cm × 163 cm or 1.1 m^2 in all new or renovated facilities. For each 11.4 kg increase in final weight, stall width should be 2.54 cm and the length must be revised upward to 178 cm (Smith et al., 1990).

Objectives

This study is an attempt to assess the lying behaviour, postural and other behavioral development of Holstein veal calves kept under commercial conditions in individual stalls or group pens (5 calves/pen) at 5, 10, and 17 weeks of age.

A field study was conducted at a commercial research facility in Northern Indiana during the fall-winter of 1988-89. Eighty Holstein male calves were monitored, 40 were raised in individual stalls (61 cm × 152 cm or 0.9 m^2) and 40 in eight penned groups of five calves (1.3 m^2/ calf) with feeding partitions. Stalls and pens were constructed from oak and had slatted wooden floors. The barn was well ventilated by time-programmed and thermostatically controlled fans set at 21 °C. Continuous lighting was available and measured to be 10 lx.

Results and discussion

Calves were observed at 20-minute intervals for 72 recordings in 24-hour periods at 5 and 10 weeks and for 12 hours at 17 weeks old. In addition, ease of locomotion was assessed early on the morning of trucking to the slaughterhouse.

Initially, each calf was categorized for lying or standing; then one of 40 postural or behavioral activities were noted. Later, values were grouped into similar directed activities for ease of analysis, showing the percentage time spent in each. All results are shown as percentage frequency (Tables 1–6).

Other behaviours noted but not tabulated were: anticipating feeding (drinking), and restlessness after feeding (about 4%); ruminating; head on slats; head on another; feet stretched out; bawling; coughing; eliminative; and sexual behaviour. These occurred in similar or low frequencies (%).

For each calf going to market, we established a predetermined scoring system from 1-5 with 1 = calf could not walk; lifted onto the truck to 5 = walked or ran unassisted onto the truck. No shock prods (goads) were used. One calf rated a 3 and the rest rated 4. With experienced handlers, well-maintained stalls and pens, all calves moved freely; however, stall calves attempted to leave the alley to join other calves standing in stalls. Group penned calves exhibited more interest and directed their attention to more sniffing and investigation of the

Table 1. Resting positions in veal calves--stalls vs. pens (by weeks of age).

	Stalls (%)			Pens (%)		
	5	10	17	5	10	17
Erect posture, eyes open	45.0	45.5	52.6	43.0	44.5	51.0
Erect posture, eyes closed	2.0	6.0	5.0	2.5	7.0	4.0
Sternal recumbency	47.0	51.5	57.6	45.5	51.5	55.0
Head on neck	11.5	9.0	10.3	15.0	9.0	8.5
Head on flank/legs	9.1	5.0	1.5	10.2	6.0	3.5
Lying on side	1.0	1.0	0.4	1.4	1.0	0.3
Non-erect	21.6	15.0	12.2	26.6	16.0	12.3

Table 2. Grooming, investigative, idling behaviour--stalls and pens (by weeks of age).

	Stalls (%)			Pens (%)		
	5	10	17	5	10	17
Grooming self	2.3	1.0	1.0	1.5	1.6	1.0
Grooming others	2.5	1.0	0.5	2.5	0.7	0.4
Being groomed	1.5	0.5	0.3	1.4	0.1	0.0
Grooming	6.3	2.5	1.8	5.4	2.4	1.4
Investigative	3.5	0.0	0.0	5.0	0.0	0.0
Sniffing	2.0	1.0	0.0	0.8	1.0	0.0
Investigative	5.5	1.0	0.0	5.8	1.0	0.0
Idling	1.5	2.5	1.3	1.3	2.2	0.9

Table 3. Head and physical activities--stalls or pens (by weeks of age).

	Stalls (%)			Pens (%)		
	5	10	17	51	10	17
Nose up	0.5	2.5	2.3	0.71	3.0	2.5
Head out	0.6	7.0	6.3	0.71	6.0	7.5
Shaking head	0.2	1.5	2.4	0.41	2.0	2.4
Head	1.3	11.0	11.0	1.81	11.0	12.4
Rubbing	0.4	0.5	1.2	0.31	1.0	1.2
Butting	0.2	0.3	0.2	0.11	1.0	0.3
Playing	0.5	0.5	1.0	0.61	1.0	1.0
Running	0.0	0.0	0.0	0.01	0.0	0.0
Physical	1.1	1.3	2.4	1.01	3.0	2.5

Table 4. Oral activities in veal calves--stalls vs. pens (by weeks of age).

	Stalls (%)			Pens (%)		
	5	10	17	5	10	17
Chewing wood	3.2	2.0	2.3	4.0	1.4	3.5
Licking stall/pail	6.0	2.0	3.4	3.3	2.0	3.0
Tongue playing	2.0	4.0	5.0	1.4	3.5	5.0
Socking scrotum	0.0	0.0	0.0	0.0	0.1	0.0
Being sucked	0.0	0.0	0.0	0.0	0.1	0.0
Urine drinking	0.0	0.0	0.0	0.0	0.1	0.0
Oral	11.2	8.0	10.7	8.7	7.2	11.5

Table 5. Summary of postural and behavioural activities--stalls and pens (by weeks of age).

	Stalls (%)			Pens (%)		
	5	10	17	5	10	17
Lying	69.5	67.5	70.5	73.0	67.8	68.0
Sternal recumbency	47.0	51.5	57.6	45.5	51.5	55.0
Non-erect	21.6	15.0	12.2	26.6	16.0	12.3
Grooming	13.3	6.0	3.1	12.5	5.6	2.3
Head and physical	2.4	12.3	13.4	2.8	14.0	14.9
Oral	11.2	8.0	10.7	8.7	7.2	11.5

Table 6. Significance levels.

	Stalls vs. pens	With increasing age
Lying		
Sternal recumbency		**
Non-erect	* (wk 5)	*
Grooming		***
Head and physical		***
Oral		

* $P < 0.05$ ** $P < 0.01$ *** $P < 0.001$

alley floor. These observations support previous work (McFarlane et al., 1988; Sandhage et al., 1983) which showed no disorientation, impaired motor coordination or problems in calves being taken from their stalls. These impairments could be observed in calves reared on slippery floors or on worn-out slats, or if they were young, sick or anaemic.

Conclusions

1. Upright sternal posture development increased with age. This was accomplished by reduced 'sleep' and lie flat postures in both stall and group pens.
2. Grooming, investigative, and idling behaviours decreased with age and were replaced by head and body movements.
3. Frequencies of oral activity did not significantly differ between types of pen. Initially, licking stall/pail was observed; later this progressed to tongue playing (5%) . Tongue rolling has been suggested to be a 'Pavlovian type' response seen when calves think they are about to be fed (Phillips, 1989). It disappeared after feeding, when calves were resting. Webster et al. (1986) concluded that purposeless oral activity, e.g., licking or chewing the wall, or tongue rolling is not a clear indicator of stress in veal calves.
4. Behavioral trends were similar between individual stall and penned calves housed on wooden slats.
5. When removed from their stalls and group pens, all calves moved easily.

References

Agriculture Canada, 1988. Recommended code of practice for the care and handling of special fed veal calves. Publication 1821/E, Agriculture Canada, Ottawa, Ontario.

Albright, J.L., 1983. Status of animal welfare awareness of producers and direction of animal welfare research in the future. Journal of Dairy Science 66:2208–2220.

American Veterinary Medical Association, 1985. Guide for veal calf care and production. Schaumburg, Illinois.

Baker, F.H. et al., 1981. Scientific aspects of the welfare of food animals. Council for Agricultural Science and Technology Report No. 91, p. 37–38.

Boyce, J.R., 1989. Committee looks at veal calf welfare. Journal of American Veterinary Medical Association 194:1160.

Curtis, S.E., J.L. Albright, J.V. Craig, H.W. Gonyou, K.A. Houpt, J.J. McGlone & W.R. Stricklin, 1988. Guide for the care and use of agricultural animals in agricultural research and teaching. Agricultural Care Guide, One DuPont Circle, N.W. Suite 710, Washington, DC. p. 55–57.

de Wilt, J.G., 1985. Behaviour and welfare of veal calves in relation to husbandry systems. Thesis, Wageningen. Institute of Agricultural Engineering,Wageningen. 138 pp.

de Wilt, J.G., 1986. Behaviour of veal calves in individual crates and group pens. Netherlands Journal of Agricultural Science 34:233–235.

McFarlane, J.M., G.L. Morris, S.E. Curtis, J. Simon & J.J. McGlone, 1988. Some indicators of welfare of crated veal calves on three dietary iron regimens. J. Anim. Sci. 66:317–325.

Phillips, T., 1989. Answering allegations about veal. Large Animal Veterinarian 44(1):26–28.

Sandhage, M.E., J.L. Albright, L.M. Micinski VanDame & S.L. Walker, 1983. Veal calf behaviour in standard wooden crates. Journal of Dairy Science 66(Suppl.) 1):137.

Smith, C.K., J.L. Albright, M. Barkley, D. Briner, E. Fleck, J.C. Gawthrop, W.M. Grover, V.W. Hays, H. Polzin, D. Shields, T.B. Smith, R.L. Sommers & J.M. Wagner, 1990. American Veal Association, Inc. Guide For the Care and Production of Veal Calves. Third edition. Naperville, Illinois.

van Putten, G. & W.Y. Elshof, 1982. The lying behaviour of veal calves up to 220 kg: welfare and husbandry of calves. J.P. Signoret (Ed.) p. 83–104. Martinus Nijhoff, The Hague.

Webster, A.J.F., 1989. Improved husbandry systems for veal calves. In: C.J.C. Phillips (Ed.):New techniques in cattle production. Butterworths, London. p. 174–176.

Webster, A.J.F., C. Saville & D. Welchman, 1986. Improved husbandry systems for veal calves. Animal Health Trust, Farm Animal Care Trust. University of Bristol Press, Bristol.

Spatial requirements of individually housed veal calves of 175 to 300 kg

C.C. Ketelaar-de Lauwere & A.C. Smits
Institute of Agricultural Engineering, P.O. Box 43, 6700 AA Wageningen, NL

Summary

A study was carried out on individually housed veal calves of 175 to 300 kg. The first part of the study was to ascertain the exact amount of space used by calves for different postures and movements if there is no spatial restriction. In the second part the postures relevant in determining the crate sizes that are ethologically acceptable were ascertained.
Descriptors: veal calves; spatial requirements; behaviour; postures; movements; crate sizes.

Introduction

In the Netherlands, there is a trend towards keeping veal calves until they weigh 250 kg or more. For reasons of animal welfare, the government decided that the crate sizes for these heavy animals should be adjusted.

Research was carried out on the spatial requirements of individually housed veal calves of 175 to 300 kg. Ethological criteria were the most important in this research.

Material and methods

The study consisted of two parts. In the first, part the exact amount of space used by calves kept without spatial constraints in large crates of 150 cm × 250 cm was determined for different postures and movements. This part of the study yielded the crate sizes for the unrestricted adoption of these postures and the execution of these movements. In the second part, the effects of restrictions to the adoption of these postures and to the execution of these movements were determined in calves kept in crates of different sizes. The behaviour of the calves kept in the 150 cm × 250 cm crates served as a reference.

The research was carried out in three trials with fifty male calves (Dutch Friesian × Holstein Friesian) each. The animals were slaughtered when the average weight was about 300 kg. They were then 33 weeks old. Once a week, the calves were weighed, and once every fortnight, certain body sizes were recorded. The calves were individually housed in crates with a wooden slatted floor. For the experiment, 16 crates of 150 cm × 250 cm, 3 crates each of 100 cm × 200 cm and 70 cm × 170 cm and 4 crates each of 95 cm × 195 cm, 90 cm × 190 cm, 85 cm × 185 cm, 80 cm × 180 cm and 75 cm × 175 cm were used.

Behavioural observations were carried out once every fortnight from week 18 up to and including week 32. There were two types of observations:

1. Time-sampling observations: a video-recording of each calf was made at intervals of 7.5 minutes during a 24-hour period. These observations were used to measure the exact space used by the calves in the crates of 150 cm × 250 cm, and for determining which postures were adopted by the calves in the crates of different sizes.
   In the third trial, the effect of moving calves from a small crate (70 cm × 170 cm) to a large one (150 cm × 250 cm) was studied. In total, eight calves were moved (two calves per week in weeks 20, 24, 28 and 32). Time-sampling observations were carried out during the 24-hour period before, and on the first, third and fifth days after the transfer to the large

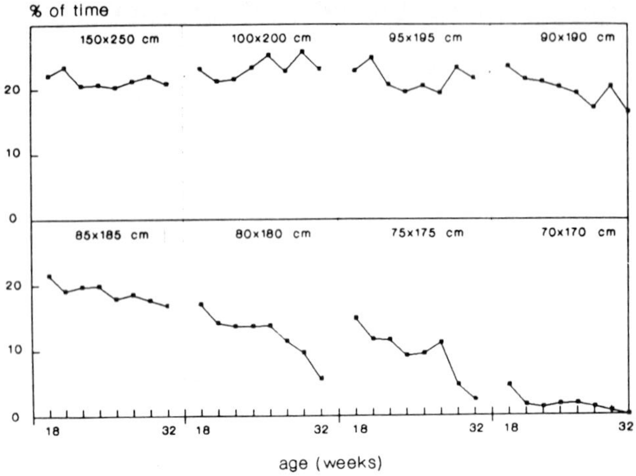

Figure 1. Lying on the brisket with front legs bent and one hind leg stretched away from the body in crates of different sizes (% of total observation time).

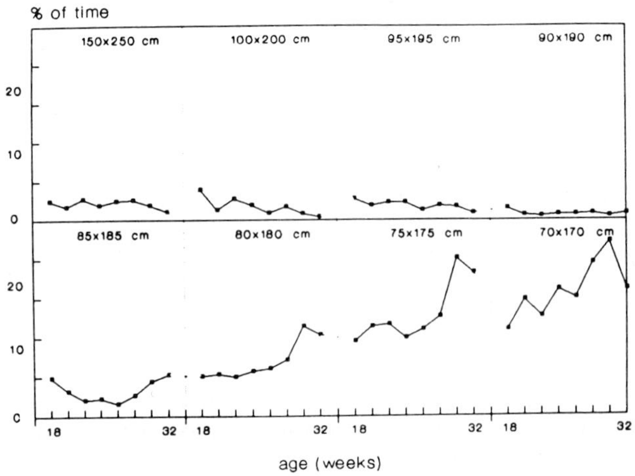

Figure 2. Lying on the brisket with front legs bent and one hind leg stretched alongside the body in crates of different sizes (% of total observation time).

crates. After one week, the calves were moved back to the small crates and their behaviour on the following day was observed.

2. Continuous observations: a video-recording was made of one calf during an 8-hour period. The way of getting up and lying down was studied. In total, six calves were studied in this way for each crate size.

Results

From the exact space used when adopting certain body postures and when performing certain movements, it appeared that certain body postures put requirements mainly on the crate width and others on the crate length. The crate width was needed for lying with one or two hind legs stretched away from the body. The crate length depended on the space needed to lie with one or two front legs stretched, and on the space needed for standing with the head stretched forward. A crate size could be determined for all these body postures (Ketelaar-de Lauwere & Smits, 1989). Which body posture was chosen as a criterion for the crate width and for the crate length depended on the results of the behavioral observations.

The behavioral studies on the calves in crates of different sizes showed that lying postures that required little space were seen less often in large crates than in small ones. Conversely lying postures that required much space were seen less often in small crates. These differences became more marked as the calves grew older and their weights increased. In large crates, no change was seen in the frequencies of the lying postures that required little or much space as the calves grew older. In small crates, the frequency of lying postures that required little space increased as the calves grew older, and the frequency of lying postures that required much space decreased. Furthermore, those lying postures requiring little space, which were rarely seen in the large crates, increased in frequency in the small crates as the calves grew older. In these lying postures, the calves lay with one hind leg stretched alongside the body. Figure 1 shows the time spent lying with one hind leg stretched away from the body, a posture requiring much space, and Figure 2 shows the time spent lying with one hind leg stretched alongside the body.

Calves moved from a small crate to a large one showed a strong increase in the frequency of lying with one hind leg stretched away from the body, and a strong decrease, usually even a cessation, of lying with one hind leg stretched alongside the body.

After moving the calf back to the small crate, the reverse happened: the frequency of the first mentioned lying posture decreased strongly and that of the second increased strongly (Table 1).

The continuous observations indicated that, when getting up or lying down, the calves in the small crates more often interrupted the movement, more often bent their heads sideways or up during the movement and more often hit their heads against the partitions of the crates during the movement. Moreover they needed more time to get up and lie down. These calves tended to get up on their forelegs first (like horses). The calves in the large crates slipped more often on the floor during getting up and lying down.

Discussion and conclusions

The results show that the body postures calves adopt are dependent on the size of the crates. This becomes more marked as the calves get older and heavier because the crate size becomes more restrictive. Most striking here are the lying postures requiring little space, which are not seen in the large crates. It seems that these 'special' lying postures are a substitute for the lying postures requiring much space. This contention is supported by the behaviour of the calves moved from a small crate to a large one and back. When in large crates, these calves adopted lying postures that required much space as frequently as other calves that had been accommodated in the large crates all the time. Moreover the 'special' lying postures almost disappeared in the large crates. It seems that the lying postures with one hind leg stretched alongside the body are not 'natural' for the calves because they do not lie in these postures when unconstrained. De Wilt (1985) found that individually housed veal calves at eight weeks of age could not easily lie with both hind legs stretched away from the body in crates 70 cm wide. From

Table 1. Average frequency of lying with one hind leg stretched away from the body (Lying Posture 1) and with one hind leg stretched alongside the body (Lying Posture 2) before and after transfer from a small crate to a large one ('before' and 'after', respectively) and after moving back to a small crate ('return') (% of total observation time).

Week	Calf	Lying Posture 1 before	after	return	Lying Posture 2 before	after	return
20	1	2.2	43.9	29.7	29.1	0.2	2.2
	2	3.4	18.6	9.3	15.1	0.3	1.6
24	3	0.0	32.2	3.3	23.1	0.3	34.1
	4	0.5	21.2	0.5	22.0	0.0	36.2
28	5	1.1	28.7	0.0	39.2	0.3	43.8
	6	1.7	24.1	0.0	21.0	1.2	32.0
32	7	0.0	34.5	0.0	30.9	0.7	29.0
	8	0.0	19.5	0.0	48.6	1.9	49.3

Table 2. Recommended crate sizes for veal calves of 175 to 300 kg.

Age (weeks)	Mean weight (kg)	Crate width (cm)	Crate length (cm)
18	175.2	80	170
20	189.3	85	170
22	206.1	85	170
24	227.2	90	175
26	246.4	90	180
28	262.6	90	185
30	283.0	90	185
32	301.0	95	190

16 weeks of age, lying with one hind leg stretched alongside the body was still the only way for the calves to stretch their hind legs.

The results of the continuous observations indicated that calves in crates often had difficulty getting up or lying down. These difficulties in large crates seemed to be related to the slipperiness of the floor. The difficulties shown by calves in small crates were most obvious and seemed to originate from the restricted space. According to Bogner (1981), it is necessary to keep calves on a non-slip floor. Therefore, wooden slats would be unacceptable. Graf et al. (1976) observed that a slippery floor could cause severe difficulties for calves when getting up or lying down.

The behavioral observations revealed which postures were important in determining acceptable crate sizes. The ultimate criterion chosen for the crate width was the space calves need for lying on the brisket with the front legs bent and one hind leg stretched away from the body. This lying posture was seen very often (32.8% of total lying time; Ketelaar-de Lauwere & Smits, 1989) and it was replaced by an 'unnatural' lying posture when there was no longer enough space. Moreover, thermoregulation was eased when a calf was able to stretch at least one hind leg away from the body (van Putten & Elshof, 1975). The criterion chosen for recommending the crate length was the space calves need when they stand in the crate with a straight back and the head stretched forward. It was the longest posture for which the exact amount of space could be measured (Ketelaar-de Lauwere & Smits, 1989). This made it possible to guarantee that there would be some extra space at the front of the crates, so that the calves could get up and lie down (Kaemmer & Schnitzer, 1975). Table 2 gives the crate sizes.

References

Bogner, H., 1981. Einige Mindestforderungen für die Haltung und Mast von Kälbern, wie sie sich aus tierschutzbezogenen Untersuchungen ableiten lassen. Tierzüchter 9:376–378.

de Wilt, J.G., 1985. Behaviour and welfare of veal calves in relation to husbandry systems. Thesis, Agricultural University, Wageningen.

Graf, B., R. Wegman & M. Rist, 1976. Das Verhalten von Mastkälbern bei verschiedenen Haltungsformen. Schweizerische Landwirtschaftliche Monatshefte 54:333–355.

Kaemmer, P. & U. Schnitzer, 1975. Die Stallbeurteilung am Beispiel des Ausruheverhaltens von Milchkühen. In: Die Beurteilung von Liegeboxen. KTBL-Schrift, Darmstadt.

Ketelaar-de Lauwere, C.C. & A.C. Smits, 1989. Onderzoek naar de uit ethologisch oogpunt minimaal gewenste boxmaten voor vleeskalveren met een gewicht van 175 tot 300 kg. De hoeveelheid ingenomen ruimte en gedragsaspecten. IMAG-rapport 110, Wageningen.

van Putten, G. & W.J. Elshof, 1975. Platzanspruch eines Mastkalbes. In: Haltungssysteme und Verhaltensanpassung. KTBL-Schrift, Darmstadt. p. 56–63.

Development of abnormal oral behaviour in group-housed veal calves: effects of roughage supply

J. Kooijman *, H.K. Wierenga * & P.R. Wiepkema **
* *Research Institute for Animal Production 'Schoonoord', P.O. Box 501, 3700 AM Zeist, NL*
** *Department of Animal Husbandry, Ethology Section, Agricultural University Wageningen, P.O. Box 338, 6700 AM Wageningen, NL*

Summary

During 2 trials, the development of two abnormal patterns of oral behaviour in group-housed veal calves was studied. The effect of roughage intake on these abnormal behaviours was investigated.

Based on the results of these trials it was concluded that feeding roughage to veal calves in addition to milk replacer reduces the development and occurrence of abnormal oral behaviour. Both the quantity and the structure of the roughage appeared to determine the extent of its effect on the calves' behaviour.

Descriptors: veal calves; abnormal oral behaviour; roughage supply.

Introduction

In today's veal calf production, the animals frequently show behaviour considered to be abnormal. As a consequence of the milk replacer feeding method, the calves develop redirected sucking behaviour like preputial sucking or sucking on other parts of penmates' bodies (Mees & Metz, 1984; de Wilt, 1985). Besides this redirected sucking behaviour other abnormal oral behaviour patterns, such as excessive sucking/biting/licking (manipulation) on inanimate objects and tongue-playing, are performed (Sambraus et al., 1984; de Wilt, 1985). Although many researchers have speculated about the causes and function of this abnormal behaviour (Sambraus et al., 1984; Wiepkema, 1987), an acceptable explanation is still lacking. However, the occurrence of these behaviour patterns is currently considered to be indicative of impoverished well-being (Wiepkema et al., 1983).

For the production of white veal, calves either receive no roughage at all or very little throughout the fattening period. Consequently, the calves are not (or hardly) able to chew on roughage and to ruminate (van Putten, 1982). As calves would under (semi-)natural conditions develop into ruminating animals, the absence of roughage may contribute to the development of abnormal oral behaviour. Indeed, several authors (van Putten & Elshof, 1978; Papendieck, 1979; Sambraus et al., 1984) have reported a relationship between roughage deprivation and the occurrence of abnormal oral behaviour. Especially the manipulation of inanimate objects and tongue-playing have been mentioned in this context.

The aim of this research was to investigate the development of abnormal oral behaviour patterns of milk-fed veal calves. Furthermore, the effect of roughage supply on the occurrence of these behaviours was studied.

Material and methods

For this investigation, a barn with 18 pens for 5 calves each was available. The pens had wooden slatted floors and self-closing feeding gates. There was a wooden trough for roughage in front of each pen.

The calves were fed milk replacer in buckets twice daily (7:00-7:30 h and 16:00-16:30 h) according to a commercial scheme. The barn was illuminated from 6:00 h until 20:30 h.

In 2 trials, of 31 and 26 weeks duration, respectively, the following treatments were compared:

A. milk replacer only (control);
B. milk replacer + ad libitum straw pellets;
C. milk replacer + ad libitum hay;
D. milk replacer + 200 g straw pellets per calf per day;
E. milk replacer + 1 kg maize silage per calf per day.

Three pens were used for each treatment. In the first trial, only Treatments A, B and C were tested; in the second trial, all 5 treatments were tested. In the first trial, only 80% of the milk replacer ration was given with the hay diet compared to the other two diets.

The calves' age averaged one week on arrival at the experimental farm. After one week of habituation, observations of behaviour were carried out for 30-min periods starting after milk intake. The behaviour of the 5 calves in a pen was recorded every 10 seconds. The following behaviour patterns were recorded:

Manipulating objects	The calf bit, sucked, licked or in some other way made oral contact with inanimate objects.
Tongue-playing	The calf made strange, swinging, corkscrew-like movements with his tongue in or outside his mouth.

The fattening period of the first and second trial was divided into 8 and 7 periods of 24 days, respectively. In each 24-day period every pen was observed 4 times. In the second trial no observations were carried out during the third and fifth period.

Observations during 24 hours were carried out to determine the time spent on roughage intake. The behaviour of the 5 calves in a pen was recorded every 5 minutes. For labour reasons, 24-hour observations were carried out in the first trial only during periods 6, 7 and 8; and in the second trial only during periods 1, 2 and 4. During each of these 24-day periods each pen was observed once. The 24-hour observations were confined to Treatments B and C.

Assuming that one interval score represents an average of 10 seconds, the measured behaviour scores of the half-hour observations were calculated as percentages of the observation time spent on different behaviour patterns. An analysis of variance was carried out within each period on the means per pen for each behaviour pattern. For the 24-hour observations, the mean percentages of intervals in which roughage intake was recored were compared by means of the Student's *t* test for each period.

Results

The occurrence of the manipulation of inanimate objects and tongue-playing behaviour during the successive 24-day periods of the 2 trials is shown in Figure 1. During the first trial the highest incidence of both behaviour patterns occurred when the calves were fed milk replacer only (control treatment). In the same treatment the occurrence of these abnormal oral behaviour patterns increased throughout the fattening period. The lowest incidence was recorded when the calves received additional hay ad libitum.

The highest incidence of manipulating objects and tongue-playing during the second trial occurred in the control treatment and when the calves received just 200 g of straw pellets in addition to the milk replacer. In both treatments, the incidence increased throughout the

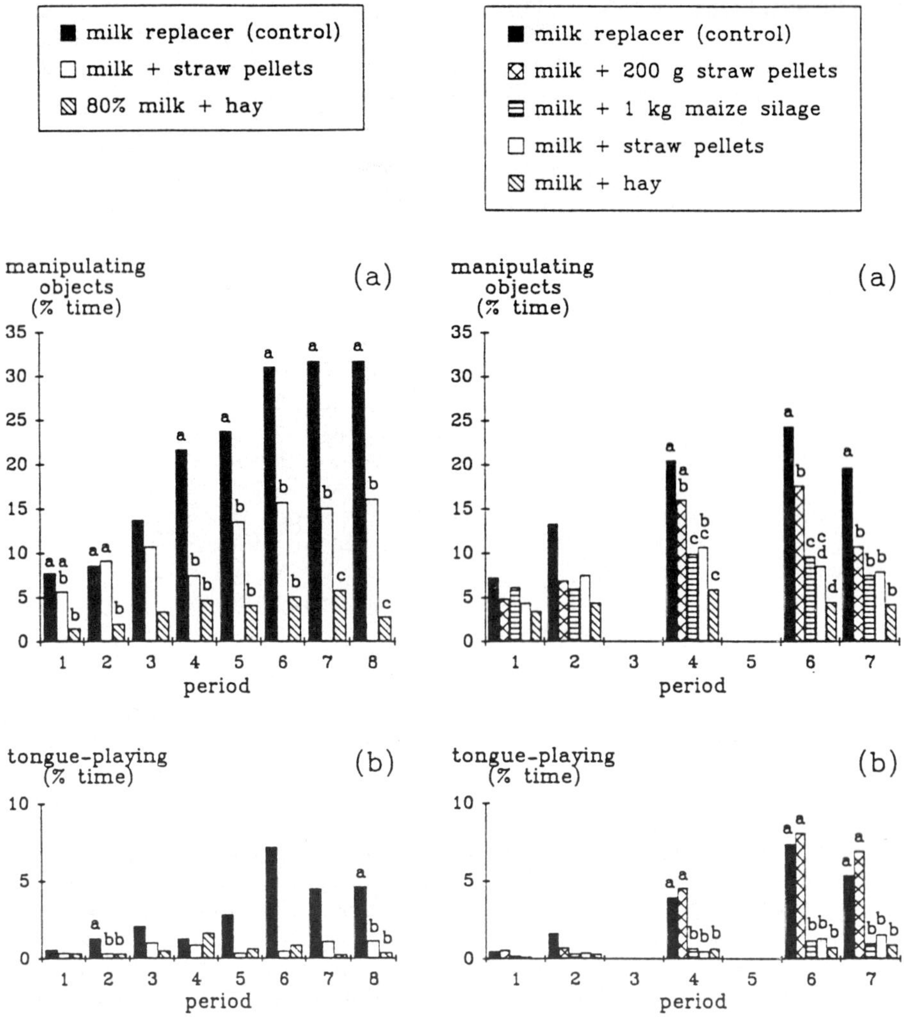

Figure 1. Occurrence of (a) manipulating objects and (b) tongue-playing during the first (left part) and second (right part) trial, as percentages of the observation time (= 30 mins). Each period lasted 24 days. Different symbols (a, b, c, d) indicate significant differences ($P \leq 0.05$) between treatments within each period.

fattening period. As in the first trial, the lowest incidence of both abnormal oral behaviour patterns was recorded when the calves were fed hay ad libitum in addition to the milk replacer.

Time per 24 hours spent on roughage intake tended to increase throughout the fattening period (Figure 2). Furthermore, the hay-fed calves spent significantly more time on roughage intake than the ad libitum straw-pellets-fed animals in each 24-day period. In the first trial, straw pellets and hay intake increased from practically zero in week 1 to more than 2.5 kg per animal per day in week 31. In the second trial, straw pellets and hay intake increased from zero in week 1 to 1.1 and 1.6 kg per calf per day in week 26, respectively.

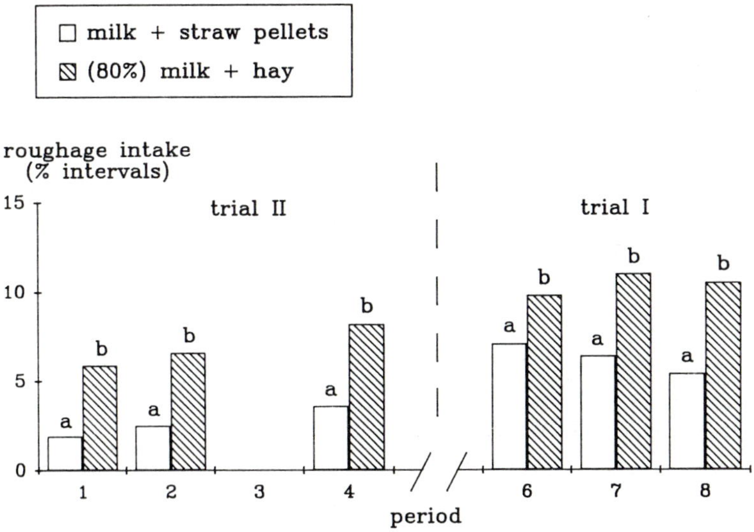

Figure 2. Time spent on roughage intake during 24 hours, as percentages of the total number of observed intervals. Further explanation: see Figure 1 and text.

Discussion

The results show that deprivation of roughage results in an increased display of excessive manipulation of inanimate objects and tongue-playing behaviour by the calves. Other authors (e.g. van Putten & Elshof, 1978; Papendieck, 1979; Sambraus et al., 1984) have also reported a correlation between a lack of structured feedstuffs and the display of these abnormal oral behaviour patterns. Apparently the animals have a strong (innate) need for roughage which increases throughout the fattening period. This increased need is inferred from both the increasing amount of roughage eaten and the increasing time spent on roughage intake throughout the fattening period. Not being able to fulfil this need results in a higher tendency to develop and perform abnormal oral behaviour. Individual factors may determine to what extent the calves express these abnormal behaviour patterns.

An important aspect is the amount of roughage supplied; supplying just 200 g of straw pellets did not appear to satisfy the calves' need. Satisfaction of the need for roughage also depends on its structure. The results indicate that when high-quality roughage such as hay (containing a large amount of fibre) was supplied, less manipulating objects and tongue-playing behaviour was recorded than when low-quality roughage (straw pellets or maize silage) was supplied. The jaw and tongue activity required for both the intake and the rumination of roughage may be important factors in explaining the influence of feeding roughage.

Assuming abnormal behaviour is indicative of impoverished well-being (Wiepkema et al., 1983), it can be concluded that roughage supply results in a significant improvement of veal calves' well-being. The amount of roughage supplied is crucial in this respect. Supplying 200 g of straw pellets per calf per day was not sufficient to meet the calves' needs in our trials.

Acknowledgements

This research was partly funded by the Fonds Welzijn Landbouwhuisdieren (Farm Animals Welfare Fund).

References

de Wilt, J.G., 1985. Behaviour and welfare of veal calves in relation to husbandry systems. Thesis, Wageningen Agricultural University. IMAG, Wageningen. 137 pp.

Mees, A.M.F. & J.H.M. Metz, 1984. Saugverhalten von Kälbern - Bedürfnis und Befriedigung bei verschiedenen Tränkesystemen. KTBL-Schrift 299:82–91.

Papendieck, T., 1979. Ethologische Reaktionen von Kälbern auf Rohfasermangel bei Haltung auf Lattenrosten. Landbauforschung Völkenrode, Sonderheft 48:62–66.

Sambraus, H.H., M. Kirchner & B. Graf, 1984. Verhaltensstörungen bei intensiv gehaltenen Mastbullen. Deutsche tierärztliche Wochenschrift 91(2):55–60.

van Putten, G. & W.J. Elshof, 1978. Zusatzfütterung von Stroh an Mastkälber. KTBL-Schrift 240:210–219.

van Putten, G., 1982. Welfare in veal calf units. Veterinary Record 111:437–440.

Wiepkema, P.R., D.M. Broom, I.J.H. Duncan & G. van Putten, 1983. Abnormal behaviour in farm animals. Commission of the European Communities, Brussels.

Wiepkema, P.R., 1987. Behavioural aspects of stress. In: P.R. Wiepkema & P.W.M. van Adrichem (Ed.): Biology of stress in farm animals: an integrative approach. Martinus Nijhoff, Dordrecht. p. 113–133.

3. Housing and management

Group housing of veal calves

A.C. Smits * & J.G. de Wilt **
* *Institute of Agricultural Engineering, Wageningen, NL*
** *Directorate for Animal production and Dairying, The Hague, NL*

Summary

Stimulated by public criticism of the welfare of veal calves in individual crates, the feasibility of housing veal calves in small groups was investigated. The calves were kept in groups of five, fed from open buckets, provided with some roughage and tethered during the first 6–10 weeks after arrival. This group-housing system was compared with individual housing in crates, in terms of behaviour, health, production, labour requirements and economics. The conclusion from this research into the prospects for group housing of veal calves is that group housing is a viable alternative to crates.
Descriptors: veal calves; housing system; behaviour; performance; economics.

Introduction

Individual housing of veal calves has been practised commercially in the Netherlands and other EC countries for almost fourty years with satisfactory economic results. Since the 1960s, public concern about the welfare of farm animals has led to serious criticism of the housing of veal calves in individual crates. This criticism was focused on the restriction of movement, lying postures and social contact. In order to improve the welfare of veal calves, the feasibility of group housing was studied in a joint project involving the Institute of Agricultural Engineering (IMAG), the Wageningen Agricultural University and Denkavit Nederland B.V.

In the first prototype of the group housing, the calves were loose housed in groups of five and bucket-fed twice daily. The concrete slatted floor of the group pens (3.80 m wide, 2.20 m deep) was covered with barley straw, which was also provided ad libitum in baskets. The growth, feed conversion, slaughter quality and health of the calves in this housing system were in general satisfactory and similar to those of calves in individual crates (0.65 m × 1.65 m), provided with 100 g of barley straw daily. However those calves kept in groups that showed such behaviour as preputial sucking and urine drinking produced poor growth and performance. Total labour requirement for the provision of straw and the removal of dung was about twice as high in the group pens as in the individual crates.

These drawbacks of the loose housing of veal calves on straw led to two important modifications. First, the calves were tethered to the feeding gate by neck chains during the first 6 weeks, to prevent preputial sucking. Secondly, the straw was omitted, to reduce the high labour requirements. The modified group-housing system was compared with individual housing in terms of animal welfare, production, health, labour and economic aspects.

This paper discusses the results of this comparison, which was done on experimental and ordinary farms.

Housing systems

The group pens of 2.50 m deep and 3.00 m wide were surrounded by galvanized partitions (1.10 m high). The floor consisted of wooden slats (70 mm wide and 30 mm apart) with wooden or aluminium laths, in the transverse direction, about 0.70 m apart. The calves were tethered to the feeding gate by neck chains (length 0.35 m) during the first 6 weeks after arrival. They

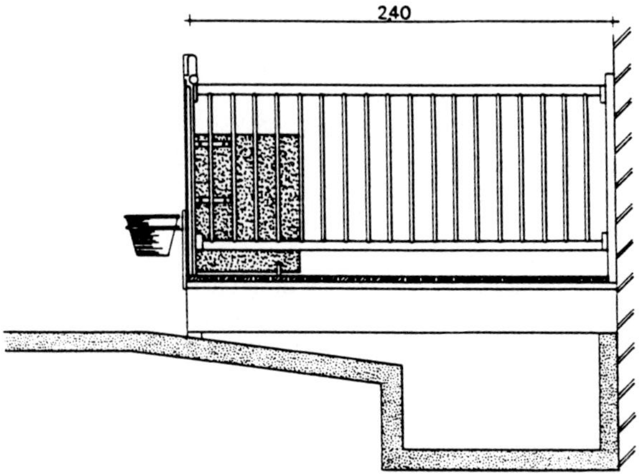

Figure 1. Side-view of a group pen for five calves with a slatted floor, in metres.

were separated by wooden partitions (0.60 m long, 0.80 m high), which were removed when the calves were untethered. A trough was positioned in the door of each group pen and about 100 g of straw cobs were supplied daily (Figure 1).

The crates (about 1.65 m long and 0.60–0.70 m wide) consisted of two side-walls (1.10 m high), a partly rippled front and back, and a slatted floor (slats 70 mm wide and 30 mm apart).

Almost all the calves were male and about one week old on arrival. Most calves were black and white, a few were red and white.

Behaviour

In one unit of the experimental farm, 7 group pens and 12 individual crates of 0.70 m wide were installed.

The lying postures of each calf were recorded once every 10 minutes during one 24-hour period in weeks 8, 12, 16 and 20 after arrival. This procedure was repeated in 4 consecutive fattening periods. This way, various lying postures and activities of calves in group pens and in individual crates could be compared (de Wilt, 1985).

The results shown in Figure 2 focus on the position of the head during lying. These results are part of a more comprehensive study of various lying postures and activities of calves in group pens and individual crates (de Wilt, 1985).

The incidence of lying with the head turned back on the shoulder or belly decreased more rapidly over time in the crates than in the group pens (Figure 2A). The occurrence of lying with the head reclined forwards on the floor increased in the crates but not in the group pens (Figure 2B). Supporting the head otherwise, for instance against the side-walls in the crates or on congeners (or the fence) in the group pens was least common in the crates (Figure 2C). Lying with the head upright, which is complementary to lying with the head supported was more frequent in the group pens because the incidence of this activity in the crates increased over time (Figure 2D).

These results indicate that lying with the head turned backwards is thwarted by the side-walls of the crates. The head is held upright more often. Apart from possible discomfort, the impairment of lying with the head supported may affect sleep (Ruckebusch & Bell, 1970).

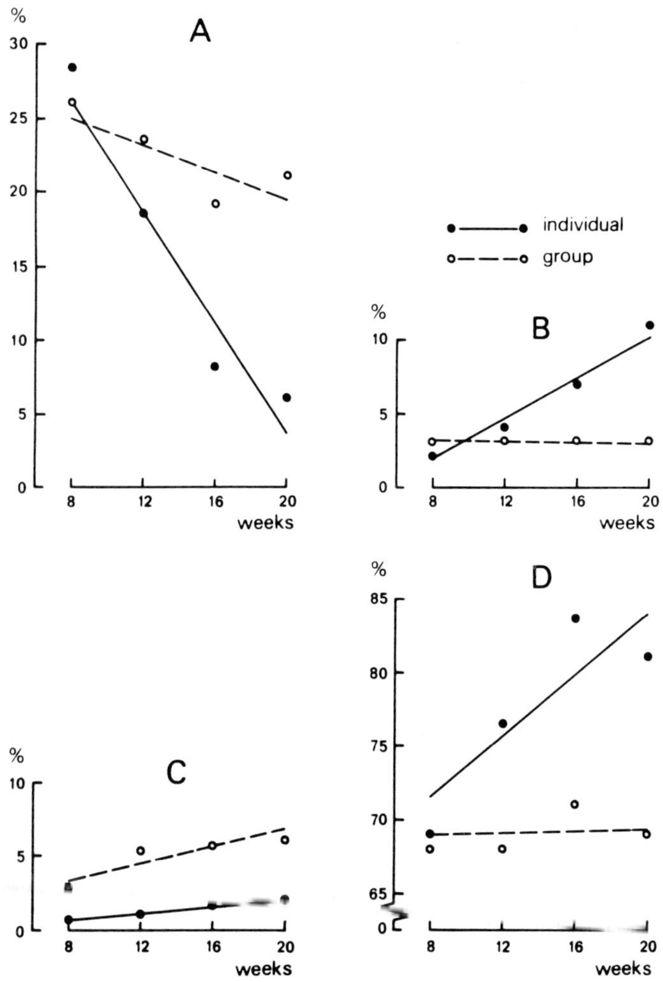

Figure 2. Lying with the head turned backwards (A), reclined forwards (B), on congeners (C) or upright (D) in individual crates and group pens with slatted floors from week 8 to 20. The lines show the overall average and trend of these lying postures in both housing systems, while dots indicate averages per week. The values indicated are percentages of total lying.

From the observations on lying behaviour, comfort behaviour, social activities, exploration and stereotyped behaviour, it was concluded that compared with individual housing in crates, this system of group housing is a major step in the direction of improved calf welfare (de Wilt, 1985).

Table 1. Treatments of enteric and respiratory disorders and other illnesses in individual crates and group pens, as a percentage of total number of calves.

	Housing system	
	individual	group
Number of calves	6809	8953
Enteric disorders (%)	14.2	18.8
Respiratory disorders (%)	39.7	53.2
Other (%)	9.4	15.0

Table 2. Dead calves and culls as a percentage of total number of calves, in individual crates and group pens.

	Housing system	
	individual	group
Number of calves	6809	8953
Deaths (%)	3.01	3.95
Culls (%)	1.42	2.18

Table 3. Production results in individual crates and group pens of slaughtered calves.

	Housing system	
	individual	group
Number of calves	6587	8477
Initials weight (kg)	41.4	41.5
Liveweight at slaughter (kg)	221.9	227.0
Length of fattening period (days)	171	176
Growth (g/d)	1058	1053
Feed conversion	1.843	1.826

Table 4. Slaughter quality of calves in individual crates and group pens, rated from 0 (very bad) to 10 (excellent).

	Housing system	
	individual	group
Number of calves	6240	7963
Meat colour	6.8	6.7
Fat colour	7.3	7.3
Meatiness	6.9	6.9
Fat cover	7.0	7.2

Health and production

The health and production of calves in group pens and in individual crates was compared on twelve farms during five consecutive fattening periods. (Smits & Ham, 1988). Each farm had both housing systems; the feeding schedules were identical.

In order to prevent preputial sucking and to make health management easier, the group-housed calves were tethered with neck chains between short partitions during the first 8–10 weeks after arrival.

The farmers recorded any illnesses, health treatments, culls and deaths. Slaughter quality (meat and fat colour, meatiness and fat cover) was established visually on a scale from 0 (very bad) to 10 (excellent).

Treatment of enteric and respiratory disorders and other illnesses, including trauma and navel infections, was more frequent in the group pens than in the individual crates (Table 1). The incidence of respiratory treatments was relatively high in both housing systems. Preputial sucking was seldom observed and was not regarded as a problem. The higher frequency of diseases in the group pen was also reflected in the mortality and culling rate (Table 2). After eliminating the dead and culled animals from the calculation of the production results, daily growth and feed conversion were similar in the two housing systems (Table 3). The liveweight of calves at the slaughterhouse was somewhat higher in the group pens than in the crates, because the group-housed calves had a slightly longer fattening period. The slaughter quality of the calves in the two housing systems was similar, except for a small difference in fat cover in favour of the group-housed calves (Table 4), perhaps because group-housed calves were more active.

There were remarkable differences between individual farms in the comparison of both housing systems; on some farms mortality, culling rate and production results were better in the group-housing system.

The early detection of illnesses and the adequate treatment of individual calves in group housing require extra attention from the farmers. In the group-housing system, there were no meaningful differences between the first and the later fattening periods in terms of health and production of the calves.

Labour requirements and economics

The activities of the farmers, such as stalling, feeding, health control and medical treatment, tethering and untethering the calves, and cleaning the pens, were recorded in both housing systems. There were differences between housing systems in the time required for these various activities but the total labour requirements were similar in the group and the individual housing (Giesen, 1984).

The economic consequences in terms of labour income were calculated by Kasper et al. (1989) for individual crates of different widths and group pens with different types of feeding gates. The calculations were also made for a combined system, in which the calves were first housed in crates and then transferred to group pens at 10–12 weeks after arrival (Figure 3).

Since it is much more common to convert existing stalls than to build completely new stalls, the calculations refer to the consequences of conversion. If crate size increases, the number of calves per stall section will decrease and this leads to a reduction in return from labour. The differences in return from labour between the three types of group housing represent variations in housing costs, which are mainly the result of the type of feeding gates.

It can be established that group housing is an economically acceptable alternative to individual housing in crates of 0.70 m wide or more. Furthermore, the combination of at first individual housing and later group housing may be more profitable than individual or group housing alone.

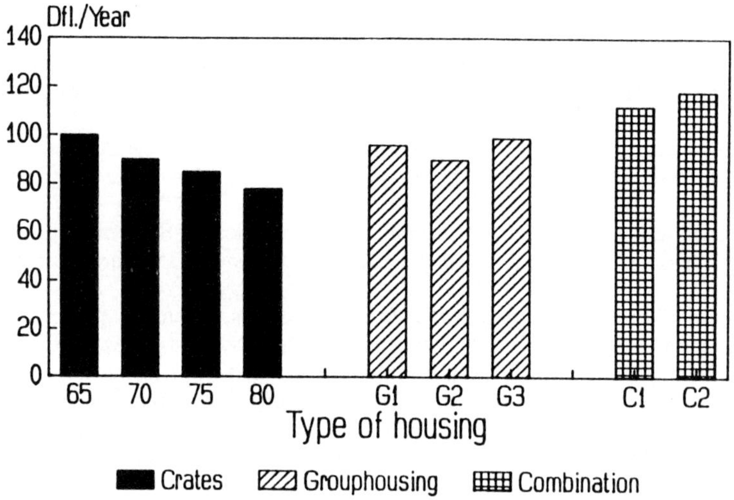

Figure 3. Farmer's return from labour for three different housing systems. From the left, crates of different width (0.65, 0.70, 0.75 and 0.80 m), three types of group housing and a combination of crates and group pens.

Conclusions

The housing of veal calves in groups of five fed from open buckets, provided with some roughage and tethered during the first 6–10 weeks is an important step towards improved calf welfare. The higher mortality and culling rate in the group pens may be reduced by better disease control and better housing and climate, especially in the first weeks after arrival. The individual recognition of calves in group pens needs further improvement, as do the feeding equipment, the ventilation system and the floor type. Finally, the feasibility of housing calves individually in rearing units during the first weeks followed by loose housing in group pens needs further investigation.

References

de Wilt, J.G., 1985. Behaviour and welfare of veal calves in relation to husbandry systems. Thesis, Agricultural University. IMAG, Wageningen.

Giesen, J.H.J., 1984. Arbeidsbesteding in de vleeskalverhouderij. IMAG nota 108, Wageningen.

Kasper, G.J., J.H.J. Giesen & A.C. Smits, 1989. Vleeskalverenhuisvesting in economisch perspectief. IMAG rapport 106, Wageningen.

Ruckebusch, Y. & F.R. Bell, 1970. Etude électropolygraphique et comportementale des états de veille et de sommeil chez la vache (Bos taurus). Annaleste Recherches Véterinaires 1:41–62.

Smits, A.C. & P.J.M. Ham, 1988. Praktijkonderzoek groepshuisvesting vleeskalveren. IMAG rapport 105, Wageningen.

Loose-housing experience in North America

Trevor Tomkins
Milk Specialties Company, P.O. Box 278, Dundee, Illinois 60118 US

Summary

In the early 1980s a number of companies introduced and promoted loose-housing veal-production systems in the US. The system consisted of a European-designed automatic milk-feeding machine, and an ad libitum feeding program. Holstein bull calves of about 45 to 50 kg initial weight were raised in groups of 40 to 50 with one automatic feeder per group. The length of the fattening period was 14 to 16 weeks. Calf performance was extremely variable in comparison with individual bucket feeding systems and profitability was lower. The main reasons for the poor performance related to higher morbidity and mortality, higher feed/gain ratios and wider variation in weight at slaughter age. (Mortality > 10% vs. < 6%; feed/gain 1.75–1:95:1 vs. 1.65–1:75:1; slaughter weight at 15 weeks 120–180 kg vs. 150– 175 kg.)
Descriptors: loose housing; veal; automatic feeder; milk feeding.

Introduction

Following the apparent commercial success in the late 1970s of some loose-housed veal production systems in Europe, and in particular England (Paxman, 1981; Webster & Saville 1981; Webster et al., undated), a number of companies involved in the veal industry in North America introduced loose housing to the United States in the early 1980s. The system was promoted and sold on the basis of increased growth rates with ad libitum feeding, reduced labour compared with conventional bucket feeding, low capital investment and of being more humane.

The system consisted of a European-designed automatic milk-feeding machine which reconstituted dry powder and an ad libitum feeding program. Milk was available through a nipple. Holstein bull calves of about 45 to 50 kg initial weight were raised in groups of 40 to 50 with one automatic feeder per group. The length of the fattening period was 14 to 16 weeks.

During the period from 1982 to 1985, it is estimated that the number of veal calves being raised annually in loose housing systems in the United States rose from 0 to 200 000 (16% of US total), but from 1985 to 1990 that figure has declined to an estimated maximum of 5000 calves per annum.

Housing and environment

Some relatively low-cost operations were set up in converted dairy barns and other existing buildings, and in some cases more sophisticated purpose-built facilities were established. Calves were either kept on bedding, which was typically straw, wood shavings or corn stalks, or on wooden or concrete slats. Ventilation in the buildings ranged from natural systems with no forced-air movement to highly automated controlled-environment systems. In a few cases, particularly in the southern part of the USA, calves were kept in naturally ventilated buildings.

Many of the low-capital investment operations were clearly inadequate in many respects. Where calves were kept on bedding, the floors underneath frequently had little or no drainage, which made for excessive use of bedding material. Where inadequate bedding was used, calves were in wet conditions. In summer, the use of bedding gave rise to severe lice and fly problems.

Table 1. Feeding program for loose housed veal production.

Age (weeks)	Milk concentration (g/litre)
1 - 4	80 – 125
5 - 8	125 – 160
9 - 12	160 – 200
13 - 16	200

Table 2. Performance of veal calves on loose housing compared with conventional veal production systems. Data from producers' records 1983–1985.

	Loose housing	Conventional
Time on feed (weeks)	14 – 16	14 – 16
Start weight (kg)	45 – 50	45 – 50
Finish weight (kg)	120 – 180	150 – 175
Feed/gain	1.75 – 1.95	1.65 – 1.75
Mortality (%)	> 10	< 6.0

Where slatted floors were used, these were frequently very wet and slippery during the first few weeks. This made management difficult, when trying to restrain and isolate a calf for treatment.

Nutrition

Holstein bull calves were typically purchased at sale barns, then sorted at a central location, and shipped to veal units 12 to 36 hours later.

On arrival at the units, calves were allowed to rest for 6 to 12 hours with access only to water or electrolyte solutions through a nipple assembly or open trough.

Calves were usually introduced to milk from the automatic feeder on an individual basis to ensure that the calf would suck from a nipple. In many cases, calves would then be allowed ad libitum access to milk from that stage. In other cases, calves would be restrictedly fed on the nipple on an individual basis and only given milk replacer ad libitum after 2 to 3 days. Milk was given at a temperature of about 40 °C. The milk feeding program consisted of about 40 kg starter feed (21% crude protein (CP), 16% fat) and 185 kg finishing feed (16% to 18% CP, 18% fat). The starter feed contained added elemental iron. Milk solids were increased from 8% at the outset to 18–20% at the finish. The feeding program is summarized in Table 1.

The milk replacers generally contained lower percentages of skim-milk powder and higher percentages of whey proteins.

Health

The general experience was that calf health was considerably poorer in loose-housing systems. At the outset, scouring and diarrhoea were major problems, particularly where calves were given access to ad libitum quantities of low-solids milk replacer. These scouring problems were not easy to treat on an individual basis. Frequently, calves became severely dehydrated. Hygiene was often very poor, and probably gave rise to disease transmittal. Scouring problems generally lasted 2 to 3 weeks.

Between three and five weeks, calves that had not started well on the system were often susceptible to pneumonia. These problems were exacerbated when ventilation was inadequate.

During the later stages of the growth cycle, a variety of problems were common, some specifically related to loose housing. Bloat, related to overconsumption, enterotoxaemia and abomasal ulcers were common.

In many cases, stereotyped behaviour developed in early stages, including preputial sucking and urine drinking.

It was common where calves were bedded on straw for them to start consuming significant amounts and to develop active rumens. Where this happened, circumstantial evidence suggested less problems associated with enteric disorders, but lower milk intakes, provided the straw consumed was clean.

Performance

From the outset, calf performance was extremely variable in comparison to conventional bucket-feeding systems (where calves are kept in individual pens and fed twice daily) and profitability was lower as shown in Table 2. This data is extracted from producer records available during the period 1983–1985.

The main reasons for the poor performance are related to higher morbidity and mortality, higher feed/gain ratios and wider variation in weight at slaughter age.

The management of iron supplementation was generally more difficult than for calves raised in individual pens and this gave rise to wider variation in meat colour and affected financial returns. In general, the meat-packing industry did not like loose-housed calves because of greater variation in size, colour and conformation.

Conclusions

The loose-housing concept as introduced in the early 1980s has not been as successful in the US as it had reportedly been in Europe and this related to a number of factors. Poor calf performance, higher morbidity and mortality, higher feed/gain ratios and wide variation in weight at slaughter age.

In addition, management of the loose housed calf was considered more difficult and claims that the system was labour-saving were ill-founded. Additional costs, such as bedding, further detracted from profitability. Mechanical unreliability of equipment also gave rise to managerial and nutritial problems, which contributed to greater expenses.

A recent attempt has been made by Milk Specialties Company to re-examine loose housing with a computer-controlled system for individually fed transponder-equipped calves. This has not demonstrated that some of the problems with loose housing can be overcome where calves are in large groups of 25 calves or more.

References

Paxman, P., 1981. Quantock loose house system. In: Alternatives to intensive husbandry systems: Proceedings of symposium, Wye College (University of London) Publ. Universities Federation of Animal Welfare (UFAW 1981).

Webster, A.J.F. & C. Saville, 1981. Rearing of veal calves. In: Alternatives to intensive husbandry systems: Proceedings of symposium, Wye College (University of London) p.86–94, Publ. Universities Federation of Animal Welfare (UFAW 1981).

Webster, A.J.F., C. Saville & D. de B. Welchman, 1986. Improved husbandry for veal calves Published by University of Bristol Press. Bristol, England.

Automated feeding of milk replacer and health control of group-housed veal calves

K. Maatje * & J. Verhoeff **
* *Research Institute for Animal Production 'Schoonoord', P.O. Box 501, 3700 AM Zeist, NL*
** *Department of Herd Health and Reproduction, Faculty of Veterinary Medicine, University of Utrecht, P.O. Box 80151, 3508 TD Utrecht, NL*

Summary

Comparative investigations were carried out between bucket-fed veal calves in individual crates and group-housed calves with automated and rationed teat feeding of milk replacer. Rations in both systems were identical. The calves in group pens were identified electronically in the feeding station. Behaviour, growth rate, feed intake, health and slaughter quality were determined.

The growth rate and feed intake of the group-housed calves were lower than those of the calves in individual crates. The feed conversions were equal. There were no differences in carcass conformation. The colour of the meat of the group-housed calves was more favourable. Records of the feeding behaviour of the group-housed calves were studied to see if sick animals could be identified at an early stage.
Descriptors: veal calves; group housing; automated feeding.

Introduction

The main reason for the development of group-housing systems is that they are assumed to be more attractive to the nature of veal calves than individual housing in crates. The calves have more freedom of movement and more social contact in group housing (de Wilt, 1985).

Automated feeding systems for milk replacer were already available in the late 1960s. However the ad libitum intake of milk replacer had too many disadvantages (N.N., 1976). Since the development of electronic identification systems, individually rationed automated feeding has become possible. Automated feeding from teat dispensers has some clear advantages over bucket feeding. First, the need for sucking is satisfied during feeding and the occurrence of cross-sucking, including preputial sucking and urine drinking is reduced.

Bucket-fed calves in group housing are tied up during the first six to eight weeks of the fattening period to prevent preputial sucking and urine drinking (Smits & Ham, 1988). With the automated feeding system, this period of tying up can be omitted. Furthermore, the supply of milk replacer can be divided over more than two meals per day, thus approaching a more natural drinking pattern and the farmer is needed less at feeding times.

Another important aim of this investigation was to see whether the automated feeding system could be used for the early detection of sick calves.

Material and methods

Five comparative trials were carried out between bucket-fed veal calves in traditional crates and group-housed calves with automated feeding on milk replacer.

In each of the trials 80 male Dutch Friesian × Holstein Friesian calves, about 7 days old, were obtained from commercial farms. Four groups of 15 calves were housed in group pens

*Table 1. Production results of group-housed calves and calves in individual crates (from three trials); *losses = deaths + culls; ** final live weight = (100/65) × cold carcass weight.*

	Calves in group pens	Calves in individual crates
Number of calves	180	61
Fattening period (days)	161	161
Losses (%)*	5.6	0
Initial weight (kg)	41.5	41.0
Final liveweight (kg)**	218.1	225.4
Daily gain (g)	1097	1145
Feed conversion	1.78	1.76

Table 2. Carcass quality of calves fattened in group pens and in individual crates. Scales: a: 1 = maximum, 10 = minimum; b: 1 = maximum, 6 = minimum; c: 1 = pale, 8 = dark;

	Calves in group pens	Calves in individual crates
Number of calves	170	61
Fleshiness[a]	6.76	6.76
Fat covering[b]	1.90	2.38
Meat colour[c]	3.40	4.27
Haemoglobin in blood at slaughter (g/100 ml)	8.40	9.08

and 20 calves were housed in individual crates. In one trial, the groups were changed to one group of 30 calves and two groups of 15 calves. The crates and the group pens were situated in the same unit. For each calf 1.5 m^2 of floor space was available in the group pens. The crates were 70 cm wide and 170 cm long.

Group-housed calves were recognized by a computerized identification system. A signal then made the teat appear and deliver 0.5 l of milk replacer. This procedure was repeated until the calf had received its meal. During the later experiments, preparation of the milk replacer was centralized and it was pumped into a pipeline connected to the feeding station. The milk replacer was distributed by a flow meter. The daily rations were divided into four meals for the calves in group pens, and into two meals for the calves in individual crates. In addition to the milk replacer, the calves in both systems received straw pellets.

During the fattening period, data were collected about behaviour, haemoglobin in blood, liveweight (every two weeks), feed intake and health. After slaughter, the carcass quality and the condition of the lungs and rumen were assessed visually.

Feed intake, drinking rate and rewarded and unrewarded visits to the feeding station were recorded automatically. A veterinary surgeon physically examined the animals twice a week. Both the automatically recorded data and the veterinary observations were studied to see wether sick animals could be detected at an early stage.

Results

Behaviour observations

In the groups of 15 animals, the calves quickly became accustomed to the automated feeding system. After 4 days more than 90% of the calves went to the drinking machine of their own accord. In the larger groups of 30 animals, it took 14 days before the calves were familiar with the system.

Preputial sucking and urine drinking were rare. Only two animals out of 300 in 5 trials were removed from the group housing because urine drinking was disturbing their growth rate. Tongue playing was observed less in the group-housed calves than in those in individual crates. Occasionally, tail biting was observed in the group-housed calves.

Production results

Table 1 presents the production results from both sets of calves. Data from three comparable trials were collated. The results show that the final live weight and growth rate of the group-housed calves was less than that of the calves in individual crates. The differences were consistent in the three trials ($P < 0.05$). The feed conversion of the calves in group pens and the calves in crates were similar. However during the trial with the centralized system for preparation of milk replacer, there was a small statistically insignificant difference between the feed conversions: 1.81 for the calves in group pens and 1.75 for the calves in crates.

The results in Table 2 show that fleshiness was, on average, equal for the two housing systems. However the group-housed animals were less fat and the colour of the meat was better (paler). The differences in meat colour were significant ($P < 0.05$).

Animal health

The incidence of illnesses is presented in Table 3. It was insignificantly higher in the group pens than in the individual crates. Only the number of cases of buccal diphtheria was lower in the group-housed calves.

Results from post mortem examinations (Table 4) show that group-housed animals suffered more from respiratory diseases than those reared in individual crates. This concurs with the observations of the animals' health. Ruminal hyperkeratosis was diagnosed more frequently in the group-housed calves but did not significantly affect their carcass weight at slaughter in either housing system.

Health control

Table 5 presents number and percentage of treated and untreated cases of illness with distinct changes in feeding behaviour. With the exception of feed intake, the running average per calf was calculated for each parameter. The actual parameter was compared to the running average for each day. The animal was assumed to have been detected when a certain empirical threshold was exceeded or when the feed intake differed from the total ration per day. Table 5 shows that the drinking rate was the most sensitive parameter in a sick, or sickening, animal. If all the above parameters are taken into account, 80% of the cases of illness were accompanied by a change in the animal's feeding behaviour. The percentage of false-positive detections was rather high (15.3%).

Table 3. Incidence of illnesses per calf for calves housed in group pens and in individual crates during the fattening period.

	Calves in group pens	Calves in individual crates
Number of calves	170	61
Respiratory disease (%)	0.67	0.38
Diarrhoea (%)	0.39	0.22
Buccal diphtheria (%)	0.16	0.21
Other (%)	0.14	0.02
Total (%)	1.36	0.83

Table 4. Abnormalities found by examination of the lungs and rumen after slaughter.

	Calves in group pens	Calves in individual crates
Number of calves	170	61
Lungs and pleura (%)	24	17
Hyperkeratosis of the rumen (%)	14	10

Table 5. Number of treated and untreated cases of illness accompanied by changes in feeding behaviour.

	Treated cases (57)		Untreated cases (187)	
	n	%	*n*	%
Drinking rate (DR)	27	47	86	46
Feed intake (FI)	22	39	67	36
Rewarded visits (RV)	15	26	56	30
Unrewarded visits (UV)	19	33	48	26
DR or FI	34	60	111	59
DR or FI or RV	39	68	129	69
DR or FI or UV	44	77	128	68
All criteria	47	83	144	77

Conclusions

- The average final liveweight of the group-housed calves was 6–8 kg less than that of the calves housed in individual crates. This is due to a lower intake of milk replacer. The feed conversion of the calves in group pens and in individual crates was about the same.
- Carcass quality was equal for both systems but the meat colour of the group-housed calves was significantly better (paler) than that of the calves in individual crates.

- The incidence of illness was higher in the group pens than in individual crates, probably because of the greater pressure of infection within the group.
- Computerized recording of feeding behaviour can be useful in the detection of sick animals.
- Automated feeding of milk replacer in groups of 20–30 veal calves prevents the occurrence of preputial sucking and could allow adequate production and health. Further development is necessary.

References

de Wilt, J.G., 1985. Behaviour and welfare of veal calves in relation to husbandry systems. Thesis, Agricultural University. IMAG Wageningen.

N.N., 1976. Automatisering bij de voedering van vleeskalveren. Proefstation voor de Rundveehouderij, report no. 45 (in Dutch).

Smits, A.C. & P.J.M. Ham, 1988. Praktijkonderzoek groepshuisvesting vleeskalveren. IMAG report no. 105 (in Dutch).

Effects of computerized milk feeder on behaviour and welfare of calves[1]

V. Ferrante *, E. Canali **, M. Verga * & C. Carenzi *
* *Istituto di Zootecnica Veterinaria, Via Celoria 10, 20133 Milano, IT*
** *I.D.V.G.A. C N R, Via Celoria 10, 20133 Milano, IT*

Summary

The feeding system has an important influence on welfare and production in group-reared calves. Computer-controlled machines could solve many of the problems associated with group feeding, such as control of individual milk intake and cross-sucking. The aim of this research was to evaluate the effects of a computerized milk feeder on the behaviour and welfare of calves. Sucking duration, frequency and intervals between successive bouts of milk intake were collected for each calf continuously using a videorecorder system for two weeks. The majority of the sucking sessions lasted less than 3 min; they seemed to be regularly distributed over 24 h. The average sucking frequency was 17 times and the average sucking time was 26 min 30 s over 24 h; the duration of sucking periods and intervals between sucking sessions did not change throughout the 2 weeks.
Descriptors: calves; computerized teat dispenser; sucking behaviour.

Introduction

Various studies demonstrate the influence of housing and feeding systems on behaviour and on physiological response of reared calves (Friend & Dellmeier, 1988; de Wilt, 1985). Group-housing calves, bucket feeding or the use of a teat dispenser seem a good solution to the problems associated with calves raised in individual boxes, but it could give rise to other problems linked with feeding, i.e. cross-sucking as well as difficulty in the evaluation of individual milk intake and disease control (Mees & Metz, 1984; van Putten, 1982). Computerized feeding machines could solve many of the problems associated with group feeding and seem to give superior feeding efficiency and to reduce the chance of illness (Webster et al., 1986). The aim of this research was to evaluate the effects of a computerized milk-feeding system in terms of sucking behaviour of calves.

Material and methods

Five male Friesian calves raised in individual boxes were bucket-fed twice a day until the age of two weeks. They were then fed by a computerized system in a group pen (5 m × 6 m), on concrete floor with straw bed, and were observed for two weeks with a videorecorder. The recorder began filming every time a subject entered the feeding station. The computerized feeding machine was programmed to distribute a portion equivalent to 500 g in a period of about two hours (5 g/min), up to a maximum of 2500 g each time. During the following 24 h, the animals could eat up to half the total amount of the milk ration not consumed the previous

1 This work was carried out in collaboration with the Experimental Farm 'V. Tadini', Gariga di Podenzano, Piacenza Italy.

day. The milk-replacer powder contained 29% protein. The amount of milk fed daily to each calf was 7 l the first week and 8 l the second week, the powder concentration was 100 g/l. Individual body weights were recorded at 15 and 30 days of age. Number of sucking sessions, intervals between sucking sessions and their distribution throughout the 24 h were evaluated as cumulative frequencies. Moreover, successful and unsuccessful attempts to enter the station when already occupied by another calf, were recorded. Data were analyzed by least-square-means analysis of variance.

Results

The average number of sucking sessions and the average duration of sucking sessions per day were 17 ± 10 times and 26 min 30 s ± 15 min 30 s respectively. Average daily gain was 603 g ± 103 g, while the average amount of milk fed was 6.8 ± 0.4 l. The majority of sucking periods (75%) lasted less than 3 minutes (Figure 1).

The distribution of intervals between sucking sessions is shown in Figure 2. The most frequent intervals were between 5 min and 10 s (17%) and between 90 min and 120 s (9%). While in the other intervals, the beginning of a new sucking session was random, according to Hammell et al. (1988).

The distribution of the sucking sessions throughout the 24 h is shown in Figure 3. Sucking sessions are distributed during the 24 h but two points of higher activity are evident in the late morning and in the afternoon.

Figure 4 shows the duration of sucking sessions over the 24 h, which agrees with the diurnal fluctations in the time spent on meals found by Hammell et al. (1988). In some periods, the duration of sucking was higher than in others. Each calf seemed to have different sucking times

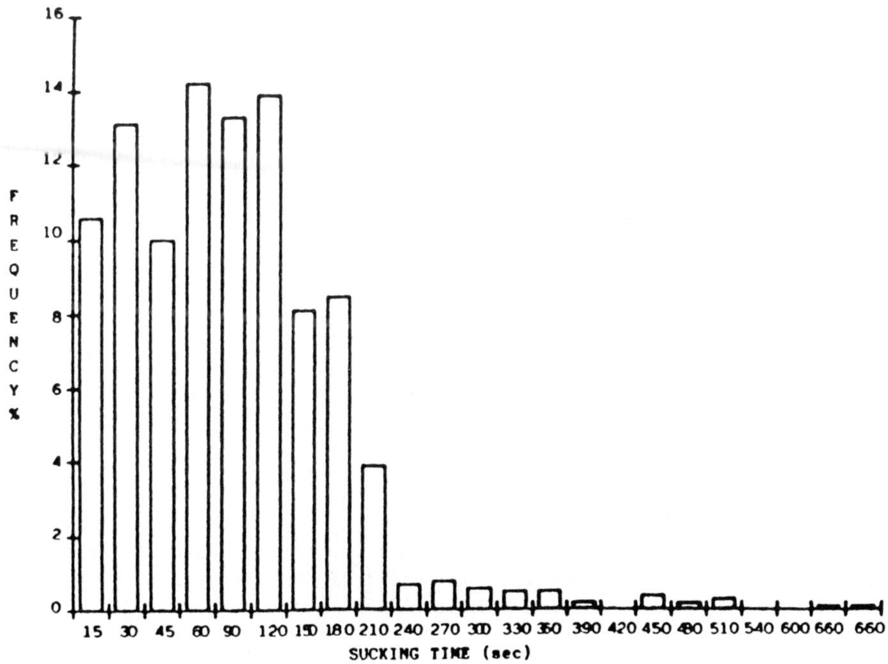

Figure 1. Distribution of the duration of sucking sessions.

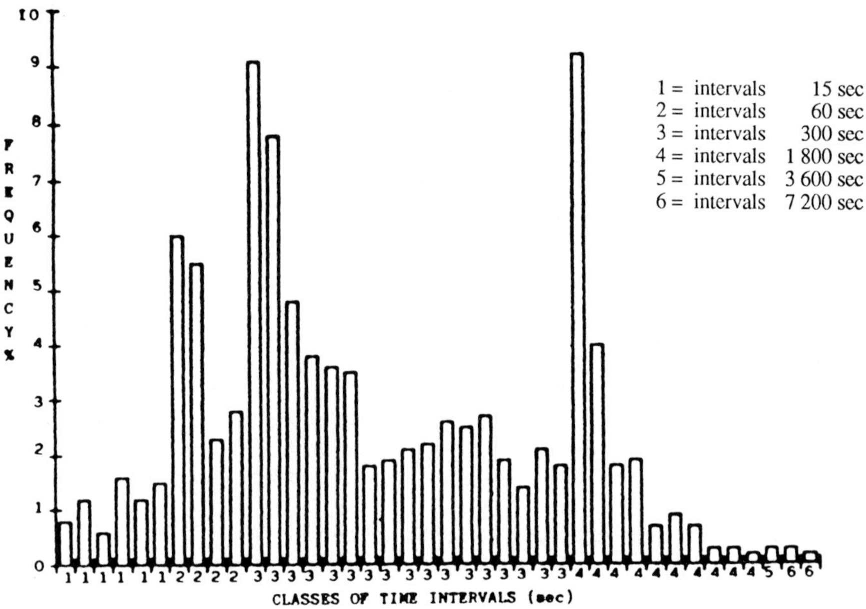

Figure 2. Distributon of the duration of the intervals between sucking sessions.

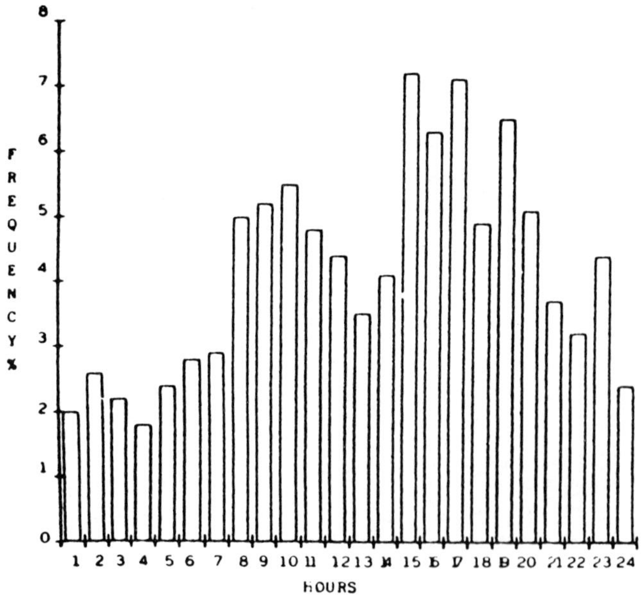

Figure 3. Distribution of the sucking sessions during the 24 h of the day.

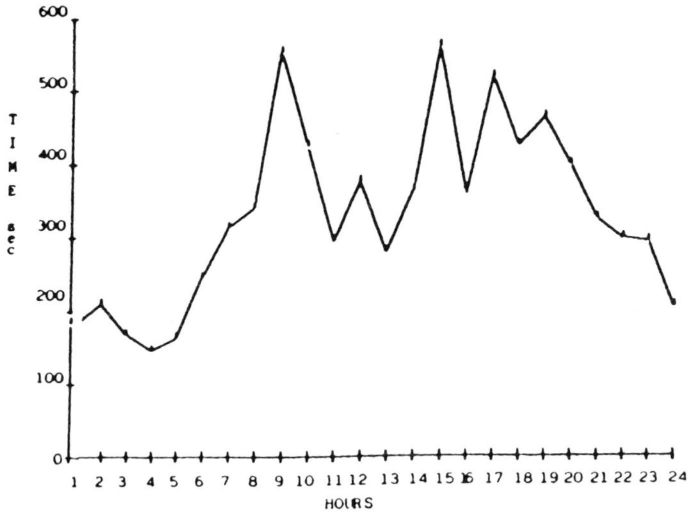

Figure 4. Daily fluctuations in time spent sucking.

and intervals between sucking sessions. The duration of sucking periods and intervals between sucking sessions did not change throughout the 14 days ($P < 0.05$).

In 157 of 1175 sucking sessions (13%), a calf tried to enter the station when it was already occupied. In 64 cases (38%) these attempts were successful. The percentage of these agonistic interactions at the feeding station did not change during the 14 days and fights were bi-directional in agreement with Hafez & Bouissou (1975) and Canali et al. (1986).

Conclusions

The computerized feeding machine allows a distribution of sucking sessions over 24 hours. The duration of sucking periods and intervals between sucking sessions did not change throughout the 2 weeks.

Since calves normally receive 4–10 feedings from their dams (Webster et al., 1986), it seems that this system could permit sucking rhythms closer to the natural ones as the average number of sucking sessions was 17 ± 10 times.

However further studies are needed on cross-sucking, agonistic interactions at the feeding station and on production traits.

References

Canali, E., M. Verga, M. Montagna & A. Baldi, 1986. Social interactions and induced behavioural reactions in milk-fed female calves. Applied Animal Behaviour Science 16:207–215.

de Wilt, J.G., 1985. Behaviour and welfare of veal calves in relation to husbandry systems. Thesis Agricultural University. IMAG, Wageningen , NL.

Friend, T.H. & G.R. Dellmeier, 1988. Common practices and problems related to artificially rearing calves: an ethological analysis. Applied Animal Behaviour Science 20:47–62.

Hafez, E.S.E & M.F. Bouissou, 1975. The behaviour of cattle. In: E.S.E. Hafez (Ed.): The behaviour of domestic animals. 3rd edition. Bailliere Tindall, London. p. 203–245.
Hammell, K.L., J.H.M. Metz & P. Mekking, 1988. Sucking behaviour of dairy calves fed milk ad libitum by bucket or teat. Applied Animal Behaviour Science 20:275–285.
Mees, A.M.F. & J.H.M. Metz, 1984. Saugverhalten von Kälbern - Bedürfnis und Befriedigung bei verschiedenen Tränkesystemen. KTBL-Schrift 299:82–91.
van Putten, G., 1982. Welfare in veal calf units. Veterinary Record 111:437–440.
Webster, J. 1984. Calf husbandry, health and welfare. Granada, London.
Webster, J., C. Saville & D. Welchman, 1986. Improved husbandry systems for veal calves. Report Department of Animal Husbandry, University of Bristol, School of Veterinary Sciences, Langford, Bristol.

Health of veal calves in 4 systems of individual housing during the first weeks of the fattening period

E. ter Wee *, H.K. Wierenga *, I.P. Jorna ** & A.C. Smits ***

* *Research Institute for Animal Production 'Schoonoord', P.O. Box 501, 3700 AM Zeist, NL*

** *Gardenbroeks's Veevoederfabriek 'Navobi' B.V., P.O. Box 1, 8075 ZG Elspeet, NL*

*** *Institute of Agricultural Engineering, P.O. Box 43, 6700 AA Wageningen, NL*

Summary

The aim of this study was to investigate ways of reducing disease in group-housed veal calves by different types of individual housing during the first weeks of the fattening period.

Two hundred and forty calves were assigned to one of the following housing systems in each of four consecutive experiments: tethering with a chain to the feeding rack (1), a baby-pen (2), a half-closed baby-pen (3), a crate (4).

The percentage of calves with diarrhoea did not differ between the systems. The duration of diarrhoea was significantly les in System 4 than in the other systems. The percentage of calves with a clinical symptom of a respiratory disorder and the duration of such a symptom were similar in the four systems. There were no significant differences between the systems in medical treatment. The results of the examination of the faeces for oocysts of *Cryptosporidium* were similar in the four systems.

The health of the calves in the four systems studied was similar. So veal calves can be housed in any of these four systems during the first weeks of the fattening period.

Descriptors: veal calves; health; diarrhoea; respiratory disorders; *Cryptosporidium*; housing.

Introduction

Of the veal calf population in the Netherlands, 10% is housed in groups. Group housing is an alternative to traditional housing in crates, with respect to welfare (de Wilt, 1985) and production (Smits & Ham, 1988). One problem in group housing of veal calves is the slightly higher losses than in individual housing (de Wilt, 1985; van der Mei, 1987; Smits & Ham, 1988). There are indications that diarrhoea (van der Mei, 1987) and respiratory disorders (van der Mei, 1987; Smits & Ham, 1988) are important causes of the higher losses in group housing compared to individual housing. Both diarrhoea and respiratory disorders occur mainly during the first weeks after the calves' arrival at the fattening unit (Postema, 1985). Group-housed calves are generally tethered to the feeding rack and separated by wooden partitions during the first 6 to 10 weeks.

The aim of this study was to investigate whether it was possible to improve the health of the calves by changing the type of housing during the first weeks of the fattening period. Four systems of individual housing were compared.

Material and methods

Animals

In each of four consecutive experiments, 240 black and white calves were studied. All the calves were males, except for four females in the second experiment. During the experiments 24 calves died or were culled. The results of the females, and of the dead and culled calves were excluded from the results.

Housing

For seven weeks, the calves were kept individually in one of the following ways: tethered with chains to the feeding rack(1), in a baby-pen (2), in a half-closed baby-pen (3), or in a crate (4). In general, group-housed veal calves in the Netherlands are tethered with a chain to the feeding rack as in system 1 during the first weeks of the fattening period. The chain of System 1 was 0.6 m long. Wooden partitions (0.9 m high and 0.5–0.9 m long) separated the calves. The pen of System 2 looked like a baby's playpen. This baby-pen was surrounded by a railing (0.9 m high) and a feeding rack. The half-closed baby-pen had two partly closed sides (System 3). The sides were made of wood (0.6 m high) and a railing (0.3 m high). Of the veal calves in the Netherlands, 90% are kept individually and are housed in solid-sided crates as in System 4. The sides of the crates in System 4 were 1.1 m high. The stand of System 1 was 0.6 m wide. The pens of Systems 2, 3 and 4 were 1.4 m long and 0.6 m wide.

Data collection and statistics

The health of the calves was recorded as follows:

- diarrhoea: from day 1 until day 22, the calves were examined daily for diarrhoea (yes/no). The faeces of a calf with diarrhoea deviated from the normal consistency or colour.
- respiratory disorders: laboured respiration, coughing, nasal discharge and watery eyes. The symptoms for each calf were recorded daily from day 1 until day 50. A calf had a serious respiratory disorder (SRD) if it showed a laboured respiration or coughed.
- individual medical treatments.
- faecal samples of 204 calves were screened for oocysts of *Cryptosporidium* every other day from day 1 up to 23 during first three experiments.

The results were statistically analysed except for the results of the faecal samples. The experiments were described extensively in Ter Wee et al. (1989).

Results and discussion

Diarrhoea

In this study 62% of the calves had diarrhoea (Table 1) in the first weeks after arriving at the fattening unit. The percentage of calves with diarrhoea in these experiments was similar to the 38 to 62% mentioned by Postema (1985) for veal calves housed in crates. It was expected that fewer calves in System 4 (the crates) would get diarrhoea because of the limited contact between the calves in this system than in the other systems. The mean percentage of calves with diarrhoea was lowest in System 4 and highest in System 1. However the difference in percentages of calves with diarrhoea was statistically significant only between Systems 2 and 4.

Table 1. Frequency (% of calves) and duration of diarrhoea, external symptoms of serious respiratory disorders (SRD) and medical treatment for diarrhoea and for respiratory disorders during the first seven weeks. The results are presented by system and in total. The systems are described in the text.

		System				Mean
		1	2	3	4	
Diarrhoea						
recordings	calves (%) *	65	63^{a}	62	58^{a}	62
	duration per calf (d) *	3.1^{a}	3.2^{b}	3.0^{c}	2.5^{abc}	3.0
treatment	calves (%)	53	51	48	50	51
	duration per calf (d)	3.7	3.9	3.6	3.2	3.0
Respiratory disorders						
recordings	calves SRD (%)	49	49	52	54	51
	duration SRD per calf (d)	2.2	2.3	2.5	2.5	2.4
treatment	calves (%)	33	33	39	36	35
	duration per calf (d)	5.0	4.8	4.6	5.3	4.8

* in weeks 1 to 3

^{a,b,c} difference between two systems with the same superscript is significant ($P < 0.05$).

The mean duration of diarrhoea was 3.0 days. The duration of diarrhoea was shortest in System 4. The differences between System 4 and the others were significant. Systems 1, 2 and 3 did not differ significantly ($P > 0.05$).

Of the calves 51% were treated for diarrhoea for 3.6 days. Despite differences between the systems in the recordings of diarrhoea, no differences were found in medical treatment for diarrhoea between the four systems.

Respiratory disorders

Over the whole experimental period, the four systems did not differ significantly in the percentage of calves with external symptoms of serious respiratory disorders (SRD) or in the duration of SRD (Table 1). Neither were there significant differences between the systems in the treatment for respiratory disorders.

Cryptosporidium

Of the calves, 90% shed oocysts of *Cryptosporidium* during the sampling period (Table 2). The results showed no differences between the four systems. The percentage of calves shedding oocysts was much higher than the 60% found in another experiment with 45 calves housed in crates (de Visser, 1987). The estimated duration for shedding oocysts was 7.2 days (2 × 3.6 sampling days per calf). The four systems did not differ in the duration of shedding oocysts of *Cryptosporidium*. The duration was similar to those reported for naturally (Anderson, 1981; de Visser et al., 1987) and experimentally (Pohlenz, 1987) infected calves.

Table 2. Percentage of calves shedding oocysts of Cryptosporidium during the first three weeks and duration of shedding based on sampling every other day. The systems are described in the text.

	System				Mean
	1	2	3	4	
Calves with oocysts (%)	93	84	92	90	90
Duration of shedding per calf (d)	7.6	7.0	7.8	6.6	7.2

Conclusion

The experiments showed that, except for the duration of diarrhoea, the tested systems were similar with respect to the occurrence and the medical treatment of diarrhoea and respiratory disorders. The shedding of oocysts of *Cryptosporidium* was also similar in all four sytems. The type of individual housing had only a minor influence on the occurrence of diseases. These results seem to contrast with the problems noted for group housing in earlier research (van der Mei, 1987; Smits & Ham, 1988). However van der Mei (1987) suggested that the differences in the treatment of respiratory disorders in calves in crates and in groups was not caused by a difference in the occurrence of respiratory disorders but was related to a difference in the detection of calves with respiratory disorders by the farmer.

It can be concluded that veal calves can be housed in any of the four systems studied during the first weeks of the fattening period.

References

Anderson, B.C., 1981. Patterns of shedding of Cryptosporidial oocysts in Idaho calves. JAVMA 178:982–984.

Bietz, G., 1987. Anbindehaltung von Kälbern mit Grabnerketten in Vergleich zur Einzelhaltung unter einstreulose Bedingungen. Münchner Veterinär-Medizin 42:352–357.

de Visser, N.A.P.C., H.J. Breukink, F.G. van Zijderveld & P.W. de Leeuw, 1987. Enteric infections in veal calves: a longitudinal study on 4 veal calf units. Veterinary Quarterly 9:289–296.

de Wilt, J.G., 1985. Behaviour and welfare of veal calves in relation to husbandry systems. Thesis Agricultural University, IMAG, Wageningen.

Moon, H.W. & W.J. Bemrick, 1981. Fecal transmission of calf cryptosporidia between calves and pigs. Veterinary Pathology 18:254–255.

Pohlenz, J.,1987. Die Kryptosporidiose bei Menschen und Tieren. Deutsches Tierärtzlichen Wochenschrift 94:67–70.

Postema, H.J., 1985. Veterinaire en zoötechnische aspecten van de kalverhouderij. Thesis Rijksuniversiteit, Utrecht.

Smits, A.C. & P.J.M Ham, 1988. Praktijkonderzoek groepshuisvesting vleeskalveren. IMAG-report 105, Wageningen.

ter Wee, E., H.K. Wierenga, I.P.Jorna & A.C. Smits, 1989. Gezondheid en groei van vleeskalveren gedurende de eerste zeven weken van de mestperiode in vier systemen van individuele huisvesting. IVO-report B-333, Zeist.

van der Mei, J., 1987. Een vergelijkend onderzoek naar de gezondheid en de produktie van vleeskalveren gehouden in boxen en in groepshuisvesting. IVO-report B-304, Zeist.

Applicability of natural ventilation in veal-calf houses: a theoretical approach

A. van 't Ooster
Department of Agricultural Engineering and Physics,
Agricultural University of Wageningen, Mansholtlaan 12, 6708 PA Wageningen, NL

Summary

This paper presents an evaluation of the applicability of natural ventilation in veal-calf houses based upon model calculations. The computer simulation model, NAT_VENT, describes the heat, moisture and carbon dioxide balance of the house and the natural ventilation mechanisms in detail. The theoretical backgrounds of the mechanisms for air exchange are discussed briefly, as well as thermal responses of veal calves to indoor climatic conditions. Current naturally ventilated housing systems in the Netherlands were modelled and their performance was compared. This specifically concerned sections of modern veal-calf houses holding 40–50 calves with the ridge opening either parallel or perpendicular to the crate rows or group pens. Model responses to animal density, outside air temperature, wind speed and wind direction as well as responses to opening sizes were computed and compared with required ventilation rates. Finally applicability of current opening sizes and the effects of control of opening sizes is discussed.
Descriptors: veal calves; natural ventilation; heat production.

Introduction

Natural ventilation is mostly considered as a ventilation system that holds risks for controllability of the climate in agricultural buildings. In particular ventilation capacity during hot weather conditions and its ability to keep fluctuation of the indoor climate within accepted limits are often questioned and is certainly insufficient when inadequate control systems are used. Yet know-how in this area has increased significantly during the last two decades. Physical processes acting as a driving force for natural ventilation are described well enough to be used as tools in design and control of naturally ventilated livestock houses (Bruce, 1986; Brockett & Albright, 1987; van 't Ooster & Both, 1988).

The applicability of natural ventilation in veal calf houses is evaluated by means of a computer model and reported in this paper. The computer model, NAT_VENT, describes the heat, moisture and carbon dioxide balance of a livestock building and the natural-ventilation mechanisms in detail. It can be used as a tool in judging the performance and effectiveness of ventilation opening configurations in animal houses.

Modelled building

The type of building building for the model calculations was of the most common design for calf housing in the Netherlands. It was described by Smits et al., (1986). The house has separated compartments for 40 to 50 calves in crates or group pens, connected by doors in the main path way. The crate or pen rows are perpendicular to the long axis of the building (Figure 1). Model calculations were done for a compartment in the middle section of this type of house,

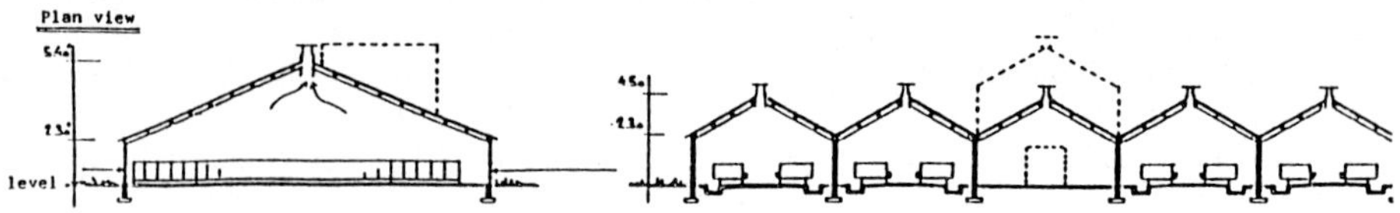

Figure 1. Section of the modelled house types, House type 1 (right) and House type 2 (left).

Table 1. Variants on the house design for model calculations.

House type	Opening sizes small	Opening sizes large
1 Ridge parallel to crate rows		
Ridge height 4.38 m; effective opening width	0.10 m	0.20 m
Side-wall openings in the doors (area per opening)	0.54 m^2	2.00 m^2
2 Ridge perpendicular to crate rows		
Ridge height 5.30 m; effective opening width	0.40 m	0.40 m
Side-wall openings in the doors (area per opening)	0.54 m^2	2.00 m^2

containing 40 calves in crates wide 70 cm. Table 1 lists 4 variants on this design, which are called House type 1 or House type 2 with small or large openings.

Ventilation requirements for veal calves

Ventilation requirements were calculated from the heat and moisture balances of the house using criteria for the indoor climate. Calculations were carried out for two extreme liveweights (LW) in the Dutch veal calf industry, 50 and 250 kg. For these two categories, ambient temperature is to be kept at least at 15 and 3 °C, respectively, and increase freely up to 23 °C in both cases (Vermorel, 1987). At high outdoor temperatures, causing indoor temperatures higher than 23 °C, it is reasonable to allow a variable temperature rise inside the house with respect to external temperature, which is dependent on the outdoor temperaturel. This temperature rise was called summer temperature tolerance and was set to 4, 3, 2.5, 2 and 1.5 °C for external temperatures of up to 21, above 21 up to 24, above 24 up to 27 °C, above 27 up to 31 and over 31 °C, respectively (CIGR, 1984). Animal heat production was calculated after Roy (1980a,b) and van Es & van Weerden (1970).

For maximum relative humidity, two criteria were evaluated. In the first criterion, advised by CIGR (1984), maximum relative humidity in the house is set at 90% minus internal temperature in °C. The second criterion is that indoors relative humidity should be kept at or below 85%. If these humidity criteria were not met, a tolerance factor for absolute humidity (i.e. absolute humidity indoors minus absolute humidityoutdoors) of 2 g/kg dry air was activated automatically. If the ventilation requirement for removal of moisture or heat from the house was very low, minimum ventilation rate required was calculated from a maximum concentration allowed for carbon dioxide, 3 l/m^3. Carbon dioxide concentration can be used as an indicator for air quality in the house. Figure 2 shows the ventilation requirement for a house compartment of 40 calves of 50 kg or 250 kg for the climate criteria stated above. This can be compared with the advice to install a ventilation capacity of at least 150 m^3/h per calf place in mechanically ventilated housing systems (Smits et al., 1986).

The required ventilation rate shows a variation of between 8.6 and 50.6 m^3/h.calf for young calves (LW = 50 kg) and between 28.5 and 161.2 m^3/h.calf for calves at slaughter weight (LW = 250 kg). If a group-housing system with standard group-housing pens (Smits et al., 1986) were applied in the same compartment, it would allow the housing of 50 calves instead of 40. Ventilation required would then range from 31.2 to 163.7 m^3/h.calf for calves at slaughter weight. The fluctuative behaviour of the curves at external temperatures above 18 °C resulted from the discontinuously decreasing summer-temperature tolerance factor and the decreasing

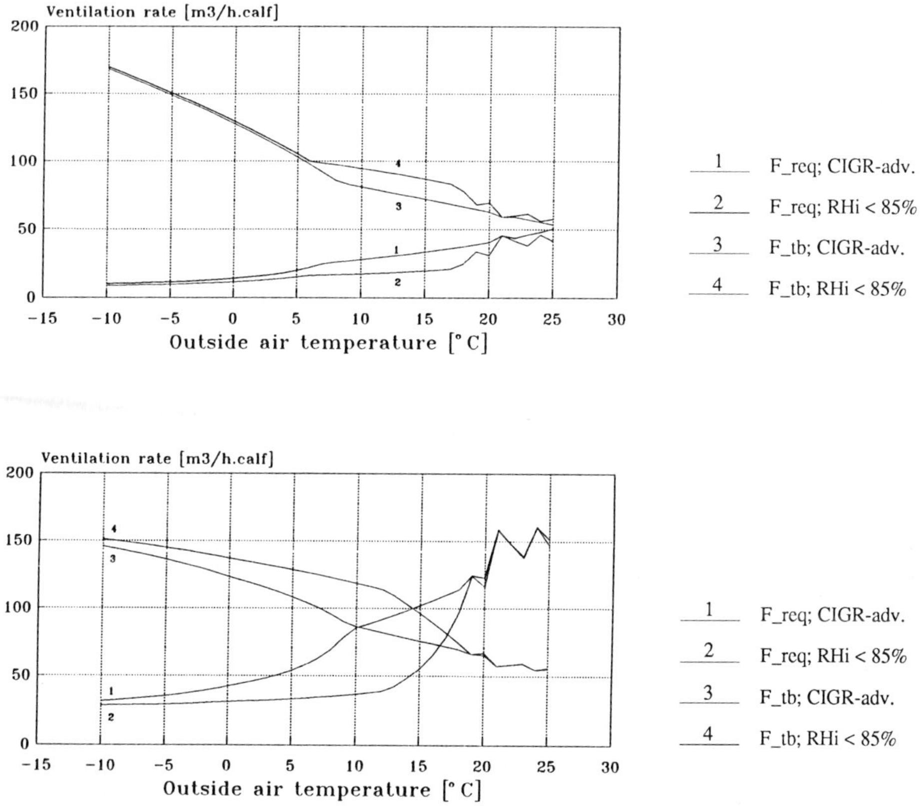

Figure 2. Required ventilation rate for young calves (LW = 50 kg) (above) and calves at slaughter weight (LW = 250 kg) (below): curves 1 & 2; Maximum ventilation by stack effect only, with indoor climate conditions as given in Figure 3: curves 3 & 4. House type 1 with small openings (fully opened).

sensible-heat production by the calves at increasing ambient temperature. These figures also show that the CIGR advice for maximum relative humidity indoors is the more stringent criterion over the full temperature range. CIGR criterion could only be kept for external temperatures up to about 7 °C, both for young and older calves (Figure 3, below). At higher outdoor temperatures, the humidity tolerance factor was activated, because the CIGR advice was not a realistic criterion for the given circumstances. The second humidity criterion was valid until the heat balance formed a more stringent criterion for the required ventilation rate. Figure 3 (above) shows the ideal indoor temperature control with use of supplemental heating for young calves at 15 °C for as long as this is possible. It also shows the free increase of indoor temperature up to 23 °C (for both liveweight categories) and the activation of the variable summer-temperature tolerances when external temperature is higher then 23 °C minus the valid summer-temperature tolerance factor of 3 °C at this temperature.

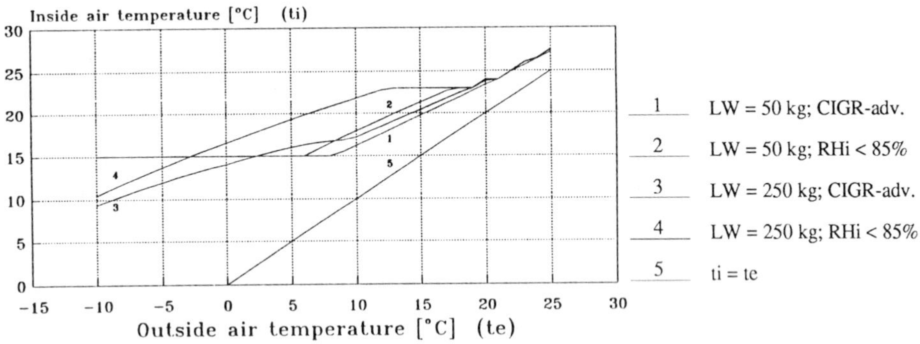

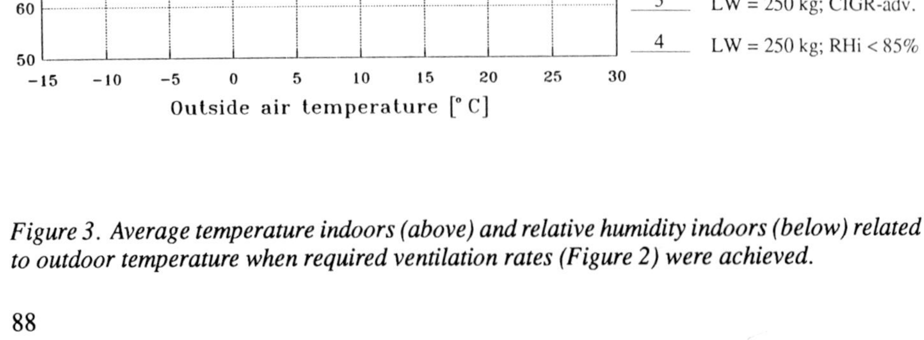

Figure 3. Average temperature indoors (above) and relative humidity indoors (below) related to outdoor temperature when required ventilation rates (Figure 2) were achieved.

Ventilation of the calf houses

The ventilation performance of the variatiants on the building design given in Table 1 were evaluated with theoretical descriptions of the air-exchange mechanisms for thermal buoyancy, wind ventilation and the combined effects of these two driving forces. The modelled theory was developed on the basis of Bruce (1986) and characterized in van 't Ooster & Both (1988).

The worst weather conditions for natural ventilation systems are generally those for which windspeed approaches zero. The ventilation system must then run on thermal buoyancy only. Almost the same can happen when a building is well sheltered from wind. Figure 2 shows the performance of the thermal-buoyancy mechanism for the two humidity criteria and the extreme live-weights, under the assumption that indoor climate is as given in Figure 3. Ventilation capacity decreases with rising external temperature, mainly because temperature difference between indoors and outdoors air decreases. For young calves, the ventilation capacity of House type 1 with small openings is just satisfactory. So a good control system could always create an acceptable climate for these calves. For calves at slaughter weight in House type 1, the criteria are violated at just over 10 °C for the CIGR advice and at about 17 °C for the criterion of fixed maximum relative humidity. If the calves at liveweight 250 kg were housed in House type 1 with large openings, the ventilation capacity, openings fully opened, would be over 150 m^3/h.calf until external temperature exceeds 20 °C. At temperatures exceeding 20 °C, the criteria cannot be maintained any more; however the capacity at 25 °C is still about 125 m^3/h.calf.

Since supplemental heating is hardly ever used in climate control in veal-calf houses, apart from the first weeks of growth, the ventilation rates given in Figure 2 are not very realistic. A better approach was obtained when the heat balance of the building and heat loss through ventilation were brought into equilibrium, thus simulating an unheated building with given opening sizes. This gave a good indication of the severity of the violation of the climatic criteria. Calculations were on House type 1 with small and large openings and on House type 2 with large openings for the higher temperature range, all house types holding 40 calves of 250 kg. Wind speed was kept low, 1 m/s. Figure 4 shows a slow decline in ventilation rate with increasing outdoor temperature for all three types. The curve marked as 1 shows ventilation rates significantly below 100 m^3/h. per calf. The other two curves are close to the minimum ventilation capacity required for mechanical-ventilation systems (150 m^3/h.calf).

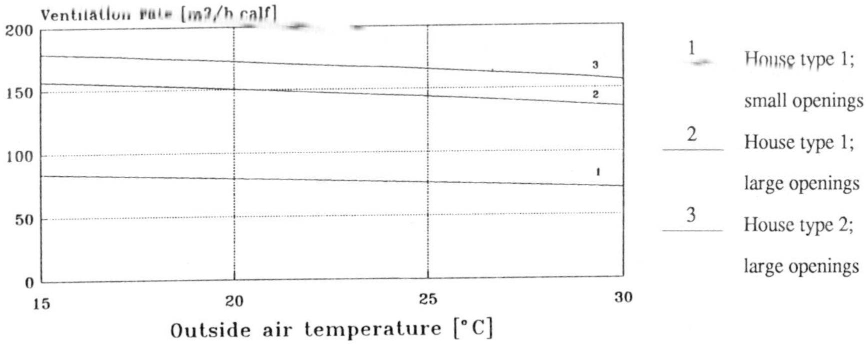

Figure 4. Ventilation rate due to thermal buoyancy and wind (1 m/s); House type 1 (small and large openings) and 2 (large openings). Openings fully open.

Very important is the reaction of the indoor climate in these cases. The intervals in which indoor temperature (t_i), relative humidity (RH_i) and the summer-temperature tolerance factor (Δt_s) varied are listed below:

1. House type 1 (small openings): $21.2 < t_i < 32.3$ °C; Δt_s: 2.3–5.1 °C; $76 < RH_i < 82\%$
2. House type 1 (large openings): $18.9 < t_i < 31.6$ °C; Δt_s: 1.6–3.2 °C; $69 < RH_i < 80\%$
3. House type 2 (large openings): $18.5 < t_i < 31.4$ °C; Δt_s: 1.4–2.8 °C; $68 < RH_i < 80\%$

Temperature and humidity criteria, apart from the CIGR advice, were obeyed in the third case and show a very bigger and a small violation of the temperature criterion only (Δt_s should be in range 2–3 °C) in the first and second case.

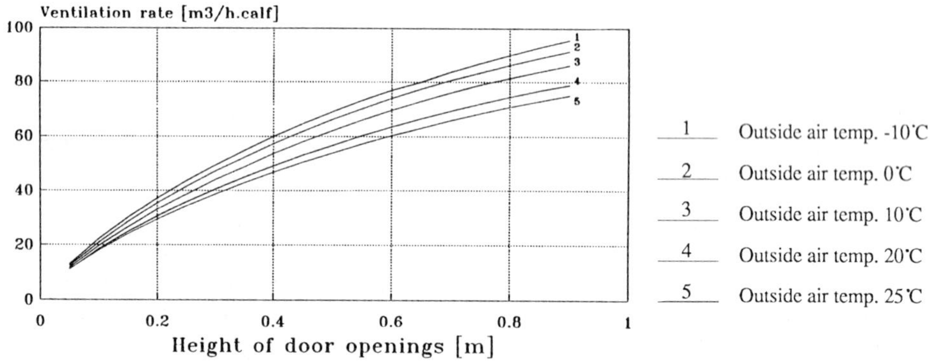

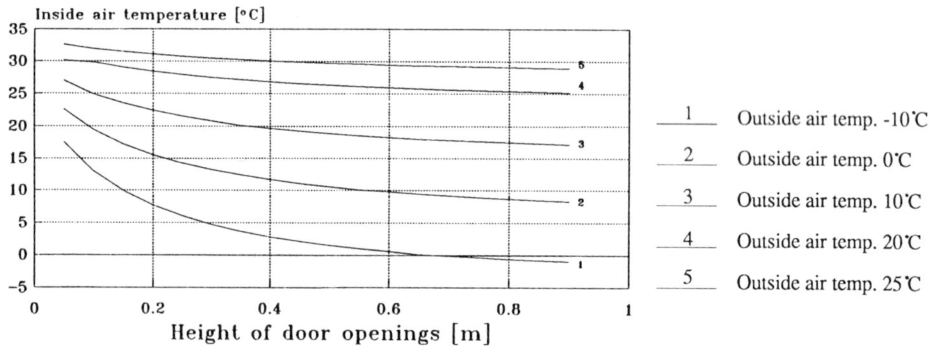

Figure 5A (above). Controllability curves of the ventilation rate for House type 1 (small openings) using the side-wall door openings as controllable openings. Thermal buoyancy only.

Figure 5B (below). Expected indoor temperature resulting from the ventilation curves in Figure 5A and animal heat production.

Controllability of the ventilation rate at very low wind speed

To indicate the controllability of the ventilation rate the effect of closing the side-wall openings was calculated for House type 1 with small openings and calves of 250 kg under windless conditions at five outdoor temperatures. Figure 5A gives the relation between the height of the side-wall openings and the ventilation rate. The ridge was kept fully open during the process. The ventilation rate can be controlled over a wide range by adjusting the side-wall openings only. Figure 5B gives the indoor temperature resulting from the ventilation curves in Figure 5A. Indoor temperature can be significantly affected at low outdoor temperatures. This range is decreasing with increasing outdoor temperature, mainly by a decreasing sensible heat production from the animals. Under the given climatic conditions, absence of wind, the criteria set for internal temperature could not be kept at high temperatures for the given opening configuration, as already concluded.

Combined effects of wind-driven ventilation and thermal buoyancy

The additional ventilation rate induced by airflow around the building is highly dependent on the wind-pressure pattern and on whether pressure differences over openings due to thermal buoyancy and wind ventilation reinforce or neutralize one another. To study these combined effects, data on wind-pressure coefficients for the two building shapes considered were derived from Moran (1980), ASHREA (1989) and Pringle (1981) and listed in Table 2.

Figure 6 shows an example of wind-pressure coefficients with poor respectively significant effect on the ventilation rate with increasing windspeed. Total ventilation rate due to the combined effects of thermal buoyancy and wind does not equal the sum of the rates given by both mechanisms individually. As mentioned before, the sum of the pressure differences induced by the individual effects is responsible for the ventilation rate achieved by the combined effects. The contribution of thermal buoyancy decreases with increasing wind speed. At wind speeds above 4 m/s, wind effect only gives a moderate estimate, within 10%, of the total ventilation rate.

Table 2. Wind pressure coefficients around the calf house at different wind angles. A wind angle of 0° implies a wind normal to the long axis of the building. House placed in an open field.

Wind angle		House type 1			House type 2		
		upw. face	downw. fece	ridge	upw. face	downw. face	ridge
0	small	0.35	–0.37	–0.13	0.35	–0.37	–0.60
	large	0.40	–0.37	–0.13	0.40	–0.37	–0.60
30	small	0.35	–0.30	–0.40	0.35	–0.30	–0.58
	large	0.35	–0.37	–0.40	0.35	–0.37	–0.58
60	small	0.09	–0.15	–0.40	0.10	–0.15	–0.57
	large	0.09	–0.20	–0.40	0.10	–0.20	–0.57
90	small	–0.10	–0.10	–0.60	–0.10	–0.10	–0.13
	large	–0.10	–0.10	–0.60	–0.10	–0.10	–0.13

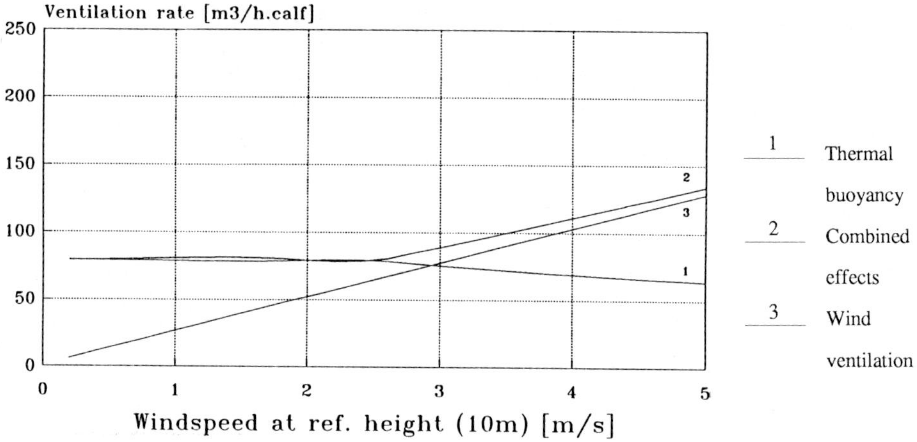

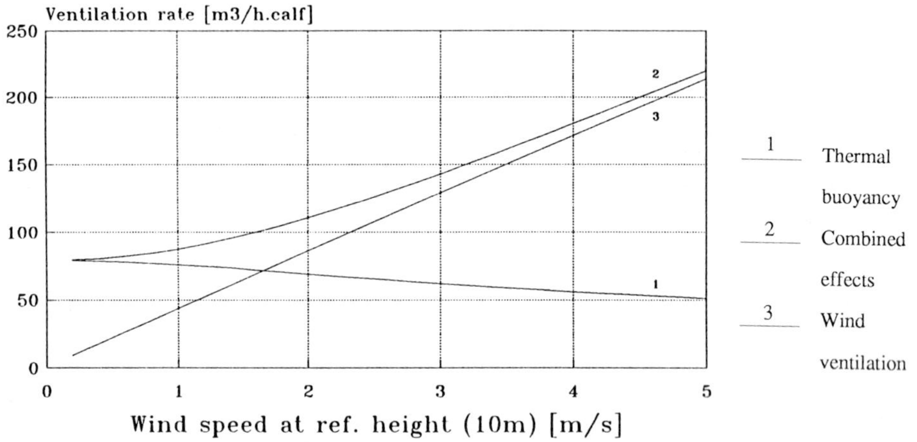

Figure 6. Combined effects, stack and wind effect as a function of wind speed with internal climatic conditions in equilibrium with the combined effect ventilation. House type 1 (above) and House type 2 (below). Small openings fully open. Wind dir. 0°.

Figure 7 indicates the effect of wind direction upon the total ventilation rate for both house types with small openings. Wind parallel to the ridge decreases the influence of the wind upon ventilation heavily. However an angle of 30° and more to the ridge improves this. House type 1 showed a clear influence of wind direction upon the ventilation rate for wind speed over 2 m/s. This is related to the roof structure. House type 2 showed little influence of wind direction upon the ventilation rate for wind directions not parallel to the ridge. Apart from wind parallel to the ridge, model calculations always indicated a higher ventilation rate for House type 2. In both types, the openings were considered to be fully open. Wind speed and to a lesser extent also wind direction thus had an important influence on the control curves of the openings (Figure 5A). With wind speed, the range in which the ventilation rate can be controlled increases. Therefore it is necessary to have a good understanding of the mechanisms of wind

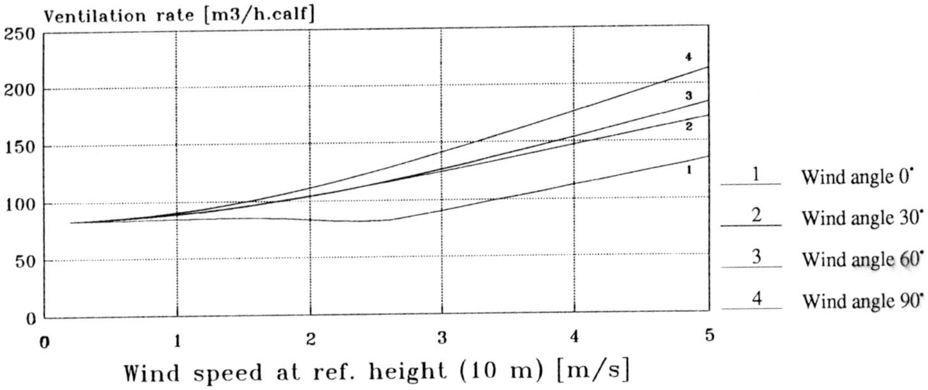

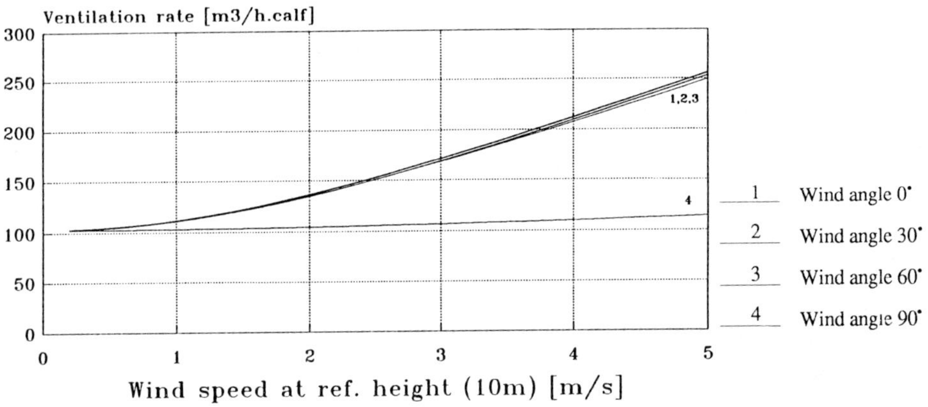

Figure 7. Combined-effect vemtilation as a function of wind speed and wind direction for House type 1 (above) and 2 (below) with small openings fully open.

ventilation and thermal buoyancy and their combined influence to achieve an appropriate control of the ventilation rate. Controllability of all openings is also preferable for this goal. A computer model containing a proper description of these mechanisms can contribute to a better control of the natural ventilation process.

Conclusions

- The computer simulation model NAT_VENT is an important tool to estimate and compare the ventilation performance of different house designs.
- The model results indicate fairly well the applicability of natural ventilation in calf houses, but validation with measured data is necessary.

- After validation, the model can be implemented in control systems for control of opening sizes.
- 3–4 cm^2/kg live weight inlet area and 2.9–3.3 cm^2/kg live weight outlet area seems to be satisfactory for house designs (Figure 1). Relatively low air speed in inlet openings (to reduce the chance of draughts at animal level) can be achieved by choosing an inlet area of 5 cm^2/kg liveweight and an outlet area of 2.5 cm^2/kg liveweight. In a group housing system or a compartment with a smaller crate size, maximum openings should be larger.
- Under poor wind conditions higher temperature rises than indicated by the summer-temperature tolerance factors have to be allowed in the described house types with small-openings (Table 1).
- In areas where very low wind speeds are rare, a well controlled natural ventilation system can be a good alternative to a mechanical ventilation with a capacity of 150 m^3/h.calf, with respect to the airexchange rate of the building. However topological aspects and built environment can be of great influence and also air mixture inside the house needs attention.
- Under Dutch climatic conditions, relative humidity can be kept within accepted limits during warm days (65–89%). To keep the CIGR advice in hot weather is impossible with any ventilation system if outside air humidity is not very low.

References

ASHRAE (American Society of Heating Refrigerating and Air Conditioning Engineers), 1989. Airflow around buildings. In: R.A. Parsons (Ed.): Handbook of fundamentals. SI Edition. Chapter 14. ASHRAE, Atlanta, Georgia, US.

Brockett, B.L. & L.D. Albright, 1987. Natural ventilation in single airspaced buildings. Journal of Agric. Engng. Res. 37:141-154.

Bruce, J.M., 1986. Theory of natural ventilation due to thermal buoyancy and wind. In: Proceedings CIGR Section II seminar on pig housing, rabbit and small birds species housing, Rennes, FR.

CIGR (Commission International du Genie Rurale), 1984. Report of working group on climatization of animal houses, Section II (Farm Buildings). Scottish Farm Buildings Investigation Unit, Aberdeen, GB.

Moran, P., 1980. Full-scale experiments to acquire wind loading data for use in the design of agricultural buildings. Journal of Agric. Engng. Res. 25:287–297.

Pringle, R.T., 1981. A design guide to mechanically ventilated livestock housing. Bulletin no. 26. North of Scotland College of Agriculture, Aberdeen, GB.

Roy, J.H.B., 1980a. The calf and its physical environment In: The Calf. 4th edition. Butterworths, London. p. 147–198.

Roy, J.H.B., 1980b. The nutrient requirements of the calf-appetite, water, energy and protein. In: The Calf. 4th edition. Butterworths, London. p. 221–252.

Smits, A.C., T. v.d. Brink & G.A. Toren, 1986. Huisvesting. In: H.J.C.M. Sturkenboom & L. Pelser (Ed.): Praktische vleeskalverhouderij. PR, Lelystad, NL. p. 23–36.

van Es, A.J.H. & E.J. van Weerden, 1970. Kan de groei van gezonde mestkalveren worden voorspeld uit het opgenomen rantsoen? Landbouwkundig Tijdschrift 82:109–114.

van 't Ooster, A. & A.J. Both, 1988. Towards a better understanding of relations between building design and natural ventilation in livestock houses. In: E. Barber (Ed.): Proceedings of the third international livestock environment symposium. ASAE, St. Joseph, Michigan, US. p. 8–21.

Vermorel, M., 1987. Effects of climatic conditions on energy metabolism and performance of calves. In: M.W.A. Verstegen; A.M. Henken (Ed.): Energy metabolism in farm animals: effects of housing, stress and disease. Nijhoff, Dordrecht, NL. p. 180–198.

Purification of veal calf manure

P.J.W. ten Have *, H.C. Willers * & W.G.J.M. van Tongeren **
* *Institute of Agricultural Engineering IMAG, P.O. Box 43, 6700 AA Wageningen, NL*
** *Netherlands Organization for Applied Scientific Research TNO, Division of Technology for Society, P.O. Box 342, 7300 AH Apeldoorn, NL*

Summary

In the Netherlands, veal calf manure has been purified in centralized activated-sludge plants since 1976. Apart from BOD reduction, N is removed by (de)nitrification. Shortly, simultaneous precipitation of P is practised in two new plants. In order to remove soluble material from veal calf manure and effluent from biological treatment, experiments with reverse osmosis have been carried out. Data are presented on the performance and the costs of both the biological and the membrane systems.
Descriptors: veal calf manure; nitrification; phosphate; reverse osmosis.

Introduction

Since the 1960s it has been difficult to dispose of the manure produced in areas with a high concentration of veal calves. Many of the units which feed calves for veal or beef production, do have little cultivated land. This problem has resulted in the practice of treating veal calf slurry in biological purification plants (VCPP) since the end of the 1960s (ten Have, 1971). Initially, this was done on the farms. But the results were poor, and this caused the Veluwe Water Authority to forbid on-farm purification in 1972. In its area about half the Dutch veal calf production is located. As an alternative, the Manure Bank of Gelderland erected a central purification plant (Elspeet Old) with an annual capacity of 18 000 m^3 in 1976. The capacity of this plant has meanwhile been raised to approx. 27 000 m^3 per annum. In 1985 a new VCPP (Elspeet New) was commissioned at a distance of 5 km from the first one. The new plant has a much larger annual capacity of about 80 000 m^3. A third plant, which is nearly identical to Elspeet New VCPP, was erected in Putten in 1986. Plans exist to raise the total processing capacity in the Veluwe area to approx. 650 000 m^3 per year. In 1989 a volume of 183 000 m^3 manure with an average of 1.55% dry matter was purified.

This paper describes several aspects of treatment and performance. Furthermore, the concentration of veal calf manure and effluent from a VCPP by reverse osmosis is discussed.

Biological treatment

The procedure in the plant is as follows: the slurry is trucked in by tanker and unloaded into a reception tank. A screen retains any parts over 15 mm. Then, the slurry is stored in a balancing tank. Once daily the batch of slurry to be treated is pumped into an aeration tank, in a period of about an hour. Aeration is started to provide oxygen to the bacteria in the activated sludge and to mix the content of the tank. Some 21 hours later, aeration is switched off to allow the sludge to settle. After a 2-h sedimentation period, the supernatant liquor is drained for another hour. Next, the cycle is repeated. Through a balancing tank, the effluent is drained off gradually into the water authority's sewerage system.

Sludge from the aeration tank is regularly pumped into a sludge tank to be concentrated, by gravity, to a dry matter (DM) content of about 5%. The supernatant liquor from this stage

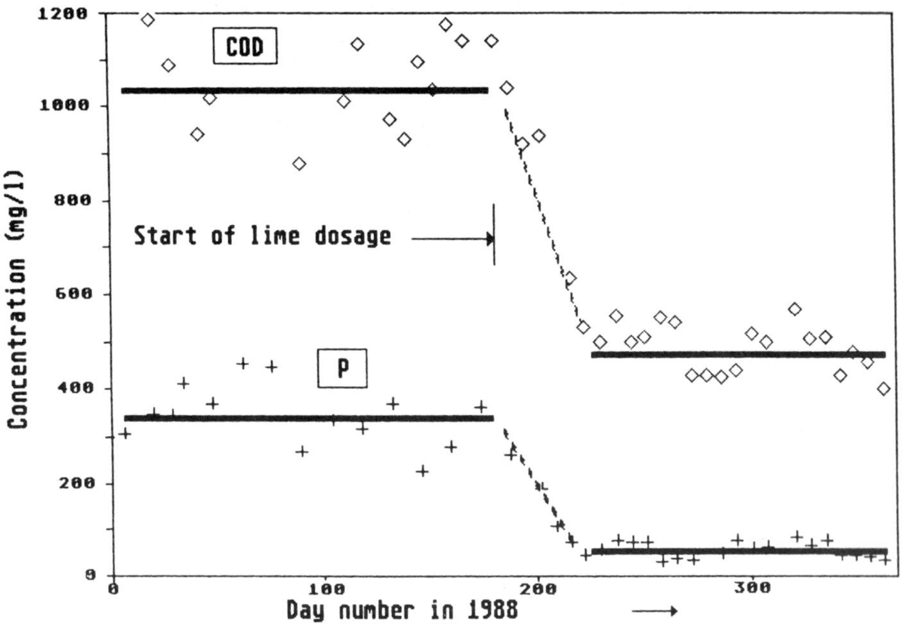

Figure 1. Changes in the concentration of COD and P after the start of the dosage of $Ca(OH)_2$.

Table 1. Elspeet New purification plant, average results in 1987.

		Influent	Effluent	Reduction %
COD	mg/l	13 000	850	93.5
BOD	mg/l	6 500	91	98.6
Nkj	mg/l	2 800	99	96.5
NH_4-N	mg/l	ca 2 400	62	97.4
NO_2-N	mg/l	0	24	
NO_3-N	mg/l	0	102	
N total	mg/l	2 800	225	92.0
P total	mg/l	520	380	27.0
K	mg/l	3 100	3 100	0.0
Cl	mg/l		2 600	
Dry matter	mg/l	15 000	ca 10 000	
Ash	mg/l	9 500	ca 8 000	
pH		7.6	7.7	

is returned to the aeration tank. Tankers remove the surplus sludge to be used in agriculture as a fertilizer.

Reduction of the BOD (biological oxygen demand) and removal of the NH_4 by nitrification and denitrification are the primary aims of a veal calf slurry purification plant. Besides, two of the three existing plants have adopted the practice of dosing $Ca(OH)_2$ (lime) into the aeration tank in order to further reduce the phosphate content. The hydraulic retention time in the two latest VCPPs is about 20 days; the load is about 0.05 kg BOD/(kg sludged) and 0.02 kg NH_4-N/(kg sludged) at an average sludge concentration of 7 kg/m^3.

Table 1 shows the average manure composition before and after treatment in Elspeet New VCPP in 1987. The reduction performance is also stated. The COD (chemical oxygen demand) and N reductions were more than 90%. Of phosphate only a minor part was removed, and soluble salts, such as K, were not removed at all.

In the course of 1988 adding 2–2.5 kg $Ca(OH)_2$ per m^3 manure was initiated. As can be seen from Figure 1, this did not only cause the P concentration to fall considerably, but also the COD. Flocculation of organic matter is likely to have caused this.

In futural plants a number of changes of the present-day procedure are possible. In the first place, the discontinuous process can be changed into a continuous one, with separate tanks for nitrification, denitrification and sedimentation. This change will contribute to a better handling of certain problems caused by inadequate denitrification due to too low a COD:N ratio in the manure. Furthermore, the odour emission will be conceivably lower compared with that experienced when slurry is pumped into the aeration tank in a very short period. Finally, it will be easier to link the plant to a continuous sludge dewatering system.

A second possible change is further dewatering the surplus sludge. It was found in experiments that it is not easy to dewater this sludge (DHV Consultants, 1988). Two types of poly-electrolyte (pe) are required to condition sludge formed in simultaneous phosphate precipitation, and the dosage of these are rather high, namely some 5 g cationic pe/kg sludge and some 2 g anionic pe/kg sludge. The use of a belt press did not prove to be possible. Provisionally, decanting centrifuges have been chosen, because they provide the possibility of thickening the sludge up to 10–11% DM without much supervision. A product of this DM content can still be pumped. There is a need for further thickening because the direct application to cultivated land is likely to suffer from the competition from other manure types. It is expected that a major portion of this sludge is to be processed in manure processing plants, which generally implies that it is to be subjected to thermal dehydration. The main market for the final product will have to be found abroad. Thermal dehydration being expensive, it is considered to adopt dewatering in existing and new VCPPs with centrifuges. In that situation, of each tonne of veal calf slurry with 1.5% DM, some 0.1 t of sludge with 10% DM will need further processing. Without transport and overload costs, treatment is estimated to cost about NLG 14 per tonne of slurry at a capacity of 180 000 m^3/yr. The investment will be about NLG 9 million.

Reverse osmosis

Experiments have been carried out to concentrate veal calf slurry and effluent from the above biological purification by reverse osmosis (RO). Tubular membranes, manufactured by PCI (type AFC 99), have been used because these are not very susceptible to clogging and can be cleaned easily. The experiments with veal calf slurry were tentative and did not last longer than 1–3 days each. With effluent, prolonged research was carried out which took about five months (van Tongeren, 1990).

First of all, most of the suspended solids have to be removed from veal calf slurry, e.g. by sedimentation. The NH_4 content of the permeate, the liquid separated by reverse osmosis, is considerably reduced as a result of lowering the pH value (with HCl or H_2SO_4) to about pH = 6. The maximum achievable concentration factor is about 3.5, which implies that the volume of the concentrate is about 30% of the volume of the settled manure. Under these conditions,

Table 2. Application of reverse osmosis on settled veal calf manure; [1] Concentration factor 3.5; [2] Acidification with H_2SO_4.

		Raw manure	Settled manure	Permeate reverse osmosis
COD	mg/l	15 000	10 000	200
BOD	mg/l	7 500	5 000	50
Nkj	mg/l	2 800	2 700	100
NH_4-N	mg/l	2 400	2 400	100
P total	mg/l	600	270	3
K	mg/l	3 000	3 000	100
Cl	mg/l	2 500	2 500	150
pH		7.8	7.8 → 6[2]	6.5
Relative volume[1]	%	100	90	65

Table 3. Application of reverse osmosis on effluent from biological treatment; [1] Concentration factor 5.5; [2] Acidification with H_2SO_4.

		Effluent	Permeate reverse osmosis
COD	mg/l	600	15
BOD	mg/l	100	< 5
Nkj	mg/l	80	5
NH_4-N	mg/l	50	5
P total	mg/l	40	0.2
K	mg/l	3 000	100
Cl	mg/l	2 500	100
pH		8 → 6[2]	6
Relative volume[1]	%	100	82

a flux across the membrane of 10–12.5 l/(m^2 h) can be taken into account. The total concentration factor, due to sedimentation and reverse osmosis, is about 3.

Table 2 gives the composition of raw manure, settled manure and permeate, as may be expected under practical conditions. Despite the acidification, the BOD and Nkj concentrations in the permeate are still rather high so that water authorities in the Netherlands are not likely to accept it to be directly discharged into open water. So, the permeate will have to be disposed of into the sewer or be spread on land.

The mixture of sediment and concentrate is in fact nothing else than veal calf slurry which has been thickened three times. The lower moisture content makes it more attractive as a fertilizer and reduces the transport costs per unit quantity of the original slurry. The same applies to the processing costs in a drying plant.

Acidification to pH = 6 (about 1.5 l H_2SO_4/m^3) also has favourable effects on the effluent from a VCPP, in this case on the flux. Under practical conditions, about. 30 l/(m^2 h) may be taken into account, with a maximum concentration factor of 5.5. It is possible to continue the processing for 1–2 weeks without intermediate cleaning. Table 3 shows how the effluent composition improves as a result of membrane filtration.

The application of reverse osmosis on effluent from biological treatment improves the quality of the final effluent, but on the other hand increases the volume of substances for which an outlet has to be found by about 0.15 m^3/m^3 manure.

The costs of reverse osmosis are estimated at 5–15 NLG/m^3 water removed. The lowest costs apply to effluent treatment in the VCPP, the highest to farm-scale concentrating of slurry. These costs do not include those of marketing and processing the concentrate and permeate.

Final remarks

Biological treatment of veal calf slurry in the Netherlands is based on relatively many years' experience. It can be claimed to have proved its practical value.

Plants still to be erected or expanded, will see changes which are partly based on the experience acquired, such as the splitting up of a number of functions (nitrification, denitrification, sedimentation). Other changes and expansions are caused by ever-increasing environmental demands. Examples of these are the measures taken to control odour emissions and solutions for the problems of how to find environmentally sound destinations for the minerals contained in manure. It may be expected that in a few years' time, also salts and other residual substances will have to be removed from the effluent. It has been proven that the latter can be achieved by means of reverse osmosis, but evaporation seems to be another proper technique for this.

References

DHV Consultants, 1988. Slibontwateringsonderzoek op de kalvergierzuiveringsinrichting aan de Knardersteeg te Putten. Report 1-4061-19-01.

ten Have, P., 1971. Aerobic biological breakdown of farm waste. In: Livestock waste management and pollution abatement. Proc. Int. Symp. Livestock Wastes, Columbus, Ohio. p. 275–278.

van Tongeren, W.G.J.M., 1990. Omgekeerde osmose van effluent van een biologische kalvergierzuivering (duurproeven). TNO report No 90-098.

4. Health and care

Control of infectious disease in housed veal calves

A.J.F. Webster
Department of Animal Husbandry, University of Bristol, Langford, Bristol, BS18 7DU, GB

Summary

Epidemiological surveys of mortality and morbidity in units rearing calves for veal emphasise the overriding importance of diarrhoea and pneumonia. In general the incidence is higher in calves reared in groups and receiving milk replacer ad libitum from nipple feeders. A more mechanistic approach to disease incidence suggests that diarrhoea is related to excessive consumption of milk replacer (however fed) and denial of digestible dry food. Pneumonia can be linked to over-intensive stocking and stresses of concurrent infection, e.g. enteritis. Veal calves can be reared without incurring these stresses. In these circumstances their health is as good as that of conventionally reared animals.
Descriptors: veal calves; infectious disease; housing; management.

Introduction

The health status of veal calves is not satisfactory. Infectious diseases impose a severe economic burden on the veal industry through deaths, reduced performance and the costs of treatment. These problems are compounded by increasing public concern that veal calves should be reared

1. in groups, thus potentially increasing the risk of infection and
2. without regular prophylactic medication, thus potentially increasing the severity of infection.

I have argued elsewhere that I consider good health to be paramount among the 'five freedoms' that constitute acceptable welfare standards for farm animals (Webster, 1987). I believe equally strongly that a system which can only achieve 'good' health through the continuous administration of antibiotics is not good husbandry. This paper will examine factors affecting mortality and morbidity in veal calves, compare them with other artificial forms of calf rearing and explore ways in which they may be reduced without medication by attention to housing and management.

Mortality and morbidity in veal calf units

Table 1 summarizes the results of an analysis of mortality and morbidity in 4417 calves reared for veal in Holland in individual pens or group housed (van de Mei, 1987). Mortality rate in both groups was 3.7%. Culling rate, incidence of diarrhoea and losses from pneumonia were significantly higher in group housed calves but morbidity rates for pneumonia and the extent of lung consolidation were unaffected by rearing system. This suggests that group housing may exacerbate infectious diseases, especially diarrhoea and pneumonia. However, the animals designated 'group housed' were, in fact, tethered and bucket-fed to 10 weeks of age, which does not compare with calves that are unrestrained and share access to milk via a teat dispenser from the time of their arrival on the unit. This comparison is given in Table 2 which summarises the incidence of diarrhoea and pneumonia in 794 early-weaned and veal calves *on affected farms* (Webster et al., 1985a,b). Here group rearing from the time of arrival significantly increased the incidence of diarrhoea from 0–6 weeks of age in veal calves but not in early weaned calves which were also given access to dry food containing digestible fibre

Table 1. Morbidity and mortality data for 4417 calves reared for veal individually or in groups. From van der Mei (1987).

Incidence (%)	Individual pens	Group-housed[1]	Significance
Mortality	3.7	3.7	
Culling rate	2.4	3.9	*
Diarhoea, morbidity	14.6	18.6	*
Respiratory disease			
losses	0.7	1.5	*
morbidity	25.9	25.3	
severe lung consolidation	19.6	20.9	
number of treatment periods	1.6	1.5	*

[1] groups of four calves - tethered to 10 weeks of age.

Table 2. Effects of rearing system on the incidence of disease in calves on units where disease was present. From Webster et al. (1985).

Calves treated on affected farms (%)		Diarrhoea		Pneumonia	
		0–6 w	6–12 w	0–6 w	6–12 w
Early weaned	individual pens	36	nil	28	12
	group reared	34	nil	35	10
Veal calves	individual pens	32	12	54	27
	group reared	57	13	64	34

that would enter and stimulate development of the rumen. Weaning effectively eliminated the problem of diarrhoea. The incidence of pneumonia was higher in veal calves at all ages and there was a suggestion that the problem was worse in group-reared calves, although this difference was not statistically significant.

Waltner-Toews et al. (1986), at Ontario Veterinary College, have conducted a most comprehensive epidemiological analysis of effects of farm management policies on disease in calves. Data, summarised in Table 3, are for 1968 artificially-reared and early-weaned dairy calves but the morbidity rates for diarrhoea and pneumonia are broadly similar to those for veal calves. Odds ratios define the relative probability of disease for each pair of treatments under comparison. Rearing calves out of doors in individual hutches as distinct from indoors in adjacent pens produced a 4.5-fold reduction in diarrhoea and a 25-fold reduction in pneumonia! There was, however no difference between group and individual penning for housed calves. The incidence of diarrhoea was reduced by providing salt ad lib. Medication reduced the incidence of pneumonia but effects on diarrhoea were inconsistent. Ad lib access to water reduced the incidence of pneumonia but minerals increased it. These effects are inexplicable and introduce a note of caution into any attempt to interpret epidemiological data in isolation. Nevertheless, there was clear evidence that the attempt to induce passive immunity to diarrhoea in these calves by vaccinating their mothers (vaccines unspecified) was decidely counterproductive. The final observation of note is that calves which had previously been treated for diarrhoea were three times more likely subsequently to contract pneumonia.

Table 3. Effects of farm management policies on morbidity in 1968 artificially-reared dairy calves in Ontario. From Waltner-Toews et al. (1986).

	Overall	Diarrhoea	Pneumonia
Morbidity (%)	35.0	20.5	15.4
Odds ratios[1]			
hutches vs. pens	0.25	0.22	0.04
group vs. individual pens			
antimicrobials vs. none	0.30	inconsistent	0.40
milk replacer vs. whole milk	6.59	-	-
pail feeding vs. nipple feed	0.16	-	-
water ad lib vs. none			0.24
salt ad lib vs. none		0.22	
minerals ad lib vs. none			5.35
anticour vaccine to cow vs. none	5.47	4.71	4.64
treatment for scours v. none			3.00

[1] Odds ratios define the relative probability of disease for each pair of treatments under comparison. The 'overall' and 'pneumonia' columns refer to probabilities of farms experiencing significantly above-median treatment days per calf, the 'diarrhoea' column to whether or not farms experienced at least one episode of scours.

The epidemiological data given in Tables 1–3, interpreted with due caution, confirm the high incidence of diarrhoea and pneumonia in housed, artificially-reared calves, suggest that diarrhoea is affected both by the nature of access to milk powder and the provision of solid food and further suggest that diarrhoea is a predisposing factor to pneumonia. These associations require a mechanistic interpretation.

Management and enteric disease

The two most likely causes of enteric disease in calves (in the UK) after arrival on a rearing unit at 1–2 weeks of age are *Rotavirus* and *Salmonella*. Unless the origin and health status of the calves on arrival is known, these diseases are impossible to eliminate. Despite Table 3, there is now good evidence that *Rotavirus* can be controlled by vaccination of the dam (Snodgrass et al., 1982). For this to be effective, however, calves must receive not only colostrum but also a continuing supply of milk from their vaccinated dam to ensure continued supply of antibody to confer local immunity at the mucosal surface of the gut. The technique is therefore most applicable to suckler calves and of doubtful benefit to the veal industry. Continuous medication is undesirable and cannot guarantee protection, e.g. against all serotypes of *Salmonella typhimurium*. Since the control mechanisms of elimination, vaccination and medication are all, for one reason or another, suspect, the only remaining options are attention to hygiene and diet.

The incarceration of veal calves within individual crates that deny them, when young, any physical contact with their immediate neighbours, was mainly intended to reduce the spread of infection by contagion. Wathes et al. (1988) have demonstrated that *Salmonella typhimurium* can be transmitted between mice and calves at least as effectively by inhalation of aerosols as by oral administration. A recent, as yet unpublished, epidemiological study of salmonellosis in individually crated calves by Hardman & Wathes has shown that only 40% of infections can be linked to the direct proximity of an infected neighbour. This study does not apportion

Table 4. Comparison of the performance and health of Hereford × Friesian female calves at the University of Bristol. From Webster et al. (1986).

	Rearing system	
	Quantock	Access
Numbers red	82	63
Milk powder consumption (kg)	194	178
Dry food consumption (kg)	nil	30
Weight at slaughter (kg)	162	168
Liveweight gain (kg/d)	1.09	1.12
Killing out (%)	57.2	57.0
Relative incidence (%)		
deaths and culls	11	nil
enteritis	35	8.0
pneumonia	18	3.2

the transmission of the remaining 60% between aerosols and vectors but it does clearly demonstrate that individual crates confer no guarantee against the spread of enteric disease.

Tables 1–3 confirm clinical impressions that the incidence (if not the severity) of diarrhoea is higher in group-reared veal calves, especially when they drink from a common teat, but is no greater than in crates if calves are reared more conventionally, i.e. given restricted amounts of milk replacer and early access to digestible dry food so as to encourage rumen development. A major survey by Thickett et al. (1983) confirmed that the health of group-reared, teat-fed but early weaned calves in the UK was quite as good as that of individually penned, bucket-reared animals. Taken together, these observations suggest that group rearing may increase the risk of exposing calves to enteropathogens but that this may be offset by attention to diet.

Diarrhoea is nearly always a consequence of damage to the absorptive villi of the small intestine and this is nearly always (there are exceptions) caused by antigens from viruses, bacteria or food itself (or a combination of all three) which have escaped destruction in the rumen and abomasum. When abomasal pH is below 4 most microbes are killed before entry to the duodenum. The probability that antigens will escape destruction must be related to the amount of material that leaves the abomasum in a medium with a pH higher than 4. One would therefore, from first principles, predict calves to be at greatest risk from diarrhoea when

1. they take very large liquid feeds direct into the abomasum twice daily or
2. they have ad libitum access to liquid feed from an automatic dispenser, especially in the first few weeks of life when appetite control has not developed and when there is no alternative supply of palatable food. Disease surveys in veal calves would confirm this to be so.

Table 2 showed that the problem of diarrhoea in artificially-reared calves disappeared after weaning. Now anything entering the mouth first enters the rumen where it is fermented or degraded in an anaerobic medium by commensal non-pathogenic microorganisms which not only serve as the major suppliers of energy and protein to the host animal but also kill, or inhibit, orally ingested pathogens. Entrepreneurs wishing to make a more or less honest living out of the problem of diarrhoea in veal calves have sought to market probiotic microorganisms which will competitively inhibit enteropathogens. The calf rearer might think it more sensible to develop that ultimate probiotic, the rumen.

The most successful of our approaches to the development of improved husbandry systems for veal calves we named the 'Access' system (Webster et al., 1986). Calves were given computer-controlled access to a two-compartment feeding station which delivered either regulated amounts of milk via a teat or a source of palatable digestible dry feed. This was of a composition likely to encourage normal rumen development and restricted to a maximum of 400 g/day so as not to inhibit appetite for milk. Clean barley straw was provided as an additional source of long fibre. Table 4 compares 'Access' with the 'Quantock' system where calves were group-reared on ad libitum milk replacer with straw as the only source of dry food. The Access system produced similar growth rates at lower feed costs. More important in this context, rates for mortality, enteritis and digestive disorders were greatly reduced, to a level that would be considered respectable in a conventional calf-rearing enterprise. The number of calves involved in our University-based 'Access' trials was small, mainly because the price of calves has made almost all forms of veal production uneconomic in Britain. It did however, provide sufficient confirmation of first principles to permit us to conclude that 'no calf should be deprived access to palatable, digestible dry food after two weeks of age'. Similar clauses have been incorporated into British legislation effective from 1st January 1990 and draft legislation before the Commission of the European Communities.

Management, housing and respiratory disease

The aetiology of calf respiratory disease is too complex to consider here. It is necessary, however, to recognise two distinct types. The first includes conditions such as infectious bovine rhinotracheitis (IBR) where exposure of susceptible animals to a specific pathogen carries a high probability of disease but where vaccination is effective. The second, broader category includes the family of diseases broadly referred to as calf pneumonia. This may involve one or more of a multiplicity of microorganisms, many of which can be recovered from the respiratory tracts of healthy animals - which may succumb to disease at a later date. The presence of the organism is no guarantee of disease or indeed of immunity. Whether or not this equilibrium stage proceeds to disease appears to require some disturbance which increases the magnitude of the challenge from the environment or reduces the resistance of the host.

Since elimination of calf pneumonia in commercial veal units is impossible, medication is expensive and unpopular and vaccination has, to date, been rather unpromising, control becomes once again a question of housing and management. This has two primary objectives:

1. to minimise the challenge to the lower respiratory tract from primary pathogens and other substances likely to irritate epithelial surfaces,
2. to avoid stresses that may impair the immune and other defences of the respiratory tract.

Challenge - air quality in veal houses

A calf may inhale

1. primary pathogens (viruses and bacteria) which may colonise the respiratory tract provided that they are alive and in a fit state to do so.
2. allergens (e.g. fungal spores) which may induce inflammatory responses whether dead or alive.
3. 'nuisance' dusts and gases which may not directly damage the lung but compromise its resistance to pathogens.

The factors that influence the magnitude of the challenge are illustrated in Figure 1. The concentration C (n cm^{-3}) of any particle in air is determined by its rate of release from animals and the rest of the building (Ra and Rb, n cm^{-3} h^{-1}) and their clearance, principally by ventilation and death in situ (q_v and q_d, h^{-1}).

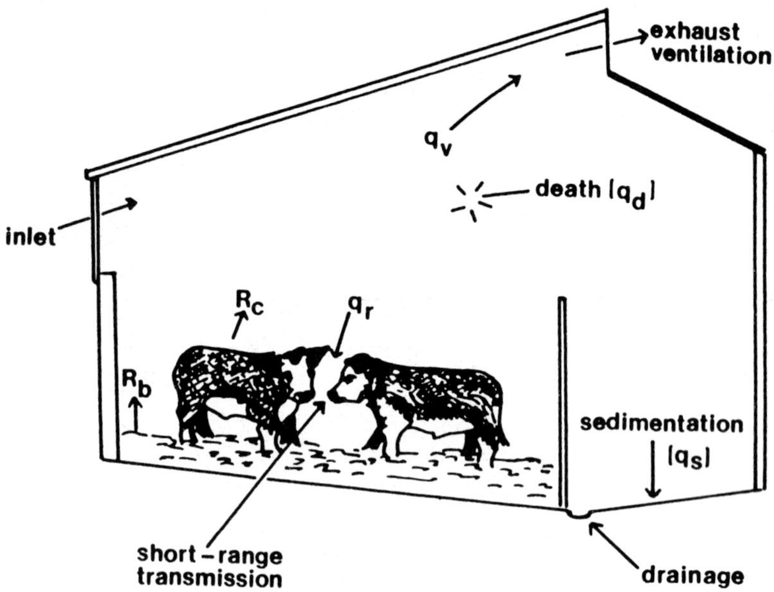

Figure 1. Pathways for release, transmission and removal of infectious organisms within the air of an animal building. Rc and Rb are release rates from animals and the building (n cm^{-3} h^{-1}), q_v, q_s and q_d (h^{-1}) are clearance rates by ventilation, sedimentation and death in situ. From Webster (1984).

The relative importance of q_v and q_d depends on the nature of the organism. Fungal spores are allergenic whether dead or alive so clearance is effected almost entirely by ventilation. It is important to remember however that the concentration of such particles (or noxious gases) in air is related to the reciprocal of ventilation rate so increases sharply at ventilation rates less than 4 air changes per hour (Webster et al., 1987).

The situation with regard to viable infectious microorganisms is more complex. Most of the organisms that can be recovered from the air are non-pathogenic although if present in sufficient numbers may contain enough endotoxin to cause damage. Very small numbers of primary pathogens are recovered from air samples even when substantial numbers of infected animals are in the building. This is because their survival time in air is very short. The life expectancy of most individual organisms within most species of respiratory tract pathogens in air is probably less than one minute, although the very few (ca 1%) which survive the initial shock of aerosolisation may live for hours. However, death in situ (q_d) is usually a much more effective mode of clearance for infectious organisms than is ventilation (q_v). Figure 2 compares the effects of changing ventilation (q_v) and stocking rate (thus Ra) on the concentration in air of microorganisms with a mean death rate (q_d) of 50 h^{-1}. Increasing air space per calf from 5–10 m^3 halves C since the animals contribute (nearly) all the microorganisms. Increasing ventilation rate from 4–40 h^{-1} reduces C by only 60%. In this example, a ten-fold increase in ventilation rate is only two-thirds as effective as halving stocking density in reducing the challenge from infectious organisms. For any specific pathogen the exact balance between q_d and q_v depends on its survival characteristics in different conditions of temperature and humidity. It should be clear however, that when respiratory infection is present within a calf house, 'improvements' to ventilation have little effect on the magnitude of the challenge from

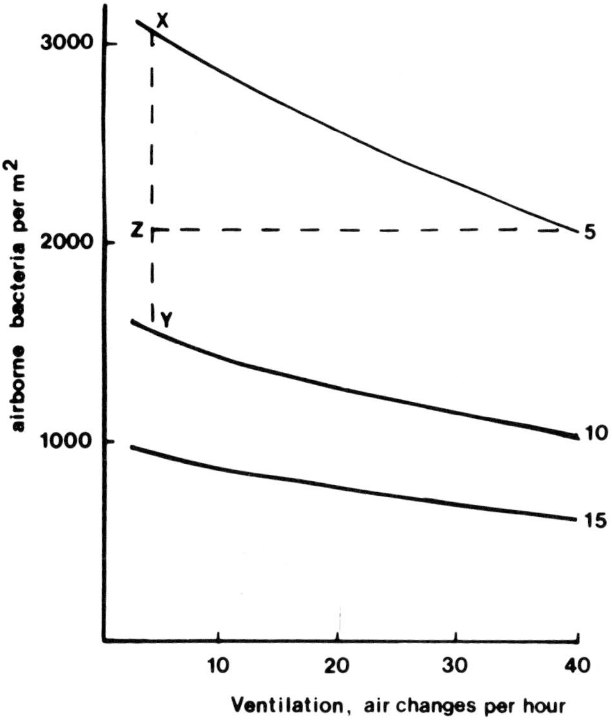

Figure 2. Effect of space allowance per calf (m^3) and ventilation rate (h^{-1}) on the concentration of airborne bacteria in a calf house. The intervals XY and XZ indicate, respectively, the effects of increasing space allowance from 5–10m^3/calf (XY) and increasing ventilation rate from 4–40 air changes per hour (XZ). From Webster (1984).

the primary pathogen and certainly cannot compensate for effects of overstocking. I have argued elsewhere that air quality in houses for veal calves reared to a weight of about 180–200 kg begins to decline very sharply when air space per calf falls below 15 m^3 (Webster, 1984). This may be considered an uneconomically generous space allowance by the industry but that does not alter the facts of the case.

The other way of reducing the challenge from pathogens is to ensure that calves are introduced in groups into buildings that have previously been depopulated, disinfected and rested. The most dangerous environment is probably that which contains recovered carriers of infection. It is also in the interests of the group to remove sick animals to hospital pens.

Resistance to infection within the respiratory tract

Undoubtedly the most effective counter to a single infectious agent is the presence of sufficient, specific antibody and the competence to recruit the full barrage of immune defence mechanisms. If effective vaccines could be produced against all the primary pathogens implicated in calf pneumonia the health, if not the welfare, of veal calves could probably be ensured in environments as polluted as those which exist in poultry houses. Meantime, however, we have no alternative but to try to minimise those stresses liable to impair resistance.

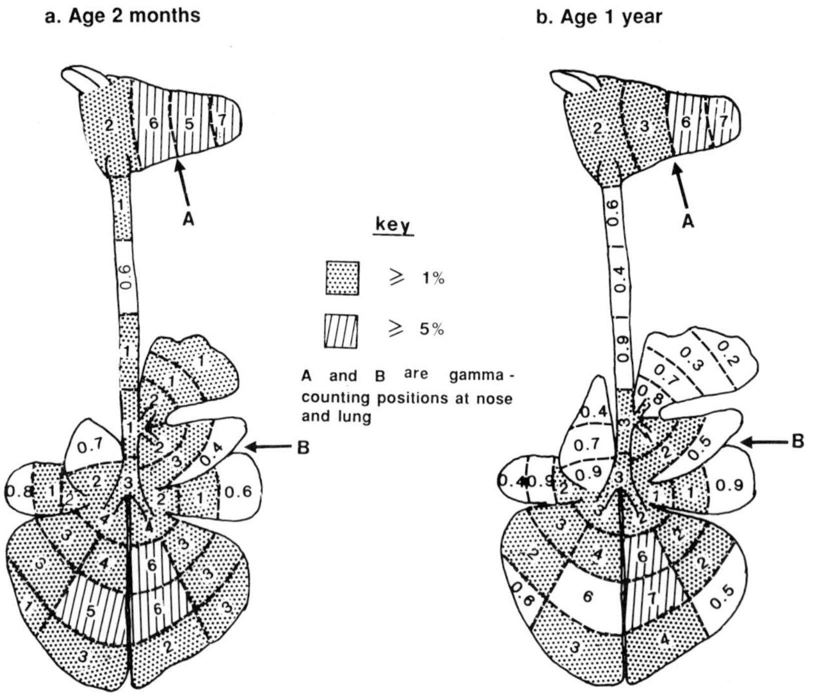

Figure 3. Regional deposition of 3μm microspheres in the calf lung at 2 months and 1 year of age. The numbers define the percentage of total deposition within each region. From Davies (1990).

Effects of the thermal environment on immune mechanisms in calves are complex and equivocal (Kelley, 1980) but it is most unlikely that changes in the physical environment within the range likely to occur in veal calf units will have any effects on the immune system. More severe stresses such as concurrent infection may well be immunosuppressive and this may explain the association between diarrhoea and subsequent pneumonia (Table 3).

There are several ways by which the environment may affect the non-specific defences of the lung. The major pathway for clearance of inhaled particles is the mucociliary escalator which mechanically removes nearly all particles depositing within the bronchioles. It is known that mucociliary clearance rates are impaired by some viral infections and this may predispose to attack by secondary bacteria. We also have evidence that mucociliary clearance is impaired in some calves below 3 months of age for reasons unrelated to infection (Davies & Webster, 1987). These individuals may be particularly susceptible to primary or secondary infection; this needs further investigation.

Perhaps the most intriguing, unexplained feature of calf pneumonia is its predeliction for the right apical lobe. This cannot have an immunological explanation. Since this lobe arises directly, and obliquely from the trachea before the bifurcation of the bronchi, many (including myself) assumed it to be a predeliction site for the deposition of respirable particles containing pathogens and/or pollutants. However, recent studies with 99m,Tc-labelled aerosols has revealed that most inhaled particles travel down the most direct route to the diaphragmatic lobes (Figure 3) which are *least* prone to pneumonia (Davies, 1990). The right apical lobe receives the least particles, which implies also the least air and because anoxia within the lung tends to *reduce* blood flow, the poorest blood supply. These relatively poor conditions of pulmonary ventilation and perfusion, greatly exacerbated by extreme confinement in individual crates that permit almost no movement, may favour the local multiplication of pathogens which enter the lung in inspired air or, more probably, by movement down the trachea against the action of the mucociliary escalator.

Conclusions

The high incidence of diarrhoea and pneumonia in veal calves remains resistant to control by vaccination and medication. It can however be related both epidemiologically and mechanistically to aspects of feeding, housing and management that are peculiar, if not unique, to veal calves. Diarrhoea can be linked to the excessive consumption of milk replacer and denial of digestible dry food. Pneumonia can be linked, in part, to over-intensive stocking and in part to the stresses of concurrent infections such as enteritis. We have demonstrated that calves can be reared for veal without incurring these stresses and that the health of such animals can be maintained without the need for constant medication. The 'Access' system is not the only way forward but it demonstrates that progress is possible if the industry is prepared to accept radical change.

References

Davies, C.P., 1990. Deposition and clearance of inhaled particles of different size in calves of two ages. Journal of Applied Physiology (in press).

Davies, C.P. & A.J.F. Webster, 1987. Deposition and clearance of monodisperse aerosols in the calf lung: effects of particle size and a mucolytic agent, bromhexine. Canadian Journal of Veterinary Research, 51:306–391.

Kelley, K.W., 1980. Stress and immune function: a bibliographic review. Ann. Rech. Vet. 11:445–478.

Snodgrass, D.R., L.K Nagy, D. Sherwood & I. Campbell, 1982. Passive immunity in calf diarrhoea. Vaccination with K99 antigen of enterotoxigenic *Escherichia coli* and *Rotavirus*. Infection and Immunity 37:586–591.

Thickett, W.S., N.H. Cuthbert, T.D.A. Brigstocke, M.A. Lindeman & P.N. Wilson, 1983. A note on the performance of and management of calves raised on cold acidified milk replacer fed ad libitum. Animal Production 36:147–150.

van der Mei, J., 1987. Health aspects of welfare research in veal calves. In: M.C. Schlichting & D. Smidt (Ed.): Welfare aspects of housing systems for veal calves and fattening bulls. CEC, EUR 10777EN.

Waltner-Toews, D., S.W. Martin & A.H. Meek, 1986. Dairy calf management. Morbidity and mortality in Ontario Holstein herds. 3. Association of management with morbidity. Preventive Veterinary Medicine 4:137–158.

Wathes, C.M., W.A.R. Zaidan, G.R. Pearson, M. Hinton, & J.N. Todd, 1988. Aerosol infection of calves and mice with *Salmonella typhimurium*. Veterinary Record 123:590-594.

Webster, A.J.F., 1984. Calf husbandry, health and welfare, Collins, London.

Webster, A.J.F., C. Saville, B.M. Church, A. Gnanasakthy & R. Moss, 1985. Some effects of different rearing systems on health, cleanliness and injury in calves. British Veterinary Journal 141:249–264.

Webster, A.J.F., C. Saville & D. de B. Welchman, 1986. Improved husbandry systems for veal calves. University of Bristol Press.

Webster, A.J.F., 1987. Meat and right, farming as if the animal mattered. Canadian Veterinary Journal 28:462–466.

Webster, A.J.F., A.F. Clarke, T.M. Madelin & C.M. Wathes, 1987. Air hygiene in stables. 1. Effects of stable design, ventilation and management on the concentration of respirable dust. Equine Veterinary Journal 19:448–453.

Ruminal drinking in veal calves

A. van Weeren-Keverling Buisman *, Th. Wensing *, H.J. Breukink * & J.M.V.M. Mouwen**
* *Department of Large Animal Medicine and Nutrition, Faculty of Veterinary Medicine, University of Utrecht, Yalelaan 16, 3584 CM Utrecht, NL*
** *Department of Pathology, Faculty of Veterinary Medicine, University of Utrecht, Yalelaan 16, 3584 CM Utrecht, NL*

Summary

Two experiments are described to enhance the knowledge of the pathogenesis of ruminal drinking (RD) in veal calves. In the first, a calf model was created in which up to half of the milk ration was given directly intraruminally. Clinical symptoms of RD were present, but less obvious than in RD patients. Pathological symptoms include villar atrophy in the proximal jejunum, and hyperkeratosis in the rumen. There was a severe drop in pH of the ruminal fluid probably caused by the production of lactate and volatile fatty acids. In the recovery period complete restoration of the jejunal and ruminal mucosa is observed.

In the second experiment, temperature sensors in the lumen of rumen and abomasum were used to record changes in temperature in the different stomach compartments, while drinking milk at 41 °C. In this way, the closure of the reticular groove while drinking could be determined indirectly. A distinct difference is found between the results of drinking or sucking of the same amount of milk, even in healthy calves!
Descriptors: veal calf; ruminal drinking; ruminal hyperkeratosis; jejunal villus atrophy; temperature sensors.

Introduction

Ruminal drinking (RD) in veal calves can be responsible for great economic losses, and is the result of a disturbance in the reflex closure of the reticular groove while drinking milk. This causes an accumulation of large amounts of milk replacer in the forestomachs. Clinical symptoms include inappetence, recurrent tympany, abdominal distension, growth retardation and clay-like faeces. These faeces have a higher dry matter content and an increased percentage of free fatty acids, compared to those of normal veal calves. Radiological studies demonstrate that the ingested milk is retained in the rumen for periods up to 48 hours. Post mortem examinations reveal cachexia, hyperkeratosis in the rumen and villus atrophy in the proximal small intestine. The syndrome appears to be related to all kinds of stress situations or may often be secondary to diseases of other organ systems (Breukink et al., 1988).

For therapy, two different approaches can be distinguished. In the first, the reconditioning therapy, the RD calf is encouraged to suck small amounts of milk from the farmer's fingers or through a floating nipple. The second technique is to wean the calf abruptly and turn it into a ruminating calf (de Visser & Breukink, 1984; Breukink et al., 1988).

In this paper, two experiments are described to enhance knowledge of the pathogenesis of RD in veal calves and to provide a method for determining the closure of the reticular groove while drinking.

Experiment 1

Material and methods

Six clinically normal male FH calves were used: 5 of them were included in the experimental group and the 6th was the control calf. In the 4th week of the fattening period, a T-shaped fistula was implanted in the proximal jejunum of 4 of the calves, including the control calf, to allow repeated biopsy sampling of the jejunal mucosa (van Weeren-Keverling Buisman et al., 1988a). In the 6th week a rumen fistula was placed in all 6 calves to administer part of the milk ration directly intraruminally, and to obtain samples of the ruminal fluid. In these samples, pH, lactate concentration, total concentration of volatile fatty acids (VFA), and the VFA pattern was determined. Besides, this fistula allowed morphological study of the rumen mucosa with an Olympus fiberscope (CF, type 1T10L, Tokyo) (van Weeren-Keverling Buisman et al., 1990a).

For the experimental calves, the fattening period was divided into three periods:

a. the pre-induction period (1st–6th week), in which the operations took place and the calves were fed twice a day from a bucket
b. the induction period (from the 7th week on), in which a gradually increasing amount of the daily milk ration, up to 50%, was infused directly into the rumen. This was continued until the faeces became clay-like
c. the recovery period, in which all milk was again given from a bucket.

Results and discussion

The clinical signs of RD which appeared in the induction period were, although present, less severe in this calf model than in RD patients. A reduction in appetite was seen in only one calf. However all calves showed ruminal tympany after intraruminal milk administration and produced clay-like faeces after 3 to 7 weeks. In one calf, diarrhoea was present for the longer part of the recovery period, which is also frequently observed in reconditioned RD patients (van Weeren-Keverling Buisman et al., 1988b).

Biopsies of the proximal jejunal mucosa revealed villus atrophy; at the end of the induction period, villi length were only 41–47% of the initial length. This was probably the result of the intraruminal milk-feeding procedure itself, as no pathogenic micro-organisms were isolated from faeces in this period (van Weeren-Keverling Buisman et al., 1990b). The exact pathogenesis of the villus atrophy still remains to be solved, but could be caused by an altered composition of the chyme or microbial flora in the small intestine.

The rumen fluid showed distinct biochemical changes after the intraruminal feeding of half of the milk ration: there was a severe drop in pH from 7.00 to about 5.00 in 60 minutes (Figure 1). This was most probably the result of a rise in lactate concentration (with a peak at 60 minutes) and a slower increase in VFA concentration (peak at 180 minutes). In the VFA pattern, there was a relative decrease in acetate, and relative increases in propionate and butyrate. Of these, butyrate especially was thought to be responsible for the observed hyperkeratosis in the rumen of the experimental calves, due to its mitogenic effect on the rumen epithelium (Sakata & Tamate, 1978). The rumen mucosa of the control calf, which was fed orally during the entire fattening period, retained its normal appearance.

In the recovery period, villi in the proximal jejunum increased to almost their original length, and the hyperkeratosis of the rumen mucosa disappeared completely. Also in RD patients, a successful recovery is described (van Weeren-Keverling Buisman et al., 1988b), although the reconditioning therapy does involve some extra time and labour.

So this RD model provides useful information about the pathogenesis of the symptoms of RD in veal calves, although the pathological changes are more distinct than the clinical symptoms.

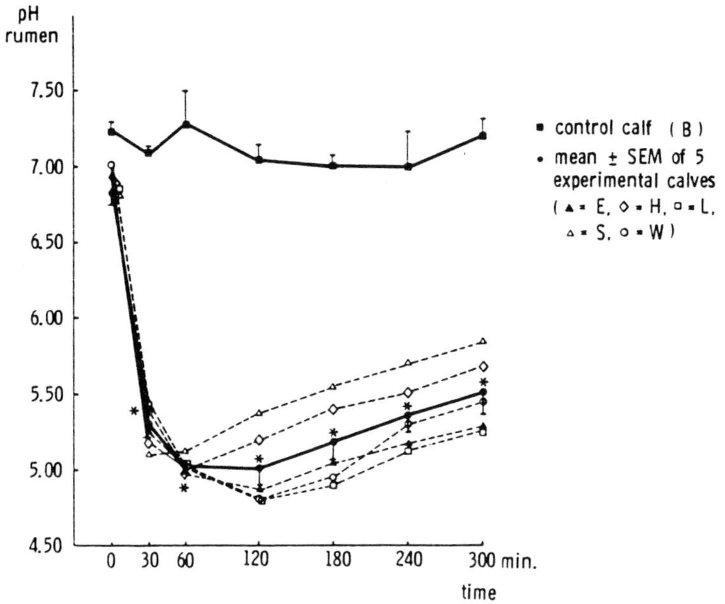

Figure 1. Rumen pH of the individual calves before (t = 0) and after drinking (control calf) or intraruminal feeding (experimental calves). Mean values (± SEM) of the experimental calves are represented by the solid line. Significant differences from values found at t = 0 are indicated by an asterisk (P < 0.001).

Experiment 2

Material and methods

Three normal veal calves between 4 and 11 weeks old were used. In two of them, two temperature sensors (self-made constructions using LM35 sensors, National Semiconductor Corporation) were introduced from the left side into the rumen and abomasum. In the third calf, the sensors were introduced from the right side. A simple electronic circuit was used to record the changes in temperature of the rumen and abomasal contents on a Brush-Gould 6 channel recorder, type 660. Recordings were made either during drinking or sucking of the normal amount of milk replacer at 41 °C (van Weeren-Keverling Buisman et al., 1990c).

Results and discussion

In the two calves operated from the left side, the temperature sensors functioned well for 1–2 months but, in the third calf, the rumen sensor became dislodged after one week. This was probably the result of the position of the connecting wire of this sensor beneath the abomasal corpus, which became rather distended after intake of 6–8.5 litres of milk replacer.

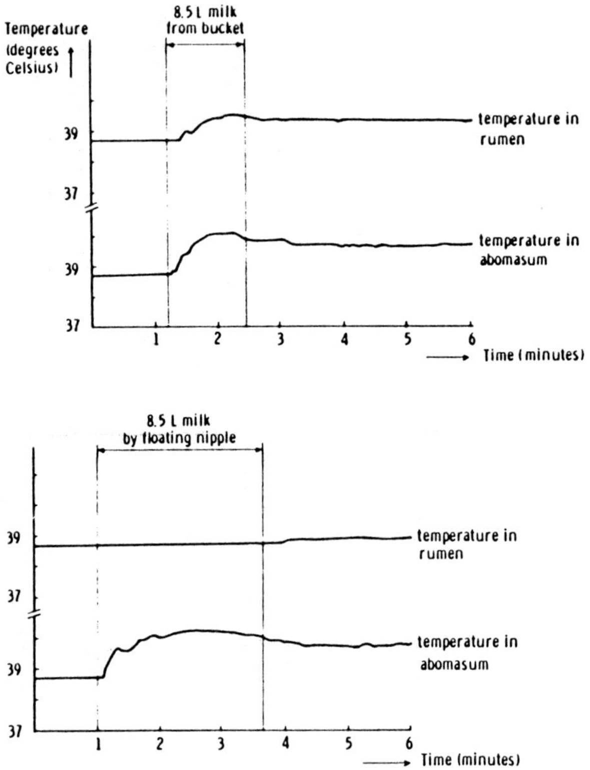

Figure 2. Temperature recordings in rumen and abomasum while drinking 8.5 litres milk from a bucket in 74 s (above) and the same amount of milk in 157 s through a floating nipple (below). Data from one healthy veal calf, 13 weeks old.

Recordings

During the intake of 8.5 litres of milk replacer at 41 °C, a clear difference was observed in the recordings of either drinking from an open bucket or sucking through a floating nipple. In the last case, only the sensor in the abomasum recorded a temperature increase of about 1.5 °C, while drinking from an open bucket resulted in a similar temperature increase in the abomasum, but also to a lesser extent in the rumen. Besides, the intake time was almost half as long as with sucking (Figure 2). It seems that a substantial part of the milk ration enters the rumen with this method of feeding, even in healthy veal calves.

Using temperature sensors seems to be a promising technique to evaluate the effect of different feeding systems or management measures for the prevention of ruminal drinking in veal calves.

References

Breukink, H.J., Th. Wensing, A. van Weeren-Keverling Buisman, E.G. van Bruinessen-Kapsenberg & N.A.P.C. de Visser, 1988. Consequences of failure of the reticular groove reflex in veal calves fed milk replacer. The Veterinary Quarterly 10(2):126–135.

de Visser, N.A.P.C. & H.J. Breukink, 1984. Pensdrinkers en kleischijters. Tijdschrift voor Diergeneeskunde 109:800–804.

Sakata, T. & H. Tamate, 1978. Rumen epithelial cell proliferation accelerated by rapid increase in intraruminal butyrate. Journal of Dairy Science 61:1109–1113.

van Weeren-Keverling Buisman, A., H.J. Breukink & M.A. van der Velden, 1988a. A simple and effective biopsy technique of the small intestine in the calf. Journal of Veterinary Medicine A 35:152–156.

van Weeren-Keverling Buisman, A., E.N. Noordhuizen-Stassen, H.J. Breukink, Th. Wensing & J.M.V.M. Mouwen, 1988b. Villus atrophy in ruminal drinking calves and mucosal restoration after reconditioning. The Veterinary Quarterly 10(3):164–171.

van Weeren-Keverling Buisman, A., Th. Wensing, T.S.G.A.M. van den Ingh & H.J. Breukink, 1990a. Intraruminal administration of milk in the calf as a model for ruminal drinking: Clinical aspects and biochemical and morphological changes in the rumen. Journal of Animal Physiology and Animal Nutrition 63:255–266.

van Weeren-Keverling Buisman, A., J.M.V.M. Mouwen, Th. Wensing & H.J. Breukink, 1990b. Intraruminal administration of milk in the calf as a model for ruminal drinking: Morphological and enzymatical changes in the jejunal mucosa. Veterinary Research Communications 14:129–140.

van Weeren-Keverling Buisman, A., R. Kuiper, P. Gootjes, M.A. van der Velden & H.J. Breukink, 1990c. Temperature sensors in the rumen and abomasum to study reticular groove reflex in veal calves. Journal of Animal Physiology and Animal Nutrition, in press.

Abomasal ulcers in veal calves: pathogenesis and prevention

H.J. Breukink *, Th. Wensing * & J.M.V.M. Mouwen **
* *Department of Large Animal Medicine and Nutrition, University of Utrecht, Yalelaan 16, 3584 CM Utrecht, NL*
** *Department of Pathology, Faculty of Veterinary Medicine, University of Utrecht, Yalelaan 16, 3584 CM Utrecht, NL*

Summary

Localiaztion and incidence of abomasal lesions in veal calves have been studied and the influence of roughage supply and treatment with clenbuterol on that incidence was investigated.
Descriptors: veal calves; abomasal lesions; roughage; clenbuterol.

Introduction

The incidence of abomasal ulcers in calves at weaning has been reported to be fairly high (Jubb et al., 1985). A high incidence of abomasal ulcers has also been found in veal calves (Welchman & Baust, 1987). In the pathogenesis of peptic ulcers, the disturbance of the balance factors between agressive and defensive decides whether an ulcer will develop or not. A stress-mediated increase in acid and pepsin production can cause ulcer formation in man, the pathogenic pathways of which can be various.

The stereotypic behaviour of veal calves in confinement is considered to be related to the occurrence of peptic ulceration in veal calves (Wiepkema, 1985). German workers have suggested that abomasal ulcers in veal calves may be caused by local ischaemia followed by focal necrosis as a consequence of strong contractions of the wall of the antrum pylori when large volumes of milk are consumed. (Krauser, 1987). Besides, an inadequate mucous barrier predisposes calves for mucosal defects (Lourens, 1985).

In this paper, the results of a number of studies on the pathogenesis of abomasal ulcer in veal calves will be summarized.

Material and methods

The localization of ulcers and erosions in 150 veal calves at slaughter was compared with the localization of bleeding abomasal ulcers found in 141 dairy cows.

In an experiment, different types of roughage pellets were offered twice daily to three groups of veal calves receiving a standard diet of milk replacers. A control group was fed on milk replacer only. The roughage supply was started in the 7th week with 25 g/d which was increased slowly to 100 g/d in the 16th week. At slaughter, the abomasum of all calves was inspected for the presence of mucosal lesions (erosions and ulcers). Mucin concentrations were estimated using histochemical techniques (Laurens). In a second experiment, a beta-agonist, clenbuterol, was given to some of the calves fattened in a commercial fattening unit with a history of a high incidence of abomasal lesions. The clenbuterol was provided daily with the milk replacer from the beginning of the seventh week of the fattening period until the end of the 23rd week. The abomasa of all calves were examined at slaughter for the presence of lesions (erosions and ulcers).
Erosion: superficial defect of mucosa.
Ulcers: deep defect of the mucosa and a clear loss of mucosa.

Results

In veal calves, the lesions were predominantly located in the top of the torus pylorus (Figure 1A). The localisation of ulcers and erosions in dairy cows is shown in Figure 1B. In adult dairy cows, the abomasal ulcers were located in the fundic part of the abomasum. Abomasal ulcers in adult cows frequently developed into bleeding ulcers.

The results obtained in calves supplemented with different types of roughage pellets demonstrated that the number of abomasal lesions was significantly higher *in all the receiving roughage groups* than in the control group (Figure 2).

In addition, the number of abomasal ulcers in the calves supplied with the corn-silage pellets was found to be significantly higher than in the group fed on the barley-straw pellets and the lucerne-hay pellets. The mucin concentration was lowest at the places where the mucous membrane exhibited defects. The mucus concentration varied clearly between different abomasa. No significant differences in mucus concentration could be demonstrated between the groups (Figure 3).

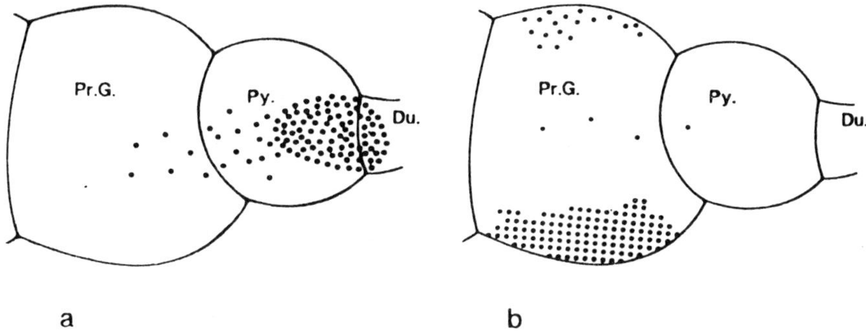

Figure 1. The localisation of abomasal lesions in veal calves (A), and in dairy cows (B). PrG.: proper gastric gland; Py.: Pars pylorica; Du.: duodenum.

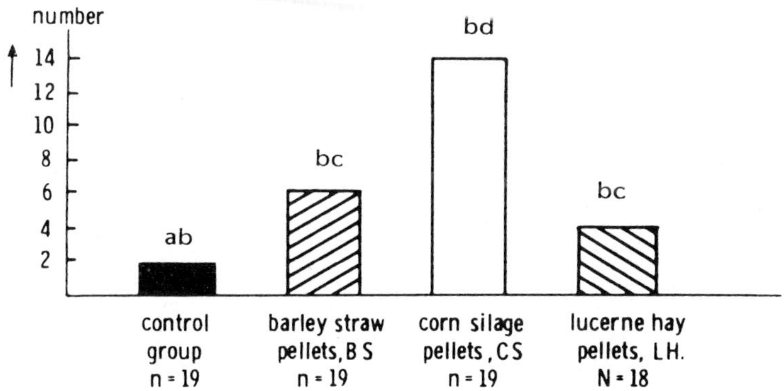

Figure 2. Number of abomasal ulcers in three groups of veal calves supplemented with different types of roughage pellets and in a control group (a/b and c/d: difference is statistically significant, t test).

The mean number of erosions in the clenbuterol-treated group was the same as in the control group (mean 3.4 erosions/calf) wheras the number of ulcers was significantly lower in the clenbuterol-treated group 0.4 and 1.4, respectively. In the clenbuterol group, no abomasa were found with more than 1 ulcer (Figure 4).

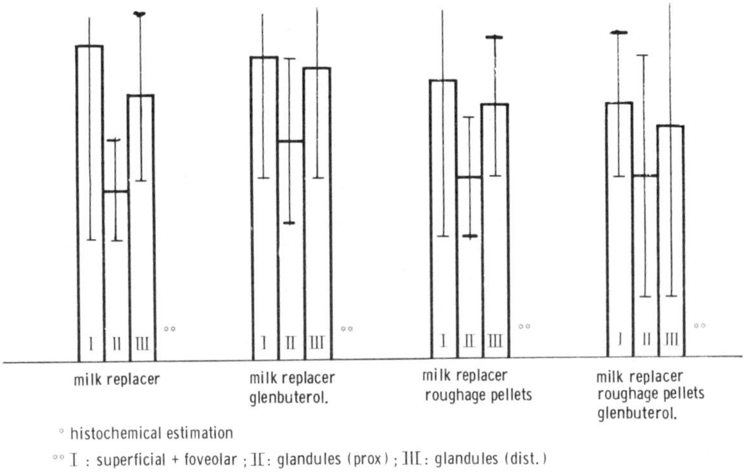

Figure 3. The mean amount of mucins (+range) abomosi in groups of veal calves fed different rations.*

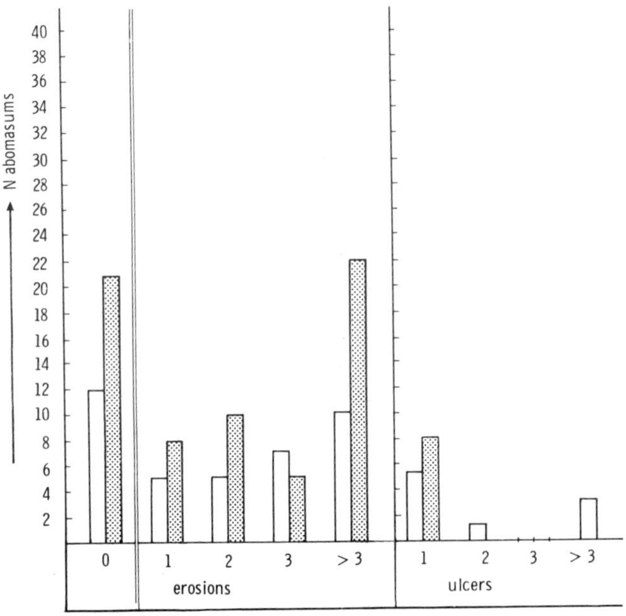

Figure 4. The number of abomasums with one or more erosions or ulcers in a group of veal calves treated with clenbuterol and in a control group.

Discussion

Stress-mediated abomasal ulcers in adult cows are mainly located in the proper gastric (fundic) gland area (Aukema, 1971). In dairy calves as well as in veal calves under conditions that stress can be involved, these ulcers were also located in the proper gastric gland area (Burato, 1984). The observation that in veal calves the lesions were almost exclusively restricted to the pyloric part of the abomasum was in agreement with the findings of others (van der Mei, 1985) and indicated that stress was not an important factor in the etiology of abomasal ulceration in veal calves.

The fact that the addition of roughage pellets to a milk replacer diet increased the incidence of abomasal ulceration has frequently been reported (Welchman & Baust, 1987). Jubb et al. (1985) reported that the incidence of abomasal ulcers and erosions in calves at weaning was fairly high (80–95%), indicating that the intake of roughage causes damage of the abomasal mucosa. The ulcers were flat, irregular defects and most healed after a while. Scarring may be evident at 6 month of age.

The literature suggests that the pathogenesis of erosions and ulcers of the pyloric mucosa of veal calves is caused by a permanent overloading of the abomasum. The filling capacity of the abomasum is limited. This results in strong peristaltic contractions in the pars pylorica with long-lasting closure of the pyloric sphincter. The resulting compression of the mucosa is most pronounced on and around the torus pylori, in the funnel-shaped pars pylorica. This may induce hypoxic lesions due to circulatory disturbances with subsequent development of erosions and ulcers (Degen, 1982; Krauser, 1987). The fact that most ulcers were located at the top of the torus pylori supported this hypothesis. Roughage particles enhance the occurence of the lesions, especially with longstanding compression of the mucosa.

Clenbuterol treatment decreased the number of ulcers as heas been found by others (Marcato, 1985), whereas the number of erosions was the same both in the treated and the control group, suggesting that clenbuterol prevents the development of erosions into ulcers. It is unlikely that clenbuterol can prevent the local damage or ischaemia which Krauser (1987) suggested to initiate abomasal mucosal damage. Perhaps a lack of mucins in the area of focal necrosis is a predisposing factor in abomasal ulceration (Wensing et al., 1986b; Pearson et al., 1987).

References

Aukema, J.J., 1971. Een onderzoek naar het voorkomen van lebmaagbloedingen ten gevolge van lebmaagulcera bij runderen uit de omgeving van Steenwijk. Thesis, University of Utrecht, NL.

Buratto, L., 1984. L'abomasite ulcerosa del vitello de latto. Obiettivi Veterinari 5:15–19.

Degen, B., 1982. Pathologisch, anatomisch und histologisch Untersuchungen zur Pathogenese der Pyolurgeschwüre bei Mastkälbern. Inaugural lecture, Berlin.

Jubb, K.V.F., P.C. Kennedy & N. Palmer, 1985. Pathology of domestic animals. 3rd ed. Academic Press. London.

Krauser, K., 1987. Untersuchungen zur Pathogenese der Pylorusulzera beim Mestkalb. Berl. Münch. Tierärtzliche Wochenschrift, 100:156–161.

Lourens, J.M., J.F. van der Wal & J.M.V.M. Mouwen, 1985. The gastric mucosal barrier and abomasal ulcers in veal calves. Tijdschrift voor Diergeneeskunde 110:755–761.

Marcato, P.S., G.J. Darsie, E.W. Salamon & A. Briganti, 1985. Ulcere abomasali nei vitelli. Alcune preliminari osservazioni sperimentali. Attidella Società Italia di Buiatrica 17:327.

Pearson, G.R., D. de B. Welchman & M. Wells, 1987. Mucosal changes associated with abomasal ulceration in veal calves. Veterinary Record 121:557–559.

van der Mei, J., 1985. Zur Einwirkung des Haltungssystems und der Beifütterung von Rauhfutter auf das Auftreten von Labmagenläsionen bei Mastkälbern. Deutsche. tierärztliche Wochenschrift. 92:463–465.
Welchman, D. de B. & G.N. Baust, 1987. A survey of abomasal ulceration in veal calves. Veterinary Record 121:586–590.
Wensing, Th., S. van Dijk, J.M.V.M. Mouwen & H.J. Breukink, 1986b. Proceedings of the 6th International Conference on Production Disease in Farm Animals, Belfast, Northern Ireland, p. 141–145.
Wiepkema, P.R., 1985. The significance of behavioural disturbances in farm animals. Tijdschrift voor Diergeneeskunde 110:12–20.

Thermal demand of one-week-old calves at two feeding levels

J.W. Schrama *, W. v.d. Hel * & M.W.A. Verstegen **

* *Department of Animal Husbandry, Agricultural University Wageningen, P.O. box 338, 6700 AH Wageningen, NL*

** *Department of Animal Nutrition, Agricultural University Wageningen, P.O. box 338, 6700 AH Wageningen, NL*

Summary

Energy balance trials were conducted in order to define the thermal demand of young calves in relation to feeding level. Heat production of ten 5-day-old Holstein-Friesian calves was measured during a 7 day period. The calves were fed at two feeding levels and were exposed to various ambient temperatures (from 18 to 6 °C). At the low feeding level (319 KJ ME $kg^{-0.75}d^{-1}$), the critical temperature of 1-week-old calves was higher compared to the high feeding level (638 KJ ME $kg^{-0.75}$ d^{-1})(15 and 12 °C resp.). Below the critical temperature, the extra heat production was estimated at 5.7 KJ $kg^{-0.75}$ $°C^{-1}$ d^{-1} and was similar at the two feeding levels.

Descriptors: calves; heat production; thermal demand; feeding level.

Introduction

Growing cattle are not very sensitive to cold conditions. Only young calves under six weeks of age are likely to suffer from the effects of cold environments (Webster, 1976). The effects of cold climates on 1 to 8 week old calves have been studied by Gonzalez-Jimenez & Blaxter (1962), Holmes & McLean (1975) and Webster et al. (1978). In these studies lower critical temperatures of 8 to 10 °C were found. These studies were all performed on calves fed at a high level, which ranged from about 750 to 950 KJ of metabolizable energy (ME) per $kg^{0.75}$ per day. These feeding levels were about 1.7 to 2.2 times maintenance requirement. In practical veal calf production, feeding schedules begin at a much lower feeding level. Usually this level is below the maintenance requirement. This is done to prevent the risk of diarrhoea during the first weeks. The lower critical temperature of homeothermic animals is dependent upon the feeding level (Robertshaw, 1981). The aim of the present experiment was to measure the effects of various low ambient temperatures on the heat production of Holstein-Friesian male calves aged 5 to 12 days at two feeding levels (one below and one above maintenance requirement). It was also decided to measure heat production at low ambient temperatures after a relatively small (4 °C) and a relatively large temperature change (9 °C).

Material and methods

Ten 5-day-old Holstein-Friesian male calves were housed for 7 days individually in an open-circuit indirect climatic respiration chamber (Verstegen et al., 1987). Six calves were kept at a low feeding level and four calves were kept at a high one. They were offered 319 and 638 KJ ME $kg^{-0.75}$ d^{-1} of a milk replacer, respectively. Ambient temperature was kept constant within each measuring period of 23 hours. Eight calves (4 of the low and 4 of the high feeding level) were exposed to an ambient temperature that changed stepwise with 3 °C between days. Ambient temperature ranged from 9 to 18 °C at the low and 6 to 15 °C at the high feeding level. Two calves of both feeding levels started and ended at the lowest ambient temperature (9 and 6 °C, respectively). They reached the highest ambient temperature (18 and 15 °C) at

Table 1. Means and statistical significance of individually measured traits during the whole experimental period of young calves at a low (FL) or high (FH) feeding level and at stepwise (TS) or fluctuating (TF) ambient temperatures between days.

Trait	Temp. changes & feeding level TS FL	FH	TF FL	s.d.	Significant effects feeding level	temp. changes
Number	4	4	2			
Initial age (d)	5	5	5			
Initial weight (kg)	42.9	45.3	36.5			
Rate of gain (g/d)	– 271	279	– 443	187	***	n.s.
HP (KJ $kg^{-0.75}$ d^{-1})	511	579	537	19.0	***	n.s.
ER (KJ $kg^{-0.75}$ d^{-1})	– 193	33	– 209	30.7	***	n.s.
ETH (KJ $kg^{-0.75}$ $°C^{-1}$ d^{-1})	– 6.0	– 5.4	-	3.7	n.s.	-

HP heat production
ER energy retention
ETH extra thermal heat production
s.d. standard deviation
*** $P < 0.01$
n.s. not significant $P > 0.10$

Table 2. Heat production (in KJ $kg^{-0.75}$ d^{-1}) and number of observations (between brackets) of young calves at various ambient temperatures by low (FL) and high (FH) feeding level and by stepwise (TS) and fluctuating (TF) changes in ambient temperature between days.

Ambient temperature (°C)	Temp. changes & feeding level TS FL		FH		TF FL	
6	-		595	(6)	-	
9	532	(6)	584	(8)	565	(6)
12	511	(8)	570	(8)	-	
15	498	(7)[1]	568	(6)	-	
18	506	(5)[1]	-		517	(8)
s.d. within temp.	20.9		27.0		22.6	

[1] 1 observation excluded from the mean, being an outlier.

day 4 of the experimental period. Two other calves at both feeding levels started and ended with the highest ambient temperature and reached the lowest ambient temperature at day 4. In addition, two calves of the low feeding level were exposed to a fluctuating ambient temperature with 9 °C between days. This scheme ranged from 9 to 18 °C and started at 18 °C on day 1 of the experiment.

Calves were weighed at the start and end of the experimental period. Gaseous exchange of CO_2 and O_2 was measured continuously. Heat production (HP) was calculated from gaseous exchange of CO_2 and O_2 in 18 min periods during each 23 hours. HP was expressed per 24 hours. Energy retention (ER) was calculated as the difference between the calculated ME content of the feed (18 KJ ME per gram milk replacer) and HP. Extra thermal heat production

(ETH) of each individual calf was calculated by regression of HP on ambient temperature. The HP of the highest ambient temperature was excluded in this calculation on the assumption that this temperature was above the lower critical temperature. Mean data per animal of the whole experimental period were statistically analysed by a two-way ANOVA model for the effects of feeding level and the effect of temperature treatment (fluctuating or stepwise changing ambient temperatures between days). Additionally mean HP per animal at each temperature were used to calculate the standard deviation between animals within temperatures.

Results and discussion

Results of traits as a mean during the experimental period are listed in Table 1. Rate of gain, heat production (HP) and energy retention (ER) were significantly influenced by feeding level, but not by the type of temperature changes (stepwise or fluctuating). Calves at the low feeding level lost weight and at the high feeding level they gained weight. The difference in rate of gain between high and low feeding level was 550 g d^{-1} ($P < 0.01$). HP was 68 KJ $kg^{-0.75}$ d^{-1} higher at the high feeding level compared to the low ($P < 0.01$). ER was clearly negative at the low and slightly positive at the high feeding level. The difference was 226 KJ $kg^{-0.75} d^{-1}$ ($P < 0.01$). Although the type of temperature changes was not significant, rate of gain, HP and ER tended to be negatively influenced by fluctuating ambient temperature. Treatments with low rate of gain had lowered ER. Change in ER between treatments, however, do not correspond well with change in rate of gain. At the low feeding level, the difference in rate of gain between stepwise and fluctuating ambient temperatures was 172 g d^{-1} and the difference in ER was only 16 KJ $kg^{-0.75}$ d^{-1}. At the high feeding level and with stepwise changing ambient temperatures, ER was 33 KJ $kg^{-0.75}$ d^{-1}. This value is low compared to the rate of gain of 279 g d^{-1}. This discrepancy might be caused by differences in deposition of protein and fat and their ratio. Furthermore ME contents of the milk replacer was calculated from its composition. It was assumed furthermore that metabolizability of feed was identical in all treatments and periods. Any deviation of these values will alter the calculated RE.

Mean data on HP at different ambient temperatures are given in Table 2. At the low feeding level and stepwise changing ambient temperature, two HP measurements of one animal were excluded from the results, being outliers. HP was elevated both at the low and high feeding level at the lowest compared to the highest ambient temperature (26 and 27 KJ $kg^{-0.75}$ d^{-1}, respectively). At the low feeding level, HP was increased at ambient temperatures below 15 °C and at the high feeding level at increased ambient temperatures below 12 °C. At the low feeding level, average HP during the experimental period was 26 KJ $kg^{-0.75}$ d^{-1} higher at fluctuating temperatures than stepwise changing ambient temperatures (not significant) (Table 1). This difference in mean HP between the two temperature treatments was mainly caused by an extra heat production with 33 KJ $kg^{-0.75}$ d^{-1} at 9 °C with fluctuating ambient temperature compared to 9 °C with stepwise changing ambient temperature.

At stepwise changing ambient temperature no significant effect of feeding level was apparent on extra thermoregulatory heat production (ETH). The mean of ETH at both feeding levels was 5.7 KJ $kg^{-0.75}$ $°C^{-1}$ d^{-1} (s.d. = 3.7 KJ $kg^{-0.75}$ $°C^{-1}$ d^{-1}). This value is in agreement with the 5.3 KJ $kg^{-0.75}$ $°C^{-1}$ d^{-1} which Webster et al. (1978) calculated from the difference in HP between 15 and 5 °C. On the other hand, based on data of the calculated thermal insulation, Webster et al.(1978) and Holmes & McLean (1974) found an increase of 18 and 13 KJ $kg^{-0.75}$ d^{-1}, respectively, in HP per °C decrease in ambient temperature below the lower critical temperature. According to Webster et al. (1978), the difference between the two methods of calculation is related to the fact that the lower critical temperature is a range of 5 °C. From the results of the present experiment, the lower critical temperature seems to be more a rather distinct value. The estimation of ETH from HP at different ambient temperatures showed a large standard deviation.This may explain the differences in results obtained by the two methods of calculation.

From the HP measurement at various ambient temperatures, it is clear (Table 2) that the lower critical temperature is increased at the low than at the high feeding level. The lower critical temperature was about 15 °C for calves at the low and about 12 °C at the high feeding level. The lower critical temperatures obtained in the present experiment are higher compared to the values of 8 to 10 °C found by Gonzalez-Jiminez & Blaxter (1962), Holmes & McLean (1975) and Webster et al. (1978). In the latter experiments, however, the feeding level was much higher than in the present study and on average the calves in our study were younger. The feeding level is important in assessing the thermal demand of young calves. It is also very important to evaluate the way in which young calves adapt to cold conditions. It is not well known, however, how calves adapt. This may be very important for energetic efficiency and occurrence of stress. From the present results it is concluded that the thermal demand (lower critical temperature) of young calves aged 1 week is about 12 to 15 °C depending on the feeding level. The extra heat production per °C below the critical temperature is estimated at 5.7 KJ $kg^{-0.75}$ $°C^{-1}$ d^{-1}, independent of feeding level.

References

Gonzalez-Jiminez, E. & K.L. Blaxter, 1962. The metabolism and thermal regulation of calves in the first month of life. British Journal of Nutrition 16:199–212.

Holmes, C.W. & N.A. McLean, 1975. Effects of air temperature and air movement on the heat produced by Young Friesian and Jersey calves, with some measurements of the effects of artificial rain. New Zealand Journal of Agricultural Research 18:277–284.

Robertshaw, D., 1981. The environmental physiology of animal production. In: J.A. Clark (Ed.): Environmental aspects of housing for animal production. Butterworths, London. p. 3–17.

Verstegen, M.W.A., W. van der Hel, H.A. Brandsma, A.M. Henken & A.M. Bransen, 1987. The Wageningen Respiration unit for animal production research: a description of the equipment. In: M.W.A. Verstegen & A.M. Henken (Ed.): Energy metabolism in farm animals. Martinus Nijhoff, Dordrecht. p. 21–62.

Webster, A.J.F., 1976. The influence of the climatic environment on metabolism in cattle. In: H. Swan & W.H. Broster (Ed.): Principles of cattle production. Butterworth, London. p. 103–120.

Webster, A.J.F., J.G. Gordon & R. McGregor, 1978. The cold tolerance of beef and dairy type calves in the first weeks of life. Animal Production 26:85–92.

5. Welfare legislation

Proposed legislation by the European Community on protection of calves

A.J. Wilson
EC Commission, Brussels, BE

Summary

A proposal for a council regulation on minimum standards for the protection of calves kept in intensive farming systems is currently under discussion in the Council of Ministers of the European Community. The proposal is based on the principles set out in the European Convention for the Protection of Animals kept for Farming Purposes. It sets out rules called for in the European Parliament's Resolution on Animal Welfare Policy of February 1987, and contributes to the completion of the internal market in the EC.
Descriptors: calves; protection; European Community.

Introduction

Public concern for the welfare of farm animals mounted during the 1960s and 1970s, and in 1976, the Council of Europe (a body quite distinct from the European Community) opened for signature the European Convention for the Protection of Animals kept for Farming Purposes. The Council of Ministers of the European Community (EC) formally approved this Convention by Council Decision 78/923/EEC (EC 1978) and the Community became a contracting party by signature of the Convention ten years later, after all EC member states had ratified it.

In the meantime, two other important developments occurred in the EC. The European Parliament was taking an increasing interest in animal welfare, and in June 1986 the Parliament's Agriculture Committee held public hearings on current issues in farm animal welfare, including intensive rearing of veal calves. Evidence was heard from producers, welfare organizations and scientists, and a report was drawn up which was the basis of the Resolution on Animal Welfare Policy, adopted by the Parliament in February 1987 (EC 1987).

The other significant initiative at Community level was the publication in 1985 of the White Paper on completing the internal market. This set out a detailed timetable for dismantling the many trade barriers which had developed over the years and which were impeding the establishment of a Common Market in Europe. Among those barriers were differences in animal welfare legislation which created unequal conditions of competition and lead to pressure for economic protection.

The European Convention

Contracting parties to the Convention are obliged to give effect to certain basic principles of farm animal welfare. For example: 'Animals shall be housed and provided with food, water and care in a manner which -- having regard to their species and to their degree of development, adaptation and domestication -- is appropriate to their physiological and ethological needs in accordance with established experience and scientific knowledge.'

General requirements were also laid down for appropriate space, freedom of movement, environmental conditions and inspection.

The 'traditional' veal production system, which had been established during the previous forty years, did not meet the criteria of the Convention in respect of space, freedom of movement or diet. A narrow crate that does not allow a calf to turn round, groom itself or even to adopt all normal lying postures cannot, however broadly the word 'appropriate' may be interpreted, be said to meet physiological and ethological needs. Moreover, the behaviour of calves during confinement in such crates, the limb weakness seen when they are released and their abnormal susceptibility to transport stress indicate that they are being caused unnecessary suffering and injury, which is also banned by the Convention.

Similarly, feeding on a diet deficient in iron and lacking any roughage is not in accordance with the Convention. It is also, incidentally, unnecessary: experience has shown that veal produced from calves fed a moderate amount of roughage is indistinguishable from 'traditional' veal, and is perfectly acceptable to the most discerning consumers.

The Parliamentary Resolution

The Resolution called on the Commission to make legislative proposals on the intensive rearing of veal calves with a view to abolishing the present system of feeding on an exclusively liquid diet and housing in narrow crates. It called for a balanced diet, including roughage and adequate iron, room to stand up, lie down, turn around and adopt a comfortable sleeping posture, and, from the age of 6 weeks, group housing.

The internal market

Some EC member states have legislation, or proposed legislation, which bans or restricts the 'traditional' veal production system, and imposes extra requirements on their farmers compared to farmers in other member states which do not have such legislation. This not only gives rise to unequal conditions of competition, but has already led to demands for imports of veal into those countries to be banned from countries with lesser welfare standards. Under Article 36 of the Treaty of Rome, a member state may justify prohibitions on trade with other member states on certain specified grounds, as long as they do not constitute arbitrary discrimination or a disguised trade restriction. Among those grounds are public morality, public policy and the protection of health and life of animals. In order to avoid distortion of competition and possible disruption of intra-Community trade, it is necessary to have Community rules on the matter. In fact, the Ministers of Agriculture of several member states have called for such legislation on a number of occasions during the last ten years.

The Commission proposal

For the reasons mentioned above, the EC Commission has made a proposal for a Council Regulation concerning minimum standards for the protection of calves kept in intensive farming systems (EC 1989), based on scientific evidence on the welfare of calves. A calf is defined as a bovine animal up to the age of 6 months, and an intensive farming system is defined as a system in which calves are housed and are fattened mainly by the feeding of milk or milk replacer without the use of their dams or nurse-cows.

It is proposed that from 1 January 1992, new production systems must provide for group housing of calves after the age of 8 weeks. Until 8 weeks, calves may be kept in individual boxes or tether stalls, but boxes and stalls must be at least 80 cm wide and 180 cm long. Group-housed calves must have at least 2 square metres of unobstructed floor space per calf.

Existing production systems must meet the same conditions from 1 January 1999, i.e. an adjustment period of 7 years.

A period of 8 weeks during which calves may be individually penned was chosen, instead of the 6 weeks mentioned in the Parliamentary Resolution, in view of research work, particularly in the Netherlands, which indicates that behavioral problems are usually resolved by 8 weeks, but may commonly persist beyond 6 weeks. When the proposal was being drawn up in working groups, by experts from the member states, there was much discussion on dimensions for individual boxes and stalls. Some wanted a specified measurement, while others considered that the size of the box should be related to the size of the calf, such as the height at the withers. It was eventually decided to propose a measurement, partly because it is easier to check, but it should be realized that this is a minimum measurement and that a later requirement in the Annex to the Regulation states that calves must always have enough room to lie down, rest, stand up and groom themselves without difficulty.

Another point called for in the Parliamentary Resolution, in relation to farm animal welfare in general, was that the Commission launch a publicity and information campaign for consumers. Our proposal envisages a broader information system, with scientific, technological and other relevant information being collected by the Commission and disseminated to the competent authorities of the member states and to 'other interested parties', who would include producers, consumers and legislators.

A lot of research and development is being carried out in calf husbandry, and it was considered necessary to keep this work under review. It is proposed to make a report to the Council and the Parliament before 1 January 1994 on the welfare of calves kept in intensive farming systems, accompanied by any appropriate proposals for adjustments to the Regulation. This review process should ensure that the Regulation takes account of technical progress and does not hinder the introduction of any new systems that could improve the welfare of calves.

Another important feature of the proposal, and one which was stressed by the European Parliament, is the requirement for inspection and enforcement. As with all such EC legislation, the primary responsibility for enforcement rests with each Member State, and the proposal contains an obligation for them to make systematic inspections of intensive farming systems in their territory, and to report the results to the Commission. In addition, there is a requirement for veterinary experts from the Commission to make inspections in the Member States, with a view to ensuring that the rules are applied in a uniform way across the Community.

Appended to the Regulation, and an integral part of it, is a technical Annex containing detailed rules on housing, care, feeding and watering. In particular, it is proposed that the food must contain sufficient iron to provide a positive state of health and well-being and growth rate, and that each calf after the age of two weeks shall have daily access to at least 100 g of dry food containing digestible fibre.

Conclusion

The proposal has yet to be debated by the European Parliament and the Council of Ministers, and the legislation that finally emerges may be somewhat different from that which is proposed. Nevertheless, it will, for the first time, lay down rules on a Community-wide basis for the welfare of intensively farmed calves.

References

Official Journal of the European Communities, Vol. L 323, 17.11.1978, p. 12.
Official Journal of the European Communities, Vol. C 76, 23.3.1987, p. 185.
Official Journal of the European Communities, Vol. C 214, 21.8.1989, p. 28

Intended welfare legislation on veal calves in West Germany

G. Baumgartner
Federal Ministry of Food, Agriculture and Forestry, P.O. Box 140 270, D-5300 Bonn 1, DE

Summary

The Federal Minister of Food, Agriculture and Forestry is empowered, wherever necessary in order to protect animals, to issue, subject to the approval of the Bundesrat, regulations specifying in detail requirements concerning the keeping of animals.

When adopting the amended Animal Welfare Act in 1986, the National Parliament asked the Federal Government

- to issue such regulations especially for the keeping of calves, pigs, poultry and fur animals, taking into account scientific evidence of ethology
- to insist on harmonization of welfare legislation in the European Community.

After intensive discussions, the Federal Minister of Agriculture elaborated a regulation with detailed requirements for calves:

- requirements for the space in pens or boxes
- group housing for veal calves older than eight weeks
- roughage for calves older than two weeks
- sufficient light, iron etc.

This draft was adopted, but preparing its own proposal, the EC Commission asked the Federal Government to postpone the implementation of this national legislation.
Descriptors: Federal Republic of Germany; Animal Welfare Act; animal welfare on veal calves; animal welfare on rearing calves

Introduction

In the Federal Republic of Germany, the production of veal calves is not as important as in France, Italy or the Netherlands. In 1988, the number of dairy cows was here about 5.2 millions, the number of veal calves produced was only about 560 000.

As in other member states of the European Community, the majority of those veal calves are kept individually in crates or boxes; boxes which were planned and constructed when it was usual to slaughter those animals at the age of about three or four months. In the meantime, following new production schemes, the animals are kept up to six months, reaching a live-weight of up to 250 kg.

From the point of view of animal welfare, it is necessary that all calves can lie down and get up easily, stretch out their legs when lying, groom themselves and turn round. But for the time being, most veal calves are hindered from doing so. This development leads to severe animal welfare problems.

For calves older than about 8 weeks, some farmers have therefore replaced individual housing systems by group housing. Group housing for veal calves beyond 8 to 10 weeks is practised in some large calf-fattening farms in northern Germany but also in some smaller farms in southern Germany.

Legal situation

The Animal Welfare Act of the Federal Republic of Germany rules that any person keeping, caring for or required to care for an animal

- must provide the animal with food, care and housing appropriate to its species, needs and behaviour;
- may not restrict the animal's freedom of movement to such an extent as to cause the animal pain or avoidable suffering or harm.

The Federal Minister of Food, Agriculture and Forestry is in charge of Animal Welfare and is empowered, wherever necessary to protect animals, to issue, subject to the approval of the Bundesrat, regulations specifying in greater detail the requirements concerning the keeping of animals.

Political situation

When adopting the amended Animal Welfare Act in 1986, the National Parliament asked the Federal Government

- to issue such regulations especially for the keeping of calves, pigs, poultry and fur animals taking into account scientific evidence of ethology
- to insist on harmonization of welfare legislation in the European Community.

Based on intensive discussions, the Federal Minister of Agriculture elaborated a regulation with detailed requirements for calves. The illegal use of hormones in veal calves, which was detected in 1988, and which in many cases was linked with poor welfare of those animals, turned this item into a problem of general public concern. At the beginning of 1989, this animal welfare regulation was adopted but, preparing its own proposal, the Commission of the European Community, based on the directive 88/182/EEC, asked the Federal Government to postpone this national legislation at least for one year. Until now, implementation has been postponed and we are urging the Council of Ministers to implement the European legislation as soon as possible. From the point of view of animal welfare, it is really necessary to improve the situation.

Main elements of the German draft

Based on the relevant scientific publications and on the discussions of the Standing Committee of the European Convention for the Protection of Animals kept for Farming Purposes and also based on a field study, prepared for the Commission of the European Community, the German draft, elaborated in 1988, follows the same line as the EC proposal of 1989. The main elements are

- requirements for the space in pens or boxes
- group housing for veal calves older than eight weeks
- roughage for calves older than two weeks
- sufficient light, iron etc.

The regulation should be applicable for veal calves up to a liveweight of 250 kg and for rearing calves up to six months of age. If calves are kept with cows, only the general rules of the Animal Welfare Act are applicable.

The design, construction and maintenance of buildings and equipment must be such that they limit the risk of disease or traumatic injuries to the calves. Floors shall not be slippery, they shall be suitable for the size and weight of the animals, and form a rigid even and stable surface. Calves must have sufficient space for their basic needs of movement and to adopt their specific sleeping postures. For the different groups of age, the minimum space requirements are defined. For new-born calves up to 2 weeks the boxes must be at least 80 cm wide and 120 cm long. Bedding of straw or comparable material must be available for those young calves. Calves older than 2 weeks may be kept on slatted floors. The requirements for the slats and the maximum width between the slats are laid down in the regulation. Calves between 2 and 8 weeks may be kept individually or in groups. Closed boxes must be at least 100 cm wide and 180 cm long. If as is usual, the trough is outside the box, the length may be reduced to 160 cm. We think that those figures are necessary, if calves between 2 and 8 weeks are to have sufficient freedom of movement and to be able to adopt species-specific sleeping postures, and freely stretch their limbs, to be able to rise and to lie down without difficulty. If those calves are kept in groups, an unrestricted floor area of at least 1,3 m^2 per calf must be available. The space of such a pen must be at least 4 m^2.

As a rule, after a transitory period, calves older than 8 weeks shall only be kept in groups. For each calf, the minimum floor area must be 1.5 m^2, the whole pen must have at least 5 m^2. If kept in groups, calves over 8 weeks of age may only be tethered or yoked for a maximum period of 1 hour during and after feeding. To rear calves on small farms, where there are not enough calves of the same age to keep them in groups, an exemption is made from the general rule of group housing. The need for social contact in those cases may be partly fulfilled by intensive care of the stock-keeper.

All calves shall receive of sufficient light (at least 50 lux) for at least 8 hours per day. Air velocity, relative humidity, levels of toxic gases and dust shall not affect adversely the health of the calves. The temperature in the lying area should be between 5 and 25 °C.

Calves must obtain sufficient available iron to maintain them in health and all calves older than 2 weeks must have access to appropriate roughage and water. Calves shall not be muzzled. For veal calves between 2 and 8 weeks, the quantity of roughage must be at least 100 g per day, for veal calves older than 8 weeks at least 200 g are necessary. Rearing calves older than 2 weeks must have access to roughage ad libitum.

Having in mind the importance of good stockmanship and regular inspection, it will be ruled that the animals shall be cared for by a sufficient number of workers with adequate theoretical and practical knowledge, and the animals must be inspected at least twice daily. If necessary, the person responsible for them shall take appropriate action. On farms with more than 50 veal calves, the results of the daily inspection must be recorded. The competent authority is empowered to check this documentation.

Final remarks

Being convinced, that these requirements are really necessary and in the interest of the animals, I hope that at EC level, the Council of Ministers will soon follow the proposal of the Commission of the European Communities, at least in principle, and adopt an EC directive. If a common solution cannot be reached, we will have to implement our national legislation.

References

Amended version of 18 August 1986 of the German Law on Animal Protection.

European Convention on the Protection of Animals kept for Farming Purposes (March 1976).

Recommendation concerning Cattle. Standing Committee of the European Convention for the Protection of Animals kept for Farming Purposes (October 1988).

Schlichting, M.C. & D. Smidt, (Ed.), 1987. Welfare aspects of housing systems for veal calves and fattening bulls. Commission of the European Community.

Verordnung zum Schutz von Kälbern bei Stallhaltung (Kälberhaltungsverordnung), Bundesratsdrucksache 634/88. Noch nicht in Kraft getreten.

French view on a veal calf legislation

P. Le Neindre * & J. Wintergerst **
* *Institut National de la Recherche Agronomique, Theix, 63122 Ceyrat, FR*
** *Ministère de l'agriculture et de la forêt, Direction générale de l'alimentation, Service vétérinaire de la santé et de la protection animales, 175, rue du Chevaleret, 75646 Paris Cedex 13, FR*

Summary

Veal calf production in France is of great importance both socially and economically, since it provides a livelihood for 8800 families. There are various production methods, and calves can be reared with their mothers or artificially. The white meat still favoured by consumers imposes constraints of a technical nature. No specific French legislation exists for veal calves and any improvement in animal welfare can be made only by reference to general texts. New legislation will have to evolve from EC recommendations and must have a firm scientific basis.
Descriptors: France; legislation; welfare; veal calves.

Veal calf production

Any specific legislation on production must take into account both social and technical factors. Veal is traditionally a luxury product: it should be light in colour, tender and have some fat. In 1988, 19% of cattle meat came from veal calves. Before 1960, most calves were reared with their mothers and slaughtered at liveweights ranging from 100 to 150 kg. Since then, milk replacer has made a new 'industrial' production process possible. Now, of the 3 100 000 veal calves produced, 350 000 are reared by their mothers, and 2 350 000 are reared on intensive farms. The mean carcass weight has risen steadily from less than 90 kg, and was 112.4 kg in 1985. This has been achieved by better feeding management and a longer rearing period.
French consumers bear two considerations in mind.

- The quality of the meat: the taste of the meat and its 'healthiness' are criticized. The use of antibiotics and growth promoters has been questioned. Since 1988, the latter have been banned.
- Rearing conditions: the separation of the calf from its dam and often from congeners, the restricted space allowed to the animal and a milk-only diet are all factors that affect public opinion about veal.

To what extent are rearing conditions affected by the demands of consumers and producers?

- Consumers pay for
  - white meat. The difference in price between white and pink meat is great (at least 3.5 FRF per kg). To produce white meat, calves must have a low iron intake, which renders them weak;
  - fat of good quality. The flavour of the meat is due mainly to the fat obtained by consumption of milk products. The whiteness of the meat is not necessary to obtain this quality but it indicates that the calf has been fed with milk.
- Producers are also subject to specific constraints and in particular they use by-products of the dairy industry:
  - Most calves on intensive veal farms are from dairy herds. They use milk products not destined for human consumption.

As a result, calves are rather weak and require close supervision. Although the mortality is low, morbidity is still quite high, the most common disorders being digestive and respiratory. To achieve the considerable growth desired, the energy, protein, mineral and vitamin content of the rations must be well balanced and there should be a large milk intake. Calves do not usually receive extra water and solid food. Housing must be designed to provide optimum temperature and humidity and as little variation in these two parameters as possible.

As calf rearing gives a small profit any new restrictions will result in the closure of many farms and an increase in size of the rest, creating a situation that will not facilitate improvement in welfare conditions.

Farmers have tried to improve their product, in particular by increasing carcass weight, but in the meantime pen size has not increased.

French regulations on animal welfare

Calf welfare could be improved by keeping consumers and producers better informed. A considerable change could be achieved, for example by informing the public that the quality of the meat does not depend on its colour. A second way could be to promote husbandry systems with better welfare conditions. In particular, calves reared by their mothers usually have better welfare conditions than artificially reared ones. This traditional production which is part of the French 'Label Rouge' scheme could be promoted more vigorously.

However, some regulations are required, since no specific legislation exists for calves and for veal calves in particular.

General legislation on welfare

Law 76-629 is the main law dealing with the protection of animals. Article 9 acknowledges that the animal has a specific status: 'an animal is a sentient being and should be kept in conditions compatible with the biological needs of its species'. Article 10 specifies that any person may own animals if he fulfils the conditions of Article 9 and if he uses the animals under the conditions provided for in Article 276 of the 'code rural', subject to the rights of others, safety and public health requirements and the provisions of the present law.

Article 276 of the 'code rural' as cited in Law 76-629 gives regulations to protect animals against bad treatment, improper use and to prevent suffering during rearing, handling, penning, transport and slaughter.

These provisions of Article 276 were published in the amended decree number 80-791 of 1 October 1980 and in particular in Clauses 1 and 2.

Clause 1: Any person, who for any reason raises, looks after or keeps domestic animals or tamed or captive animals of non-domestic species is forbidden:

1. to deprive these animals of the food and water necessary for the physiological requirements of their species and for their stage of development, adaptation and domestication;
2. to leave a sick or injured animal without care or attention;
3. to place and keep an animal in housing conditions or in an environment which, because of exiguity, climatic conditions or inadequate equipment or fitting, may be the cause of suffering, injury or accident;
4. to use, when not absolutely necessary, any device for tying or restraining, or fences or cages not adapted to the species or likely to cause injury or suffering.

If, through bad treatment or lack of care and attention, any domestic animal or tamed or captive animal of a non-domestic species is found in very poor physiological conditions the 'Commissaire de la République' can take the necessary steps to reduce the animal's suffering to a minimum. He can order the animals to be slaughtered if he thinks their state irremediable.

Clause 2: Cattle, sheep, goats and equidae must not be kept outdoors if

1. there are no installations to prevent animals suffering from the results of climatic variations;
2. the lack of man-made or natural fences or of devices for tying or restraining may cause an accident.

Animals kept, reared or fattened in high-altitude enclosures are subject to these measures only outside the normal grazing season.

The same decree (80-791) allows new clauses to be published by ministerial orders from the 'Ministre de l'Agriculture et de la Fort' and when necessary co-signed by other ministries.

The interministerial clause of 25 October 1982 is one such new clause. It defines outdoor management and housing for non-intensive rearing conditions. However the rules defined in this clause have no quantitative standards and are rather general.

With this series of legislative measures, it is possible to punish infringement by sanctions provided for under the law against bad treatment of any animal.

These regulations are supplementary to Decree 78-1085 of the text of 2 November 1978, which is the application in France of the text of the European Convention on the Protection of Farmed Animals published by the Council of Europe. The only sanctions provided for by this text for breaking the law are low fines.

New specific regulations on welfare

All new regulations will be introduced from the European framework proposals and translated into clauses to complement the decree of 1 October 1980.

This was recently done for hens in cages and is in the planning stage for calves. These steps should complete French legislation on bad treatment or improper use of animals. However the new legislation must take into account not only biological considerations but also economic requirements.

Needs of animals should be determined from well designed experiments providing clear-cut results and if possible duplicated by several independent researchers. The scientific basis of any conclusion should be firmly established before the publication of any regulation for a production process as important as production of veal calves. Certain topics still require more experiments; size of calf pen, floor characteristics, use of feeds others than milk, group and individual housing.

Current state of veal calf welfare discussions in the USA and Canada

S.E. Curtis
Animal Sciences Department, University of Illinois Urbana, IL 61801, US[1]

Summary

There is reason to believe that the US Congress will enact enabling legislation on animal welfare before the turn of the century. To determine veal calf welfare objectively, the needs of the calves and the effects of different production systems on their well-being should be documented scientifically to make a sound basis for governmental regulation of animal husbandry.

In Canada, the various parties interested in veal calf husbandry came together in 1988 to formulate recommended codes of practice for the production of special-fed veal calves. Because of these codes, there is no need for legislation at this time.
Descriptors: USA; Canada; veal calves; welfare; legislation.

Introduction

People want assurances that veal calves are being treated humanely. The valid way to determine this is to document scientifically the effects different production systems have on the various elements of the calves' well-being and then to make some rational estimate of their overall well-being, taking into account all of the calves' needs and feelings. Otherwise, one person's opinion is as good as the next one's, and progress will not be made.

We have heard of behavioral differences among calves in various production systems. Still: What does a veal calf need? Ultimately, animal welfare depends on animal feelings. We must pursue scientific research that lets the animal tell us how it feels.

Casual observations will not suffice. What do we know about how the veal calf actually feels? Where should the lines be drawn?

"Perhaps the biggest danger is that a dissatisfied public may demand changes that eliminate what is disliked but that result in either no improvement or even a reduction in the welfare of the animals involved". That is what British animal scientist Colin Spedding said not too long ago. And it seems that the measured pace at which regulation of farm animal care is developing in North America reflects this attitude.

To be sure, farm animal welfarism has been politicized in both the United States and Canada. But in both countries, the prevailing opinion among legislators is to base regulations on a sounder scientific base than is now at hand. Again, the handwriting is on the wall. Regulation of farm animal care seems inevitable. But not right away.

1 Present address: Department of Dairy and Animal Science, 324 Henning Building, Pennsylvania State University, University Park, PA 16802, USA.

The United States of America

Several sophisticated, nationwide surveys in recent months have yielded similar findings:

- roughly 3 in 4 Americans believe animals are being treated humanely on US farms and ranches, the striking exception being special-fed veal calves, for which the ratio is 1 in 2;
- 2 in 3 Americans would favour Federal Governmental regulation of animal care on farms and ranches.

In November 1988, a referendum was held in the Commonwealth of Massachusetts. A majority yea vote on the referendum would have resulted in the establishment of a state commission on farm animal care that would have licensed farms as suitable for producing animals. The referendum failed -- 29% yea, 71% nay. But it is important to recognize that this failure did not result from a debate on the merits of the proposal per se. Agricultural interests chose to play their strong suit, and their campaign slogan was: "Save the family farm, vote no". The seriousness of the Massachusetts challenge sent agriculture another clear message.

For several years, there have been state legislative initiatives on veal calf husbandry, in particular in California, New York, and Maryland. Hearings have been conducted, but so far no action has been reported from committee.

It is expected that legislative action will come first at the federal level, and that this will in effect make these activities redundant at the state level. Congressman Charles Bennett of Jacksonville, Florida, introduced H.R. 84, the Veal Calf Protection Act, in the One Hundred First Congress. Basically, this bill if enacted would direct the Department of Agriculture to set and enforce regulations for veal calf husbandry that would revolutionize our special-fed veal industry. In particular, the tether-stall means of accommodation would be outlawed altogether.

Precedent-setting hearings on H.R. 84 were conducted on Capitol Hill on 6 June 1989. At the end of a long day, Congressman Charles Stenholm, who chaired the hearing, said:

> "This is an extremely important matter. It is one of which we have felt that a hearing was needed in order to establish for the record that there are differing viewpoints. In some cases, there is scientific basis and in other cases there are personal opinions. But also, there is the realization for all of us in production agriculture that the consumer is always right. That is something we should never forget".

There is reason to believe that the Congress will enact enabling legislation before the turn of the century having to do with regulation of husbandry of all agricultural animal species. The Department of Agriculture would then embark upon a regulation-setting process, which would probably not be completed before the year 2010. This first generation of regulations will probably be crude and frankly not always in the best interests of animal well-being, given our grossly incomplete understanding of what it takes to ensure that an animal feels well, and this will then spawn the support for the research on animal welfare, and especially on the animals' conscious feelings of suffering and well-being, that is so sorely needed as the basis for governmental regulation of animal husbandry.

Canada

Agriculture Canada issued recommended codes of practice for the production of special-fed veal calves in 1988. These voluntary codes resulted from several meetings over a period of almost a year of representatives of Agriculture Canada, the agricultural industries, academe, and the Canadian Federation of Humane Societies.

In the end, the system recommended and agreed to by all parties incorporated an individual tethered accommodation. For calves weighing up to 200 kg, the minimum stall dimensions were recommended to be 65 cm × 165 cm.

The cooperative nature of the various parties interested in veal calf husbandry in Canada has been admirable. At present, all parties reportedly believe that progress on common objectives has been made. There is no legislation pending at either the provincial level or the federal in Canada at this time. People having divergent viewpoints have come together and resolved their differences, at least for the time being.

6. Marketing

Review of veal consumption in France

J.Y. Coroller * & Y. Toulhoat **
* *Scica Ouest Elevage, P.O. Box 68, 29260 Ploudaniel, FR*
** *Groupe Even, P.O. Box 67, 29260 Ploudaniel, FR*

Summary

The French are traditionally big veal eaters (greatest consumption per capita in the EC). However veal consumption is declining (8.3 kg per capita in 1963, 5.4 kg per capita in 1989). This downward trend accelerated dramatically in 1988 and 1989, when the decline in domestic consumption was 11.5% and 14.0%. This trend occurred at a time when all other meat consumption was increasing (by 30.0% between 1963 and 1978 and by 5.6% between 1978 and 1989). The main reasons for veal's declinea are

- growing scarcity of supply, and an increase in the cost of production (+13.8% in 1988, +13.0% in 1989);
- an increase in retail price (+7.9% in 1988, +15.9% in 1989) higher than that for other meats;
- a decline in the consumer's image of veal (media), making advertising difficult;
- the current image of veal being out of step with current trends towards increasing 'health and fitness';
- strong competition from other meats, especially turkey. Turkey consumption (5 kg per capita in 1989) is currently approaching that of veal.

To counteract this trend, the following steps must be taken. The high cost of veal can be justified, provided that its taste and health-giving qualities are first class. Action should be taken to disseminate information and reassure consumers about the consumption of veal (certification policy). Then, more efficient advertising campaigns than in the past could be undertaken, based on

- the well known dietetic qualities of veal, which are advantageous for low-cholesterol diets;
- culinary qualities of veal, which make menu diversification possible.

Descriptors : France; veal; consumption.

Introduction

History reminds us that veal has always been among the best of foods:

- the fatted calf slain for the prodigal son;
- the golden calf the Jews worshipped during the Exodus.

The festive nature of this meat is well vouched for. What is its status now in France?

Consumption data

Actual consumption

The French are still the biggest veal consumers in the EC, 5.4 kg carcass per capita in 1989 (Table 1).

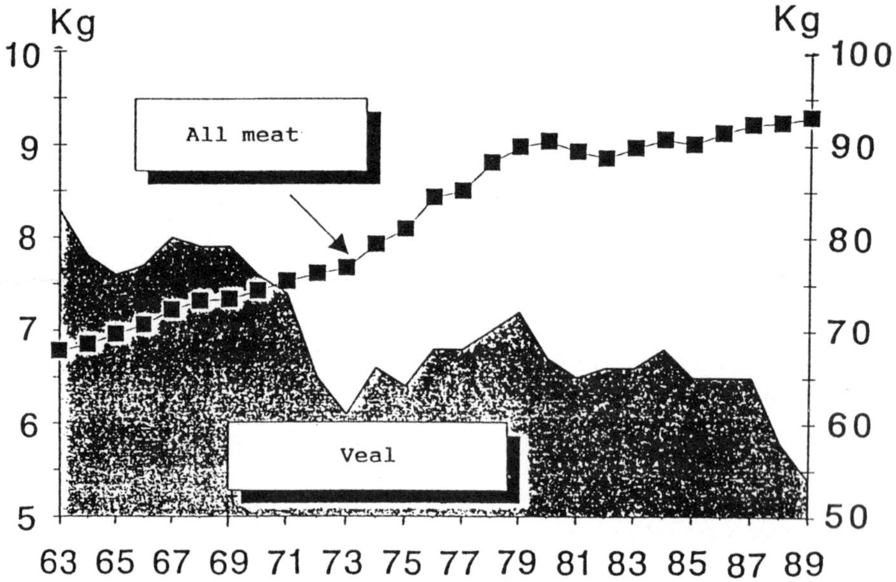

Figure 1. Average annual consumption of veal (left axis) and all meats (right axis) in France per capita. Source: SCEES.

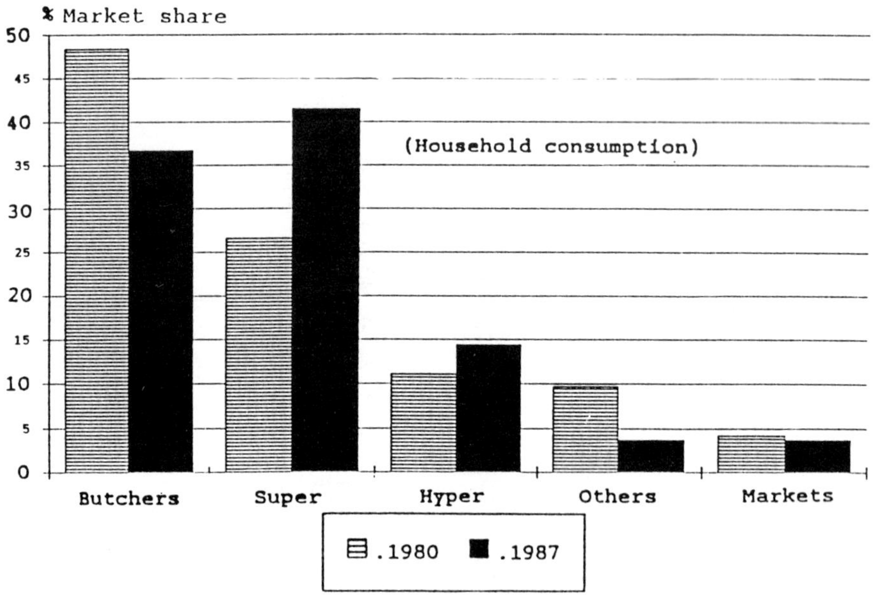

Figure 2. Veal sales by kind of retail outlet. Source: SECODIP.

Table 1. Average annual consumption per capita in the EC (kg), source: EUROSTAT, estimate in 1989.

	France	Italy	Benelux	Netherlands	West Germany
1980	6.7	3.8	3.0	1.2	1.9
1989	5.4	3.8	2.9	1.4	1.1

Veal in France: a declining market

In France, meat consumption per capita and per year increased strongly between 1963 and 1978 (about +30%) and then to a lesser extent between 1978 and 1989 (+5.6%). Since 1963, veal consumption has declined continuously, 15.7% between 1963 and 1978, 22.9% between 1978 and 1989 (Figure 1).

From accounting for 12.24% of all meat consumption in 1963, veal fell to 10.2% in 1970 and then to 5.8% in 1989. This downward trend accelerated dramatically in 1988 and 1989 and was confirmed by the SECODIP consumers' panel. According to this panel, household consumption of fresh veal fell by 14% in 1989 after falling by 11.5% in 1988.

Retail outlet trends

The share of veal purchased by households from modern retail outlets (supermarkets and hypermarkets) is rising, but butchers' trade is declining (Figure 2).

Main reasons for decline in consumption

Growing scarcity of supply plus increase in the cost of production

Veal production has experienced huge difficulties since 1988 because of changes in the economic and legislative environment;

- the ban on the use of growth promoters, and the shift to anabolic steroids;
- a modification of EC regulation 1725 (stipulating a level of skimmed milk powder and providing for a denaturation subsidy);
- the repercussions of milk quotas on young calves and markets for skimmed milk powder.

As a result, production costs increased (+13% in 1989 in France) and supply dropped steeply (Figure 3). French production of veal fell by 11% in 1989. That is more than the decline in EC production (9.6% in 1989).

Retail price of veal

In 1989, the retail price of veal increased in the same way as the production cost (Table 2). This increase was much stronger for veal than for other meats. It was also much higher than the average price index (Table 3). These price trends partly explain the decline in veal consumption in recent years (Figure 4).

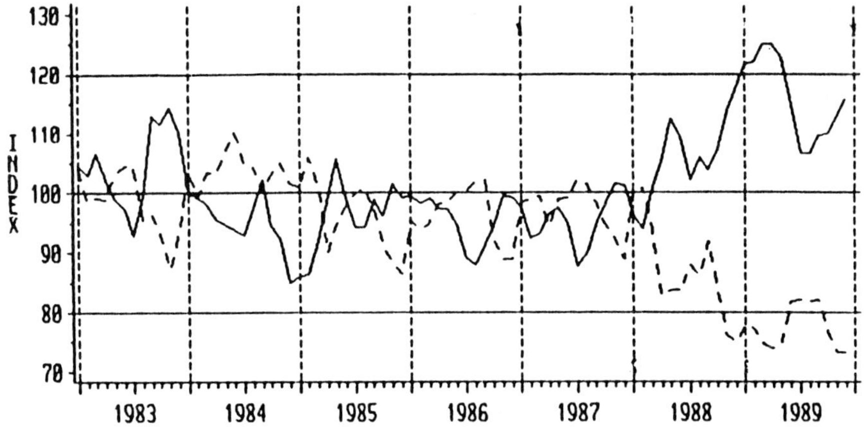

Figure 3. Comparison of trends in level (---) and cost (—) of veal production. Basis 100: average 1980. Source: SCEES - INSEE - OFIVAL.

Table 2. Comparison of trends in production cost and retail price; [1] estimate.

	1987/1986	1988/1987	1989/1988
Production cost	+ 3.3	+13.8	+ 13.0[1]
Retail price	+ 5.4	+ 7.9	+ 15.9[1]

Table 3. Comparison of trends in retail prices for various meats (%). Source: INSEE.

	1987/1986	1988/1987	1989/1988
Average price index	+ 3.1	+ 2.7	+ 3.5
Food (excl. beverages)	+ 2.1	+ 2.0	+ 4.3
Beef	+ 3.0	+ 3.4	+ 8.0
Pork	+ 1.6	+ 1.4	+ 9.4
Poultry, rabbit, game	– 1.0	– 1.0	+ 1.5

Tarnished image

Primal image of veal

We recently did a qualitative consumer study. We found that the predominant elements constituting the image of veal are

- simplicity
- fragility
- countryside.

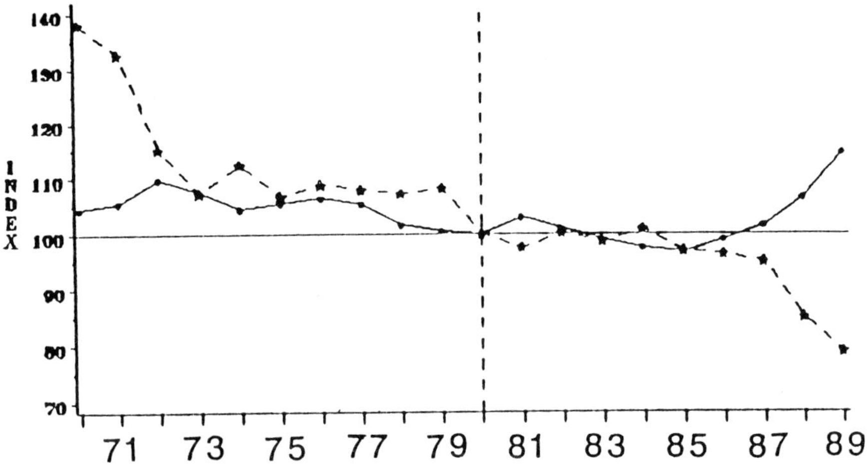

Figure 4. Relative price (—•—) and relative consumption (----) of veal. Basis 100: average 1980. Source: INSEE - SCEES, estimate in 1989.*

So, ideally, a calf sucks from its mother, lives in the fields and is housed on straw at night like its mother.

Attacks on the image of veal

Modern rearing techniques are moving veal away from this idyllic vision. They made it an easy victim of attacks by the mediat.

- In 1980 and 1984, there was a french boycott on veal consumption: the consequences were drops of 7.0% and 1.1% in consumption, respectively.
- German beta-agonist scandals (Aug. and Sept. 1988) were reported extensively in the French press.
- A recent Brigitte Bardot TV broadcast 'SOS animaux de boucherie'. This broadcast was a 'success':
  - 15% audience;
  - 100 000 Minitel calls;
  - 94% of the callers pledged they would never again buy industrial meat;
  - 44% pledged they would never again eat meat (source: RIA Jan. 22, 90).

Image at odds with 'health and fitness' trends

In today's affluent society, the consumer has the power to be more understanding and demanding when he makes his choice. Hence the consumer is increasingly concerned with the medium and long-term consequences of food on his health and fitness. Veal seems not to fulfil all requirements on this point (Figure 5).

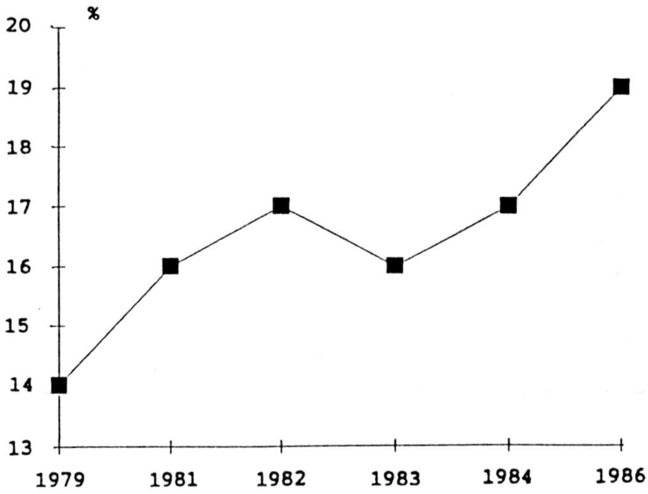

Figure 5. Proportion of French housewives not buying veal for fear that veal is less sound than other meats. Source: ARC survey for IFAA.

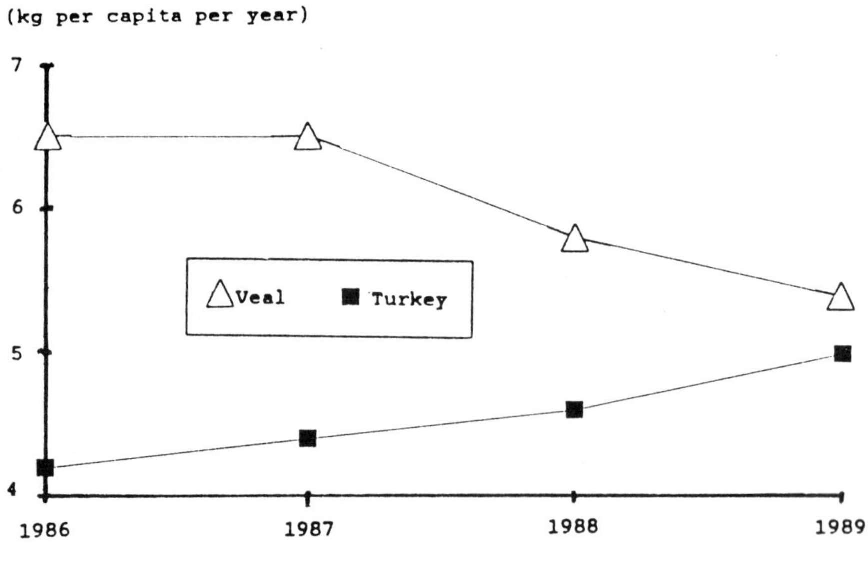

Figure 6. Comparative consumption of veal and turkey in France. Source: SCEES.

Advertising veal

Advertising is risky when production does not offer sufficient guarantees of safety.

Competition from turkeys

In order to develop this meat, turkey processors have used veal cuts and designations (i.e. 'escalopes, osso buco, rotis, blanquette'). According to SECODIP, these cuts represented 86% of household turkey consumption in 1987. This is the reason for the big market share of modern retail business in domestic consumption of turkey (73% in 1989). This more affordable meat is an easy substitute for veal. Its market is continuing to grow (Figure 6).

How to act to face the future

Breaking through restraints

High cost

It is possible to justify the high cost of veal provided that its quality is first class and meets consumer demands (colour, tenderness, flavour).

Sound health

The following points must also be observed: strict accordance with regulations, consumer information; product certification. Claims made about veal must be reassuring and not open to doubt.

Enhancing quality

Dietetic qualities of veal

In a time of media preoccupation with cholesterol, this can be a winning card. According to a CREDOC (1988) survey, 47% of the French population is concerned or deeply concerned about their diet.

Culinary qualities

Veal is easy to cook, can be used in a great variety of recipes and so makes diversification of the menu possible. It is up to veal production not to spoil these great assets by using drugs (especially beta-agonists), that alter meat quality.

Conclusion

As stated above, veal has certain assets to overcome the downturn in its consumption. It is now up to veal producers not to spoil these assets by using drugs that affect the meat and make veal calves some kind of bull calves. The consumer's image of veal should be respected.

Marketing strategies and quality control for veal in the United States

Arthur Follenweider Sr.
Swissland Packing Company, Ashkum, IL, US

Summary

To develop a successful veal-marketing plan, five factors, besides a company's personality, are of major interest: quality; supply; sales; shipping; and profit. It is important to develop and control each of these factors to create and keep a healthy marketing position. This paper describes the marketing strategies of a veal-packing company in the United States.
Descriptors: food processing; veal; United States; marketing.

Introduction

Most marketing plans have many similarities. Where they vary is in product and frequency of sales. The trick is to find and use the one that fits you best. With this in mind, let us explore some factors that are of basic importance.

The first thing is to develop a personality for your company. This is probably the most important factor in long term success or early failure.

Personality is described as 'habitual patterns and qualities of behaviour of any individual or group as expressed by physical and mental activities'. The key words are physical and mental activities. A personality that features a strong desire to work hard and have a winning attitude will succeed. A company emulates its leader.

Start with a sharply reacting organized production facility where all employees are proud to say that they work. Keep this facility hygienically clean, and repair and replace whatever needs care in the plant. This personality should be fully displayed by all employees of the company. The proper personality will have a strong and lasting effect on present and future customers.

This strong, optimistic attitude in a company's personality will also instill a positive feeling in the special-fed-calf suppliers. This feeling that they are bringing their calves to a company that can pay them and merchandize the meat products at a top price adds to their security. It is extremely important to be positive about the future. In the United States, those involved, whether they are suppliers or customers, have once again come to realize the importance of a strongly organized company.

Marketing of veal

There are five major ingredients in a successful veal-marketing plan:

1. quality
2. supply
3. shipping
4. sales
5. profit.

Quality

You cannot discuss marketing of veal in the United States without mentioning Aut Groenevelt. He started the veal industry in the US and was instrumental in the distribution and circulation of finished products throughout the retail market and restaurants in the United States. Because the starter calves were plentiful and low priced and the feed was strong, a calf of the highest imaginable quality was produced. Without this quality repeat orders would not have followed, and the programme would not have succeeded. Confirmation was outstanding, colour was beyond compare, and quality was so good that you could not touch the foreshanks together. The result was that those people who tried the product liked it and came back for more. The American public has become aware of what quality veal is and that is what they want and will accept. The Europeans are responsible for introducing this quality product and we must all live with this result. Quality is marketable.

On the subject of quality control, I can explain what we at our company do to maintain our programme. It is an extensive quality control programme and it all begins with plant sanitation. Safe food products take precedence over all other activities at our plant. The first move in the morning is for the quality control manager in conjunction with the Federal Government inspector to do a pre-inspection tour throughout the plant. Whatever does not meet proper criterion in sanitation is recleaned and sanitized. When it is ascertained that everything is go, the plant begins operation and quality grading and selection then take over.

The kill-floor workers have been highly trained to produce the best possible product. From sticking to hide removal, a distance is kept to give each worker a chance to do his job without interruption. That includes starting the dehiding and opening the calves at the right spot on the carcass. The people who operate the hide puller realize the importance of producing a hide with no cuts or blemishes.

Offal products are trimmed, cleaned, and put on trays in coolers to make the best livers, sweet breads, hearts, tongues, etc. Carcasses are railed into the first cooler after the veterinarian has inspected and passed each calf. Each carcass is trimmed and cleaned to make sure no contamination is on the calf.

Cutting and trimming according to the book on specifications is followed to the letter on the day following slaughter. This insures a fresh quality-controlled product to be shipped to the consumer. Boxes that are solid and strongly made are used. The cuts are put in vacuum bags to insure a long shelf-life by eliminating leaks in the vacuum. So-called cutting for profit by leaving more on the expensive cuts and less on the inexpensive cuts incurs large problems for the eventual consumers. Following the regulations makes lasting customers. The same is true of cheap poorly made vacuum bags and boxes.

Packing the primal and sub-primal cuts by grades, which is redone after the carcasses are chilled, ensures consistent quality. Immediately after slaughter a highly trained plant employee quality-grades each carcass. The calves are graded 'super', which pays a bonus, 'number ones', 'off color', and 'dark'. The carcasses are again graded the following day after they are chilled. This is done at the breaking area and is to verify that the proper grade was assigned to each carcass. The primals and sub-primals are again graded before boxing. If there are dark calves in the group they will be boxed as 'number two' product. This product is sold at reduced prices.

Supply and shipping

Low-cost starter calves, low-cost feed, large demand, and good profit led to increased supply. It is universally known that you must balance supply with demand. But for a long period of time, we could not get enough supply to meet the demand. In general, this situation is no longer always true. Through many factors, the supply varies from situation to situation. At this time, Holstein steers are in fashion and have thus shortened the supply of starter calves and resulting in their dramatic rise in cost. It is felt that this situation is short-lived but we are not sure that it will go away. Proper supply with proper prices are the key requirements for a strong long-range marketing plan. The secret to success in the veal business is to develop a demand for all the various primal and sub-primal cuts and the corresponding offal products that cover your supply lines.

The third factor is transport (shipping). The meat business in the United States moves the vast majority of its product by refrigerated truck. Shipping, though it is confusing, can be accomplished with accuracy and efficiency.

Sales

The sales portion of the discussion varies in method, style and direction. One way to get consumers to buy the product is advertising. This is an expensive, but potentially effective method. However, you usually never know the results of advertising. Bright colourful, informative advertisements in trade journals or in retail newspapers are suggested. Menu advertisements where possible are always very good.

Another method is telephone solicitation. This was the main method used in the meat business for many years. It was not uncommon for a salesman to sell to a customer for many years and never even know what he looked like. This will no longer suffice. To do the job properly, you must meet and discuss your product with the customer. Utilization of the mail service in the form of advertising material or informative correspondence can be effective. Another method is to promote your product in some of the many food shows throughout the United States. In my opinion, it takes a composite of all these methods to develop and sustain a strong marketing plan. It is important to note that marketing is not only essential in opening new markets but also in maintaining these markets after the initial sale.

Some products can be marketed without being concerned with continuing sales. This is not the case with veal. A programme should include trips to customers' facilities. It is equally important to have your customers or potential customers visit the production facility. Some people like to see their supplier more often than others. Some like to be more informed than others. The frequency of these activities can only be determined by the knowledge of the customers' personal desires. The customer is usually a bright and intelligent individual, who will eventually see through deception. If there is quality and supply, an informed potential customer should become a long-lasting satisfied customer.

In addition, it is always imperative to coordinate the production department with the sales department.

Profit

If the industry is to flourish in the world market it must be profitable. Profit-orientated companies make for good solid competition. One doesn't need a strong marketing programme to give away a product. A special-fed veal raiser depends on a company staying in business and paying a fair price for their calves. This cannot be accomplished without a profit-orientated market plan. The profit must be put back into your facilities to keep them updated and ready to utilize new operating methods.

Future outlook:

- Personality. This is and will be determined by the people involved.
- Quality. I feel that the quality of the finished calves will suffer through all the insecurity in the business at this time.
- Supply. I think that the United States supply will increase in the near future, probably in the fall of 1990.

- Sales. Continuing effort will slowly increase the production of various new products. These new products will have price and added value as their main theme.
- Profit. This is impossible to predict, but hopefully we will all have a profitable future.

Market-oriented production of veal: an approach from practice

S.H.M. Metz
Veevoederbedrijf ALPURO B.V., P.O. Box 1, Uddel, NL

Summary

Market-oriented production of milk-fed veal includes simultaneously market research, product development, and production and marketing of the veal. An important link in this chain is the production of the veal calves, which has to be modern, healthy and cost-effective. A system is presented for the daily control and guidance of the production of veal calves on the farms. The system includes a computer-based fast administrative processing of data on calf identity, housing, feeding, daily management and, after finishing calf fattening, also data on carcass weight and carcass quality. Deviations from standards are daily indicated by the system.

By implementing this information in farm management adequately, real progress has been made in the production of veal calves, and in meeting the consumers' demands for veal.
Descriptions: veal; veal calf; feeding; management; information; market.

Introduction

Veal is a high-quality, luxurious type of white meat, being extremely tender and highly digestible. Veal is consumed in a large number of countries. However in most of these countries, the proportion of veal in total meat consumption is very limited. Consumers' preferences for type of veal show great variation: both between and within countries, different markets exist for specific qualities of veal. Consequently a good knowledge of consumers' demands in the various markets is needed and the production of the veal calves on the farms and the slaughter and processing in the meatworks has to be directed to meet these demands.

In this paper, an outline from practice is given in what way market-oriented production of veal operates. Thereafter, a description is given of a system that is used to guide and support veal calf production on farms, as an optimization of cooperation between veal calf producers, feed company and meatworks.

Outline of market-oriented production of veal

A practical outline of market-oriented production of veal is presented in Figure 1. The production and marketing of the veal includes a chain of consecutive major processes before the meat reaches the consumer:

- breeding of young calves
- production of milk replacers
- fattening of veal calves
- slaughter of calves
- processing of veal
- distribution of veal.

Each process is linked to other processes, both by the flow of goods and the flow of information. Continuous optimization of the whole production chain is needed for veal production. In practice, this is obtained by two major mechanisms:

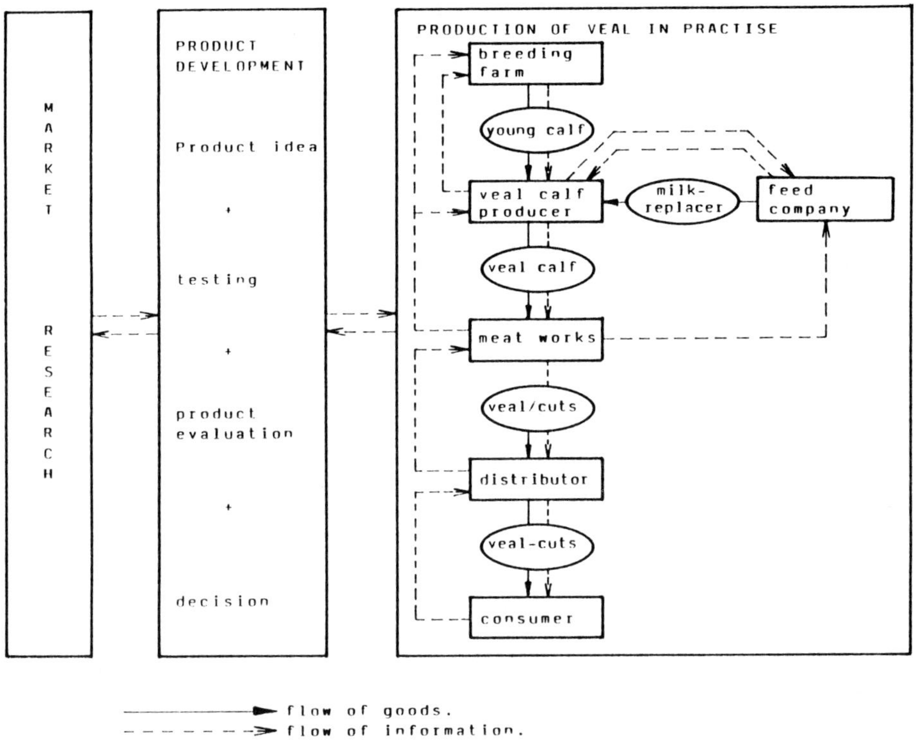

Figure 1. Outline of the market-oriented production of veal.

1. Improvement of the single processes as well as the integration between these processes by adequate flow of goods and information
2. Market research on consumers' demands for veal and development of new products to strengthen the position in present markets and to create new markets.

In the integrated production chain, the farmer (veal calf producer), the feed company and the meatworks play the central role. Together they are responsible for the quality of the veal calves at slaughter, and for the slaughter, processing and marketing of veal.

A system to control and to guide veal calf production

The fattening of calves is a process of main concern in the integrated production and marketing of veal, in the short term as well as in the long term. The quality of veal calves arriving at the meat works has to fit the needs and desires for veal. The quality of the fattened veal calves includes many aspects, like

- hygienic quality of the live animal;
- weight, type, conformation and fatness of the carcass;
- colour, taste and tenderness of the meat.

These aspects of quality are mainly determined by the farmer through type (breed, sex) of the young calves on his farm, composition of the milk replacer, housing system, feeding and management strategies. The farmer has to adapt the production of veal calves on his farm to

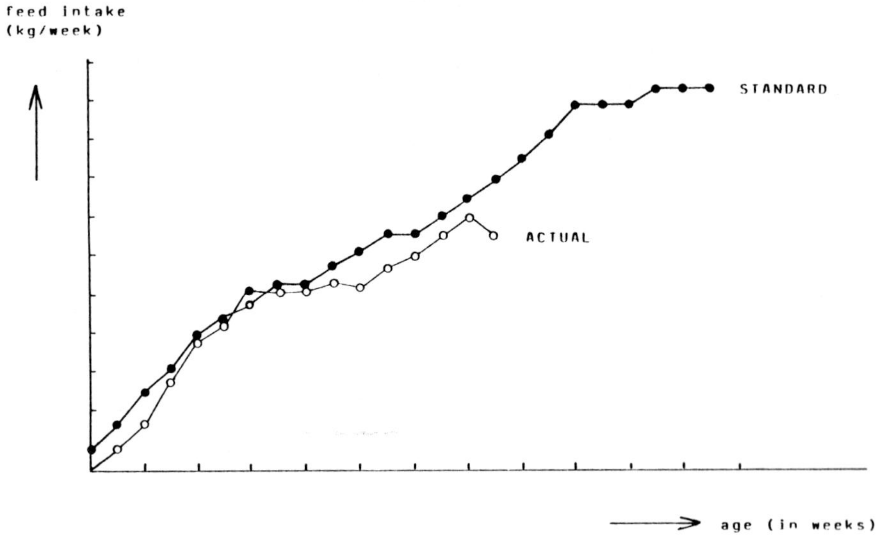

Figure 2. Actual versus standard feed intake.

Table 1. Signals for corrective measures (examples).

Week 5	use of medicine; feeding and litres correct?
Week 11	irregular feeding
Week 12	3 animals lost: total losses = 2.25%
Week 12	litres and/or feed-to-water ratio correct?

the customers' demands for quality. He needs to receive the results in carcass classification from the meat works, in order to have a useful tool for further optimization of his farm management.

Streamlining the numerous data with which the farmer has to deal, and uniting the data over a number of farms within the integrated production of veal, requires an easy-handling, standardized recording + indication + communication system. Such a system has been developed by our company. The system is a computer-based fast administrative system

- for improving the flow of information between the feed company, the veal calf producers, the meatworks and the breeding farms;
- for controlling and guiding the production of veal calves on the farms.

This system is continuously managed and controlled by the feed company and covers between 400 and 500 veal-calf producers.

The input and the output, which are specific for each farm and for each production group of calves within farms, include

- input data on
  - farm equipment (farm size, type of housing, ventilation and feeding system, hygienic control);
  - calves (breed, sex, date of input, liveweight at input),
  - desired feeding strategy;

- daily routine of feeding, management, health control, use of medicines, and losses of animals;
- slaughter data on carcass weight, carcass quality (colour + conformation + type + fatness), and quality of the 5th quarter.

• output
- overviews of the daily routine for feed intake (both momentarily and cumulative), losses of animals, use of medicine;
- deviations from the standard strategy in calf fattening (Figure 2);
- signals for corrective measures during calf fattening (Table 1);
- technical results (growth rate, feed conversion, carcass quality).

This output is generated daily, weekly or at larger intervals, depending on type or output. From the time the computer-based system became fully operational (1989), it appears to be an adequate basis for:

• continuous control of calf fattening;
• optimization of feeding and management routine on the farms;
• evaluation of the suitability of types of calves for veal production;
• continuous evaluation of feeding + management + housing strategies;
• adaptation of veal-calf production to the expected consumers' desires for veal.

These results stress the importance of the computer-based fast administrative system to support the market-oriented production of veal. It must be emphasized, however, that on the farms this data-processing system cannot replace the farm manager in optimizing veal-calf production. Although essential, it only remains a useful aid in managing the consecutive processes in market-oriented veal production.

Bio-economic model to determine the economically optimum fattening period in veal production

Gerard W.J. Giesen *, Titia E. Elema * & Ab F. Groen **
* *Department of Farm Management, Wageningen Agricultural University, Hollandseweg 1, 6706 KN Wageningen, NL*
** *Department of Animal Breeding, Wageningen Agricultural University, NL*

Summary

A bio-economic model was designed to determine the economically optimum fattening period in veal production. Given the daily feed ration and starting values for body weight and composition of a newborn calf, a biological model calculates gain in liveweight and growth of the carcass and its components by a deterministic day-to-day simulation. Starting from the calculated performances of the animal and exogeneous prices, an economic model calculates the gross margin at the end of each possible fattening period.

The model allows the calculation of influences of feed intake, biological parameters and prices on the economically optimum fattening period or optimum delivery weight and the resulting gross margin.
Descriptors: veal production; bio-economic model; optimum fattening period.

Introduction

Veal production is an important sector of Dutch agricultural production (PVV, 1988). Newborn calves are raised mainly on a diet of calf milk replacers up to a liveweight of about 238 kg in about 184 days (LEI, 1989). In 1962, veal calves were fattened up to 151 kg (van Horne & Sturkenboom, 1986). Liveweight and fattening period have increased gradually since and are still increasing.

In the Netherlands, veal production is for the greater part controlled by milk replacer producers. They arrange (by contract) when their farmers have to start fattening new calves and when they have to deliver them. Milk replacer producers repeatedly have to reconsider the economically optimum fattening period or optimum delivery weight of the animals. Of course, farmers fattening calves at their own risk are faced with the same question.

The economically optimum fattening period depends on the economic circumstances and the performance of the animals. The most important economic factors are the prices of veal, milk replacers and newborn calves. Important factors determining the performances of veal calves are breed and sex.

In this paper, a bio-economic model is presented to determine the economically optimum fattening period in veal production.

Model

The model consists of two parts: a biological and an economic model. Given the daily ration for the total possible fattening period and starting values for body weight and composition of a newborn calf, the biological model calculates gain in live-weight and growth of the carcass and its components (bone, meat, fat and waste matter) by a deterministic day-to-day simulation (Knol et al., 1990). Starting from the calculated performances of the animal and exogeneous prices, the economic model calculates the gross margin at each possible moment of delivery.

Biological model

The biological model consists of two parts: (1) the model of energy partition and growth; (2) the carcass model.

Figure 1 presents the structure of the energy partition and daily gain. For each day, the energy intake is partitioned into energy for maintenance and for protein and fat gain. Partitioning equations used depend on liveweight of the animal, feed quality and feeding level (van Es & van Weerden, 1970). From retention of energy in protein and fat tissue, growth of body components is calculated.

The model is dynamic: simulated full body weight on a certain day determines the energy requirement for maintenance, and energy partition and growth of the next day.

The daily growth of the carcass and its components (bone, meat, fat and waste matter) is also derived. The growth of bone is a fixed proportion of the fat-free gain of the body. The growth of carcass meat and fat is based on retention of energy in protein and fat in the body. Thereby, the energy retention of fat is partitioned into intramuscular and intermuscular fat.

Economic model

The economic model calculates the gross margin, which is the margin of returns over variable costs, per animal and per weekly period. Returns are calculated according to the payment system (per kg liveweight or carcass weight, possibly corrected for quality). Variable costs include costs of newborn calf, feed, interest and other variable costs. Losses as a result of failure are also included in the variable costs.

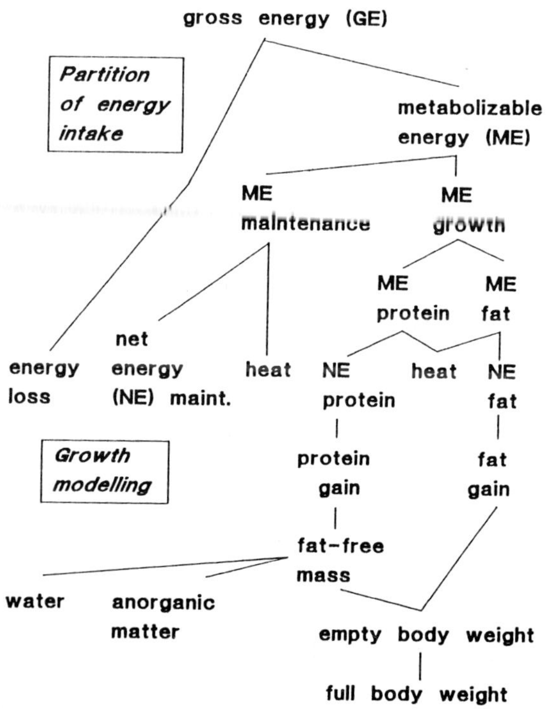

Figure 1. Schematic representation of the energy partition and growth model.

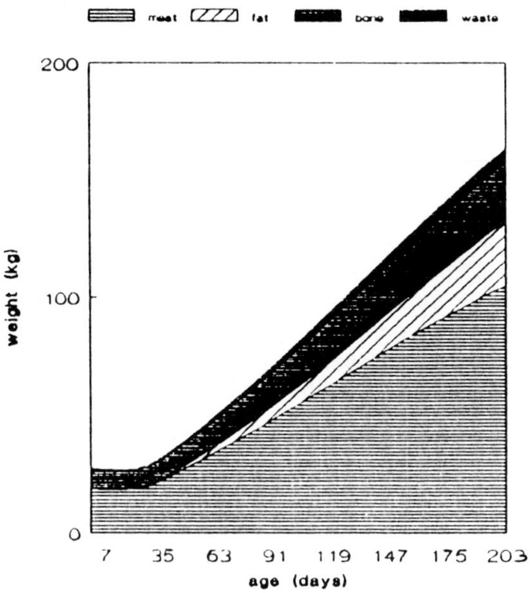

Figure 2. Simulated carcass composition.

Table 1. Simulated economic results of fattening black-and-white bull calves.

Age (days)	Live weight (kg)	Carcass weight (kg)	Gross margin (NLG) per calf place year	Gross margin (NLG) per calf
140	182.2	112.5	41.42	16.68
147	191.4	118.5	58.85	24.83
154	200.5	124.5	72.71	32.07
161	209.5	130.4	83.47	38.42
168	218.4	136.4	88.91	42.63
175	227.1	142.2	91.77	45.76
182	235.7	147.9	91.48	47.37
189	244.0	153.4	87.49	46.98
196	252.1	158.8	81.81	45.50
203	259.9	164.1	74.65	42.95

Assumed prices:

veal price NLG/kg carcass weight	9.30
price of newborn calf	400
feed prices (NLG/kg)	
starting feed	2.76
fattening feed 1 (based on skimmed milk powder)	2.72
fattening feed 2 (based on whey powder)	1.88

The economically optimum fattening period and delivery weight are reached when the 'marginal' gross margin per week of fattening a present veal calf equals the average gross margin per week of starting to fatten a new calf (Giesen et al., 1988). Average gross margin per week and therefore also gross margin per veal calf place per year are then at their maximum. If no newborn calves are available, the fattening period should be continued until the marginal gross margin per week equals zero or the gross margin of the animal is at its maximum.

Model results

In Figure 2, simulated data on carcass composition are presented for a black-and-white bull calf (birth weight 45 kg). The fattening starts at day 8 after birth. Up to the age of 65 days, calves are fed on starting feed (56 kg in total), from 65 days onwards the feeding is according to the so-called two-bags system: two feeds, one based on skimmed milk powder and the other on whey powder, are mixed. The daily ration is increased from 1.65 kg at 65 days up to 3 kg from 176 days onwards, this mainly by increasing the last feed.

In Table 1, the economic results are presented, assuming prices given at the foot of the table. Given these results, a veal calf has to be delivered at 175 days of age and 227 kg liveweight, because gross margin per veal calf place is at its maximum then. If no newborn calves are available to start up a new fattening cycle, optimum fattening period is one week longer up to 182 days or 236 kg liveweight. In that case, optimum fattening period is reached when gross margin per animal is at its maximum.

Changes in feed intake and biological parameters of the newborn calf will influence growth and carcass composition. These changes and changes in economic circumstances will influence the economic results and may have consequences for the economic optimum fattening period or delivery weight. The model presented here enables these to be calculated.

References

Giesen, G.W.J., W.H.M. Baltussen & J. Oenema, 1988. Optimalisering van het afleveren van mestvarkens. LEI-publicatie nr 3.139, Den Haag.

Knol, E., A.F. Groen & S. Korver, 1990. A biological model for veal production. Department of Animal Breeding, Wageningen Agricultural University (unpublished).

LEI, 1989. Resultaten kalfsvleesproductie 1988 (contractbedrijven). Landbouw Economisch Instituut, Den Haag.

PVV, 1988. Kalfsvleesproductie in perspectief. Produktschap voor vee en vlees, Rijswijk. PVV-rapport nr. 39.

van Es, A.J.H. & E.J. van Weerden, 1970. Kan de groei van gezonde mestkalveren worden voorspeld uit het opgenomen rantsoen? Landbouwkundig Tijdschrift 82:109–114.

van Horne, P.L.M. & H.J.C.M. Sturkenboom, 1986. Het optimale aflevergewicht van vleeskalveren. Proefstation voor de Rundveehouderij, Schapenhouderij en Paardehouderij, Lelystad. Publikatie nr. 44.

Model programme for production and marketing of free-range veal

R. A. Brown
Food Animal Concerns Trust (FACT), P.O. Box 14599, Chicago, IL 60614, US

Summary

This paper describes production and marketing of premium-priced veal from dairy calves raised in groups with access to pasture. The calves were fed acidified milk replacer and a solid ration. Limits were placed on use of medicaments. Feed costs averaged $168.50 per calf. Processing included electrical carcass stimulation. Opportunities and problems with marketing are discussed.
Descriptors: veal; free range; veterinary drugs; electro-stimulation; marketing.

Introduction

Almost all premium-priced veal in the US is produced in crates, anaemic diets without roughage, and extra-label drug usage. Therefore, a project was initiated in 1985 to determine how a premium-priced veal could be produced and sold from the same starting point (surplus bull dairy calves and milk replacers) but without immobilization of the calves, with solid as well as liquid rations, with normal iron levels, and with limited medication. This paper describes the project as it exists today.

Production practices

Production is continuous on one farm with a few calves slaughtered every week. Four to eight calves are purchased at a time directly from neighbouring dairy farms where colostrum was properly administered. Under the standards used, calves may not be purchased from an auction barn or dealer as is customary done with 'all-in, all-out' production in crates. By avoiding auction barns and dealers, calves have less exposure to pathogens, less transport stress, and are less likely to have been neglected.

Calves are housed in small groups of 8 to 12 animals in pens constructed within an old dairy barn; each pen has access to a grassy yard which the calves use frequently. The maximum recommended age difference among calves in a group is ten days. As the calves mature they are moved through a series of five pens in the barn.

Calves are trained to suck acidified milk replacer (Richard et al., 1988) with iron from nursing bottles. They are then trained to suck ad libitum from a nipple. Acidified milk replacer comes as a powder form and is mixed with cold water in a large plastic garbage can. Acidification retards spoilage so that a simple inexpensive tube and nipple can be used. Each group should have at least two nipples, which are mounted on the fencing of the pen. We recommend that feeding cans and nipples be thoroughly cleaned once a day.

Nipple feeding ad libitum appears to satisfy the calves urge to suck because cross-sucking and urine-sucking are no problem.

From the start calves have access to a grain-based ration (Boyaud, 1979) and hay. The ration initially has 20% protein. At about eight weeks of age, they are weaned to solid food only, at which time the ration has 16% protein. They reach a liveweight for maketing of 160 kg in 20 weeks.

Medication

The only routine medications used are vaccines, wormers, and coccidiostats. A competent farmer can easily identify sick calves because they are listless and do not play. When a calf becomes sick, medication is permitted under the following limitations:

1. the medication must be specifically approved for calves, ie, no extra-label usage;
2. sulfamethazine is not permitted because approval of this troublesome drug may be withdrawn;
3. withdrawal times are doubled;
4. any calf medicated within 56 days of slaughter is withdrawn from the program.

Extra-label drug usage is now a serious problem in the US when calves are raised only on liquids. It has been suspected for some time that calves denied roughage metabolize drugs more slowly. The US Food & Drug Administration confirmed this in a recent study showing that residues of sulfamethazine were higher for 'fancy veal' (liquid-fed) than for 'replacement calves' (dairy replacements receiving solid food) (Barnes et al., 1989). For instance, on Day 7, the residue in livers of treated calves was 0.39 ppm for fancy veal and 0.03 ppm for replacement calves.

On 23 May 1989, FDA met with representatives of the veal industry and ruled that liquid-fed calves would be considered a new class of animal and that no drugs at all have been approved for use with them (Stefan, 1989). In other words, all medications used in liquid-fed calves now fall in the extra-label category, a category of questionable legal status.

Farm costs

Table 1 presents the average costs per calf for the farmer in Maine this winter. Not included are building and equipment costs, electricity and water, and the farmer's time. Winters in Maine are severe, so that feed costs would be higher than those shown in summer or in warmer climates.

Table 1. Farm costs.

Calves	$ 130.00
Feed	
milk replacer	45.00
solid ration with 20% protein	13.00
solid ration with 16% protein	108.00
hay	2.50
Medication	8.00
Bedding (wood shavings)	25.00
Trucking to slaughter	6.50
Telephone, postage, misc.	9.50
Total	$ 347.50

Processing

Carcasses receive low-voltage electrical stimulation immediately after slaughter to lighten meat colour and improve tenderness (McKeith et al., undated). We consider this processing step essential if a gourmet class veal is to be produced using solid rations.

Marketing

Many (but not all) chefs, butchers and consumers prefer this free-range veal to that from confined and anaemic calves, which is softer in texture and blander in taste. Retail prices are above those for veal raised in crates at other local markets. Based on eating qualities, drug residue problems, and concern for calf welfare, we believe free-range veal produced in this way has excellent market potential.

The principal problems we have encountered relate to the fact that the necessary infrastructure for small-scale meat producers has largely disappeared in the US. Typical problems are:

1. finding adequate processing, including special cutting and Cryovac packaging, when few calves are slaughtered weekly;
2. finding trucking for small quantities at reasonable prices;
3. selling all parts of the carcass.

We have overcome these difficulties with our current customer, the Bread & Circus supermarket chain in Massachusetts, and have an economically self-sustaining demonstration project. In our opinion, the three difficulties would diminish if this product was offered in larger quantities by a meat company with its own processing and shipping facilities.

References

Barnes, C.J., & J.N. Gelata, 1989. Chemotherapeutic agents and their disposition in food producing animals. 1. Disposition of sulfamethazine in tissues and plasma of bob veal, fancy veal and replacement calves after oral administration. DVMR, Center for Veterinary Medicine, Beltsville, US.

Boyaud, D., 1979. The Shur-gain feeding program for heavy veal production. Canada Packers, Inc., Brossard, Canada.

McKeith, F.K., J.W. Savell, C.E. Murphey & G.C. Smith, undated. Enhancement of lean characteristics of veal carcasses by electrical stimulation. Meats and Muscle Biology Section, Texas A & M University, College Station, TX 77843, US.

Richard, A.L., A.J. Heinrichs & L.D. Muller, 1988. Feeding acidified milk replacer ad libitum to calves housed in group versus individual pens. Journal of Dairy Science 71:2203.

Stefan, G.E., 1989. Quality control in formula-fed veal production. Center for veterinary Medicine, Rockville, US.

7. Meat quality and growth stimulants

Biochemical muscle characteristics and veal quality traits

C. Valin
Institut National de la Recherche Agronomique - Theix, 63122 Ceyrat, FR

Summary

Production and consumption of veal are decreasing in Franc,e where traditionally veal has been highly appreciated. Many factors are responsible for this trend but it generally appears that veal is suffering either from a bad image or at least from a bad quality/price ratio. This makes it important to reconsider how the quality of this product could be improved.

When grading veal carcasses, colour has always been taken into account as a quality characteristic. The control of this quality parameter is good at the production level and at present most of the factors affecting the colour of veal are known, even if the influence of the ultimate pH is ignored during the grading operations.

The most critical point in terms of veal quality at the consumer level is tenderness. Very often veal is tougher than expected. Taking into account the age of the calf at slaughter, the physical characteristics of muscle collagen do not limit tenderness. Veal tenderness depends of course on the slaughter, chilling and ageing procedures used, but veal toughness is often associated with high weight losses during cooking because veal exhibits a low water-holding capacity because of a very low ultimate pH, sometimes associated with an accelerated rate of pH fall. Therefore the way veal is cooked is particularly important to preserve tenderness.

Introduction

Traditionally, veal consumption in France has been important: around 7 kg per capita per year. But there has been a steady decline for several years, and this has grave implications for the future.

The year 1988 was particularly difficult. Veal production within the European Community fell by 8.2%, mainly because profits were cut by an important increase in the production costs. The proportion of young calves used for veal production decreased from 27% in 1987 to 25.6 in 1988, and this change was particularly important in France where the slaughter of veal calves fell by 10% in 1988. During the same period the prohibition of anabolic agents was accompanied by much illegal use of these products and of new molecules, so helping to give a bad image to the product which was simultaneously suffering from a poor quality/price ratio. So in France where the production cost increased by 14% and the retail price by 7.8%, veal consumption fell by 10.1%. Given these trends, it is perhaps worth considering in some detail the basic veal quality traits: colour and tenderness.

Biochemical basis of veal quality traits

Veal is not cheap for the consumer. So it is vital for the product to improve quality/price ratio, otherwise the consumer will seek substitutes. Colour and tenderness are the most important quality criteria for veal. To date, colour has been important in terms of carcass grading, and tenderness has been important for consumer acceptability. Flavour is important, but has not yet appeared as a factor limiting the consumption of this type of meat.

Veal colour

The colour of veal and more precisely its luminosity depends on the amount of muscle pigment and on the ultimate pH of meat. In the usual grading system, no correction for pH is used in the colour assessment of the carcasses.

Usually, veal carcasses are classified (graded) in a four-class grading system: white, light, pink, pink and dark pink, or red. Within a class and within a carcass, there is large variability in luminosity and pigment content. The result of the large variability of luminosity within a class makes the grading difficult because many overlaps between classes are possible, as shown on Figure 1 where the data have been transformed into a three-dimensional representation of the colour of veal carcasses of the four classes. In these conditions, a small change in the ultimate pH can greatly affect the grading, which is theoretically based on the amount of muscle pigment. Indeed, in the paler muscles, such as *M. longissimus dorsi*, the luminosity is almost equally affected by the amount of pigment, the ultimate pH and the rate of pH fall.

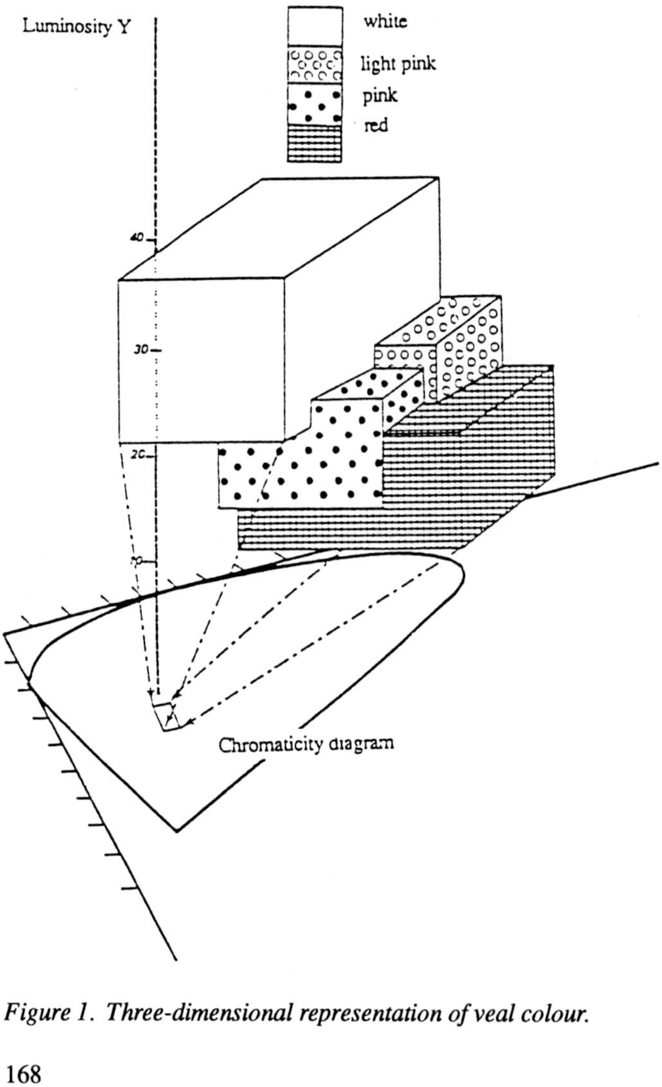

Figure 1. Three-dimensional representation of veal colour.

The control of a low amount of myoglobin in veal muscles is well achieved, as is the control of iron supplementation to improve the growth rate of the live animals without any detrimental effects on the colour (MacDougall et al., 1973). But during the last 20 years, the production methods have changed. In order to improve the growth rate and the carcass weights at slaughter, anabolic agents were used on a large scale, and the result is a slow and continuous shift towards redder carcasses. The analysis of the effects of these different factors on veal colour is summarized in Table 1 (Valin et al., 1978).

It appears that the use of anabolic agents has no significant effect on the ultimate pH and the myoglobine content. This is supported by more recent results (Renerre et al., 1989). Thus the possible muscle dependent effect of anabolic agents on the muscle typing, which is very significant in older animals such as steers (Ouali et al., 1988) and also veal (Valin et al., 1984), has no major effects on meat colour when animals are deficient in iron.

Colour is significantly affected by age or carcass weight and it appears that this effect is muscle dependent and much more important in the redder muscle of the carcass.

The effect of the milk composition on veal colour, even with an iron-deficient feeding system, is variable. For instance, in our previous experiments, the colour of veal from calves fed with starch-supplemented skim milk was redder than with a feeding based on regular fat-supplemented skim milk. It is difficult to explain this shift towards a redder colour because many factors may be involved, such as a change in gut pH, which could affect iron absorption, or a change in muscle typing.

Veal tenderness

In general, meat tenderness depends on

- the amount of collagen and the degree of collagen cross-linking
- the process of rigor mortis and the post mortem ageing both of which are affected by the thermal treatments applied to the carcasses and by the muscle typing
- the cooking procedure and the weight losses during cooking.

We will focus on the specificity of veal muscle traits which could explain the high variability in muscle tenderness frequently observed with veal. Table 2 (Kopp et al., 1984) relates the effect of sex and anabolic treatment on the characteristics of veal muscle collagen.

Both factors affect collagen, but the sex effect is much more important than the hormonal one on the amount of muscle collagen. The thermal stability is slightly lower in animals that have been treated with steriods as previously observed with calves (Valin et al., 1978) and with steers (Ouali et al., 1988). This suggests the presence of a slightly less cross-linked collagen in treated animals. But control and treated animals exhibit a high thermal solubility which shows that collagen is not really a factor limiting veal tenderness. Of course, the effect of new growth promoters such as β-agonists needs to be checked before this result is extrapolated.

The biochemical post mortem changes of veal differ compared with those of meat of older cattle. Regarding the ageing process it is well known, veal exhibits a faster rate of ageing than meat from older animals. With a slow chilling process, i.e. without any cold shortening, the ageing rate is roughly double that of meat from steers at the same storage temperature, the post mortem tenderization being achieved in four days (Figure 2).

Very often, the meat of calves treated with anabolic agents is tougher than that of control animals. This toughening always appears to be related to a decreased rate of the ageing process (Valin et al., 1978, 1984 ; Ouali et al., 1988), which probably depends on the effect of anabolic agents on the typing of muscle fibers, this effect on the typing being correlated with the extent to which the hormonal treatment has boosted the growth rate of the animals. The anabolic treatment is indeed muscle-dependent but in the longissimus dorsi the treatment shifts the typing to a slower type, which could contribute to decreasing the ageing rate (Table 3).

Table 1. Effect of anabolic agents (steroids), composition of milk and carcass weight at slaughter on ultimate pH and colour characteristics of veal. Significance: F value 5%: 4.15; 1%: 7.50**. Calves of the Norman breed, slaughtered at carcass weight of 100 to 130 kg (Valin et al., 1978).*

		F value			
		ultimate pH	haeminic iron	luminosity *Y*	redness *a*
M. longissimus dorsi					
Factor 1	anabolic agent	2.201	0.033	4.075	0.185
Factor 2	milk composition	0.362	15.568**	4.255*	9.630**
Factor 3	carcass weight	17.725**	3.786	44.813**	8.017**
M. pectoralis profundus					
Factor 1		1.435	0.369	7.656**	0.056
Factor 2		0.064	11.558**	7.153*	0.006
Factor 3		11.734**	4.283*	112.307**	4.285*

*Table 2. Effect of sex and anabolic treatment on veal muscle collagen characteristics. Significance * P < 0.05; ** P < 0.01; *** P < 0.001.*

	Sex			Anabolic treatment		
	male $n = 12$	female $n = 12$	signif.	control $n = 12$	exp. $n = 12$	signif.
Dry matter %	23.5	24.4	***	24.2	23.7	**
Collagen in DM (%)	2.1	1.8	**	1.9	2.1	n.s.
Tension (Ncm^{-2})	135	139	n.s.	148	126.8	*
ΔH of DM (J/g)	15.5	15.1	n.s.	15.0	15.6	n.s.

Table 3. Effect of anabolic agents on typing of muscle fibres in M. longissimus dorsi.

	Kopp et al., 1984			Valin et al., 1984		
	control	exp.	signif.	control	exp.	sigif.
Myosine isoenzymes (%)						
Iso I	16.2	16.2	n.s.	18.7	12.5	***
Iso II	33.7	32.0	**	34.3	36.4	*
Iso III	38.1	37.8	n.s.	32.6	34.8	*
Iso IV+V	11.9	14.0	*	14.4	16.3	**
Growth rate (g d^{-1})	1180	1210	n.s.	1369	1562	*

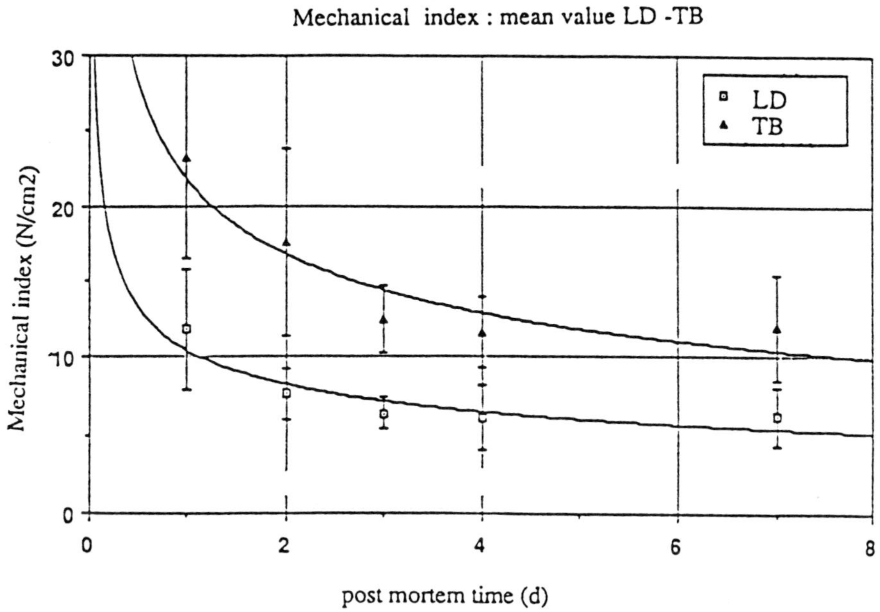

Figure 2. Post mortem tenderization in two veal muscles: M. longissimus dorsi and M. triceps brachii.

Regarding the process of rigor mortis, veal exhibits a specific behaviour characterized by a high frequency of low ultimate pH below 5.5 (Figure 3). The kinetics of the pH changes and the value of the ultimate pH varies between animals. This very low ultimate pH impairs the water-holding capacity of veal and explains the high weight losses during cooking frequently observed with veal. As we observed previously without any cold shortening, toughness of veal is closely correlated with weight losses during cooking and the drying of the cooked meat (Valin et al., 1977).

The glycolytic potential at slaughter seems to be the main cause of low ultimate pH rather than the buffering capacity which is lower in veal then in muscles of older cattle. The higher the glycolytic potential at slaughter, the lower the ultimate pH. With a glycolytic potential above 120 µmol/g the ultimate pH is less than 5.5 and whatever the muscle type, glycolytic potentials ranging between 150 µmol/g to 170 µmol/g are frequently observed in veal muscle.

In addition to low ultimate pH, some veal muscles exhibit rapid rates of post mortem pH fall which could be deleterious for the water holding capacity of meat, especially when this rapid acidification is associated with a slow chilling process as we have been able to check using electrical stimulation to get a rapid rate of pH fall associated with different rates of chilling (Table 4.).

If the protein denaturation assessed by the Hart test improves the colour by increasing the luminosity, it also increases the weight loss during cooking. In addition with higher rates of post mortem pH fall, even without electrical stimulation, it is also possible to observe an increase in the extracellular space in the muscles (Monin, personal communication). This also depletes the water-holding capacity of muscle.

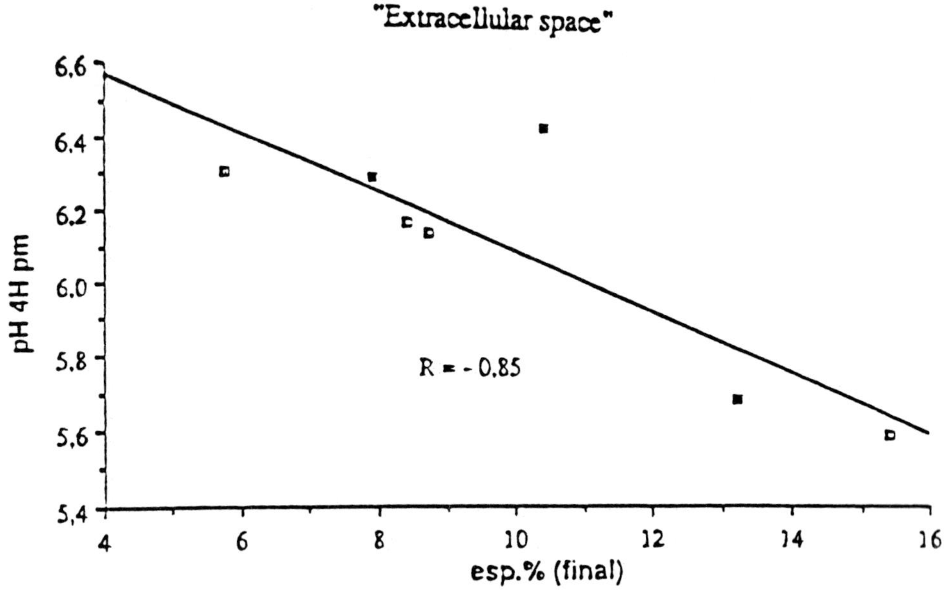

Figure 3. Relationship between extracellular space and pH 4 hours post mortem.

Table 4. Effects of chilling rate on protein denaturation of veal muscle (Legras, 1980); LD, M. longissimus dorsi; TB, M. triceps brachii; ^{a,b} $P < 0.05$.

	LD muscle		TB muscle	
	ES slow chilling	ES rapid chilling	ES slow chilling	ES rapid chilling
pH 1h	6.02	6.02	6.03	6.03
pH ult.	5.45^a	5.51^b	5.46	5.49
Luminosity	36.5^a	30.4^b	40.5^a	32.3^b
Hart test	617^a	744^b	525^a	758^b
Weight losses	24.0^a	20.0^b		

Veal flavour

Veal does not have a very strong flavour, but so far very few experiments have been done to analyse the basis of veal flavour. Undoubtedly, meat from calves fed on whole milk exhibits significantly higher organoleptic qualities than that from animals fed on lipid-supplemented skim milk (Touraille et al., 1983). In this experiment on Friesian calves displaying the same amount of intramusculair lipids pigment and ultimate pH at slaughter, the main difference was the composition of intramuscular lipids, neutral lipids and phospholipids and particularly the degrees of unsaturation. Given the influence of phospholipids in the development of meat flavours, manipulating lipid composition in veal muscle might be a way of improving the flavour of industrially produced veal.

Conclusion

Apart from poor procedures after slaughter, the characteristics of veal muscle explain many of the quality problems with this kind of meat.

The specificity of veal muscle characteristics has not yet been completely explored but, even on the basis of what is now known, it appears to us that veal quality could be improved if we succeed in controlling the glycolytic potential of veal muscle at slaughter.

References

Kopp, J., M. Bonnet, M. Zabari, J.P. Renou & C. Valin, 1984. Influences relatives du sexe et d'un traitement anabolisant sur les caractristiques du muscle long dorsal de veau. 30th Eur. Meet. Meat Res. Workers, Bristol, 3:27, 145.

Legras, P., 1980. Etude de la couleur de la viande de veau. DESS Universit de Clermont Ferrand.

MacDougall, D.B., I. Bremmer & A.C. Dalgarno, 1973. Effect of dietary iron on the colour and pigment concentration of veal. J. Sci. Food Agric., 24, 1255–1263.

Monin, personal communications.

Ouali, A., M.Zabari, J.P. Renou, C. Touraille, J. Kopp, M. Bonnet & C. Valin, 1988. Anabolic agents in beef production : effects on muscle traits and meat quality. Meat Sci., 24, 151–161.

Renerre, M., C. Touraille, P. Bordes, R. Labas, M.C. Bayle & R. Fournier, 1989. Influence of straw feeding and growth implant on veal meat quality. Meat Sci., 26, 233–244.

Touraille, C. C. Valin, B. Aurousseau & J. Sornay, 1983. Influence du mode de production sur la qualit de la viande de veau.

Valin, C. A. Lacourt, M. Renerre & C. Touraille, 1977. Influence des techniques d'levage et d'alimentation sur la qualit de la viande de veau. L'Alimentation et la Vie, 85, n1, 3–11.

Valin, C., M. Renerre, C. Touraille, J. Kopp & J. Sornay, 1978. Influence de la forme d'apport de l'nergie dans la ration et de l'utilisation des anabolisants sur la qualit des viandes de veau. Ann. Nutr. Alim., 32, 857–867.

Valin, C., C. Touraille, M. Zabari & J.P. Renou, 1984. Influence des anabolisants sur la qualit des viandes de veau. 30th Eur. Meet. Meat Res. Workers. Bristol, 4:8, 176.

Quality control during veal processing: an evaluation of modern techniques

Frans J.M. Smulders
Department of the Science of Food of Animal Origin, Faculty of Veterinary Medicine, University of Utrecht, Utrecht, NL

Summary

The eating quality of veal may be substantially affected by the way calf carcasses are processed in the abattoir. For lean light-weight carcasses, the chilling rate has been identified as a major determinant of several important sensory characteristics. Modern refrigeration systems aimed at reducing evaporative weight losses, generally include a pre-chilling period of up to 2 h at temperatures as low as –15 °C and air velocities as high as 20 m s^{-1}. The resulting rapid decline in muscle temperature causes rigor mortis to occur during a period of extensive muscle contraction, causing a phenomenon known as 'cold shortening'. The classical circumstances reported to favour this condition arise when the muscle pH is still higher than about 6.1 and muscle temperatures have already dropped below about 12 °C. Therefore, operators having adopted high refrigeration rates, have in recent years come to rely on electrical stimulation as a means of accelerating muscle glycolysis, thereby avoiding these critical conditions. Moreover, electrical stimulation is reported to make veal colour appreciably lighter, which provides added market potential. Other abattoirs, relying on moderate chilling rates, have refrained from electrical stimulation for various reasons, the major one being that cold shortening conditions rarely occur in their chill-rooms.

Recent observations strongly suggest that electrical stimulation is applied incorrectly in some abattoirs, leading to serious quality problems. There is proof too that electrical stimulation might be advantageous even at moderate chilling rates, provided a well controlled stimulation procedure is followed.

The rationale of adopting electrical stimulation for various veal-processing procedures is discussed. The effects of stimulation on the appearance (colour and water-holding capacity) and tenderness of veal were evaluated and procedures are suggested to provide maximum benefits from electrical stimulation.
Descriptors: meat quality control; chilling rate; electrical stimulation.

Introduction

The eating quality of veal is determined not only by the animal's genetic make-up or the interaction of various animal husbandry factors that are expressed in carcass meat-to-fat ratio. In the past decade it has become increasingly clear that calves with a potenial for supplying high-quality veal may still yield meat with undesirable sensory traits, as a result of ill-controlled or suboptimal processing. Although many critical factors have been identified and measures suggested to overcome problems, these suggestions have, unfortunately, not always been translated into 'Good Manufacturing Practices' by veal processors.

This paper focuses on one of the most critical processing steps in veal production, chilling, and on the measures that need to be taken to prevent adverse effects due to incorrect chilling rate.

For detailed information on the terminology and methodology of sensory analysis of meat, for quality the reader is referred to a review paper by Smulders (1986).

Post mortem muscle contraction and chilling rate

A muscle reacts to a nervous impulse by releasing Ca^{2+} from its primary cellular reservoir, the sarcoplasmic reticulum. These ions catalyse the breakdown of glycogen by activating the glycolytic enzymes present in the sarcoplasm. During this process, adenosine triphosphate (ATP) is produced, which is necessary for muscle contraction. After an ATP-requiring Ca^{2+} pump has transported Ca^{2+} back over the membrane of the sarcoplasmic reticulum, the muscle relaxes again.

After an animal has been slaughtered, glycogen is broken down, albeit anaerobically, through the enormous release of Ca^{2+} after nervous impulses at the time of slaughter. The H^+ released during glycolysis accumulate causing the pH to drop. When the muscle cell has no more ATP, the myofibrillar proteins actin and myosin can no longer freely 'interdigitate'. Once the cross-bridges between the two are irreversibly bound to form actomyosin, the muscle fibre has entered rigor mortis.

In the early 1960s, New Zealand meat scientists discovered that certain critically low temperatures caused rigor mortis in a state of severe muscle contraction (Locker, 1960; Locker & Hagyard, 1963). This phenomenon, commonly known as cold shortening, is characterized by short sarcomeres and increased toughness. A similar effect occurs at high temperatures. This heat-shortening effect will not be discussed in this paper. Figure 1 illustrates the effect of several holding temperatures on the degree of muscle contraction and tenderness.

The occurrence of cold shortening may lead to a four-fold toughening of muscles (Bouton et al., 1981). Moreover, the 'water-holding capacity' of cold-shortened meat is severely reduced. As a result, more juices are lost from the meat, causing lower (cooking) yields and less palatable meat (Honikel et al., 1981; Smulders et al., 1986). It has been calculated that cold shortening occurs when the muscle temperature drops below about 12 °C and the pH is still higher than about 6.1 (Bendall, 1973). Based on these calculations, rules of thumb have been devised. One such rule (the '10 and 10' rule) states that if a muscle is still warmer than 10 °C 10 h post mortem, one need not fear the occurrence of cold shortening (Bendall, 1973). A series of experiments in the Netherlands has substantiated, however, that cold shortening does occur in veal carcasses under many circumstances that conform to the afore-mentioned safe conditions (Smulders, 1984). Obviously, the rules of thumb need adjustment.

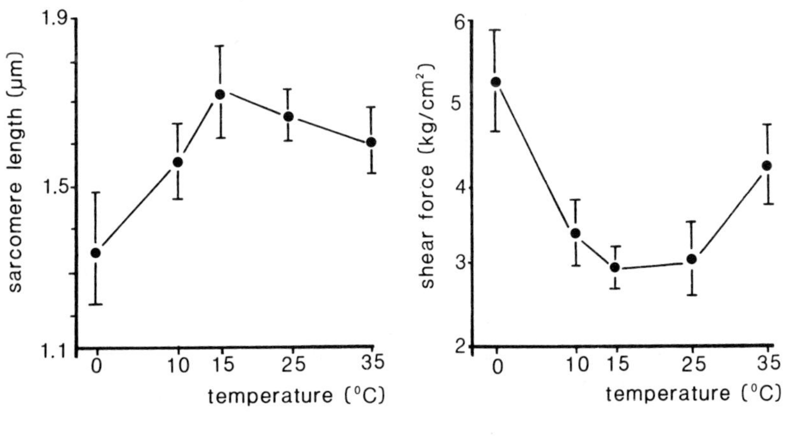

Figure 1. Effect of holding temperature on sarcomere length and shear force of longissimus muscle (Smulders et al., 1984).

Cold shortening can be prevented in two ways. The first almost obvious way is to abandon rapid carcass chilling. However this would not conform to the operator's wish to limit the residence time of carcasses. The second one, accelerating glycolysis, is achieved by electrical stimulation.

Electrical stimulation: the procedure

When a muscle cell is subjected to electrical current, certain cellular structures are damaged; as a result, Ca^{2+} is released at high concentrations, activating glycolysis. This causes a rapid pH decline. Figure 2 illustrates that electrical stimulation early post mortem, say within 5 min of slaughter, results in a faster pH fall than stimulation at a later stage. Presumably the loss of conductivity of the nervous pathways is primarily responsible for this. In practice, it means that stimulation should preferably be conducted as early as possible, but in any case within the hour, and that more electric energy is required to produce similar effects if one chooses to stimulate late (Chrystall et al., 1980).

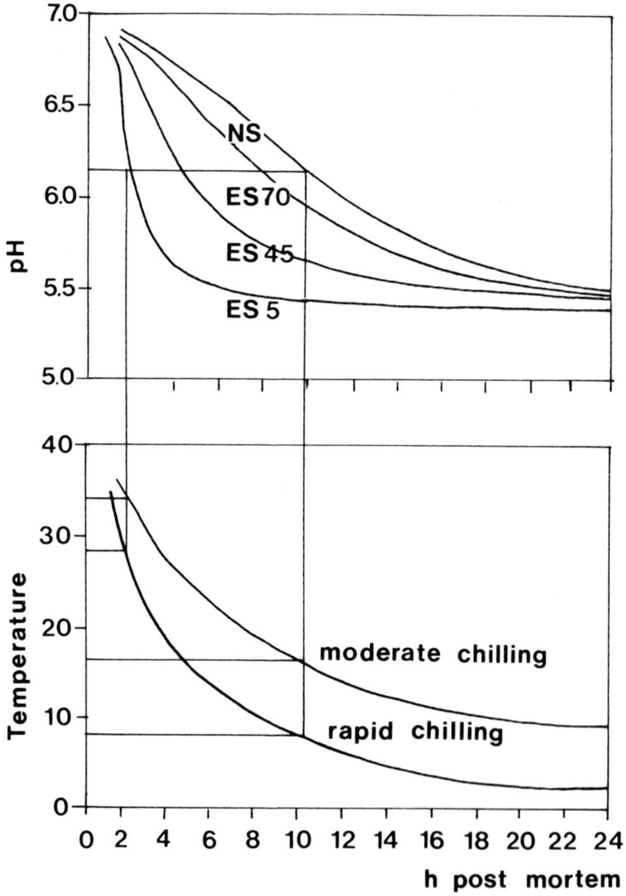

Figure 2. Effect of electrical stimulation at 5, 45, and 70 min post mortem on pH fall and the likelihood of protection against cold shortening (critical: pH > 6.1 at T < 12 °C).

Various pieces of equipment, relying on potential differences, electrical currents, frequencies, duration of stimulation, and operated manually or automatically with various types of electrodes, are available commercially (West, 1982). All of these seem able to promote a more or less rapid pH fall and may therefore be expected to protect against cold shortening. One must realize, however, that to safeguard the sensory quality of veal in modern processing lines, operators must consider more than the prevention of cold shortening alone.

Electrical stimulation: effects on sensory quality traits

Table 1 includes the effects of an extremely rapid pH decline, as effected by intensive electrical stimulation, on major sensory traits of veal longissimus muscle. Table 1 substantiates that electrical stimulation promotes an extremely rapid pH fall, effectively prevents cold shortening, and results in veal with low shear forces, which is preferred to unstimulated meat by about 80% of a taste panel. The increased tenderness after electrical stimulation does not result from prevention of cold shortening alone. Several additional mechanisms have been reported, amongst them mechanical disruption of the sarcomere structure by the vigorous contractions during stimulation (Savell et al., 1978), increased membrane damage by high temperature and low pH, leading to increased activity of proteolytic lysosomal enzymes (Dutson et al., 1980; Wu et al., 1985), and increased solubility of collagen (Judge et al., 1980). Through all these mechanisms, the tenderness of meat will be promoted, regardless of the chilling rate of the carcass.

An added advantage of stimulation is that calves with similar concentrations of plasma Hb and of muscle pigment (Haematin = Hb + Mb) yield veal with an appreciably lighter colour, as assessed both with instruments and by eye. Several mechanisms have been suggested for the increased lightness after stimulation. It may be partly related to a higher oxymyoblobin concentration due to increased penetration of oxygen into the (disrupted) fibre structure (Sleper et al., 1983). In addition, as reflected by the higher transmittance after intensive electrical stimulation (Table 1), it may be caused by an increased denaturation of the sarcoplasmic proteins and a resulting loss of water-holding capacity. This is clearly demonstrated in Table 1 by the increased drip loss in stimulated vacuum-packaged veal after 7 days of storage. It is partly because of this adverse effect that too intensive a stimulation is discouraged.

Table 1. Effect of intensive electrical stimulation (60 s, 85 V, 14 Hz) on major sensory traits of veal longissimus, as assessed 7 d post mortem (Eikelenboom & Smulders, 1986).

	Electric Stimulation	No stimulation	*P*-level
pH (40 min post mortem)	5.86	6.83	$P < 0.01$
Sarcomere length (μm)	1.7	1.3	$P < 0.01$
Shear force (kg cm^{-2})	2.8	4.9	$P < 0.01$
Panel preference (%)	79.5	20.5	$P < 0.01$
Ante mortem plasma Hb (mg/ml)	8.9	8.9	n.s.
Haematin (mg/g)	44.9	40.1	n.s.
Hunter L value (lightness)	49.8	47.9	$P < 0.01$
Transmittance (%)	38.0	27.0	$P < 0.05$
Drip loss in vacuum packing (%)	2.7	2.2	$P < 0.05$

Another important argument against 'overstimulation' became apparent in recent studies conducted in the US, in which the tenderness of meat was highest in muscles that had an intermediate glycolytic rate (i.e. a pH at 3 h post mortem in the 5.9–6.1 range); the muscle tenderness scores declined at both slower and faster rates (Marsh et al., 1987; Smulders et al., 1990). These findings once again indicated that electrical stimulation was not only useful under blast-chilling conditions where it had already been accepted as a routine procedure. Operators relying on moderate chilling might similarly benefit as long as a moderate acceleration of muscle glycolysis be targeted. However, a warning is timely for those veal processors that apply a very slow cooling rate, for this will accelerate the already high rate of pH fall to such an extent that the tissue is significantly toughened. In view of the wide variety of cooling conditions and the wide range of stimulation equipment, no precise recommendations can be made to cover all processing systems. Individual veal processors are well advised, therefore, carefully and regularly to assess the combined effects of stimulation and chilling procedure on pH fall. Monitoring the pH decline will determine whether stimulation and chilling procedures are producing optimum rate of glycolysis and warn of the need for adjustment of the stimulation duration in order to prevent overstimulation or understimulation.

Conclusions

Rapid chilling of calf carcasses may lead to serious toughening, unless carcasses are electrically stimulated. Stimulation markedly accelerates glycolysis thereby eliminating the potential for cold shortening. However, overstimulation leads to an excessively rapid decline in pH, resulting in increased drip losses and occasionally in slight toughening. Veal processors are advised, therefore, regularly to monitor muscle pH and temperature profiles.

References

Bendall, J.R., 1973. The biochemistry of rigor mortis and cold contracture. Proceedings 19th European Meeting of the Meat Research Workers, Paris: A/O 1.

Bouton, P.E., P.V. Harris & D. Ratcliff, 1981. Effect of cooking temperature and time on the shear properties of meat. Journal of Food Science. 46:1082.

Chrystall, B.B., C.E. Devine & C.L. Davey, 1980. Studies in electrical stimulation: post mortem decline in nervous response in lambs. Meat Science. 4:69.

Dutson, T.R., G.C. Smith & Z.L. Carpenter, 1980. Lysosomal enzyme distribution in electrically stimulated ovine muscle. Journal of Food Science. 45:1097.

Eikelenboom, G. & F.J.M. Smulders, 1986. The effect of electrical stimulation on veal quality. Meat Science. 16:103.

Honikel, K.O., A. Hamid, C. Fisher & R. Hamm, 1981. Influences of post mortem changes in bovine muscle on the water-holding capacity of beef: post mortem storage of muscle at various temperatures between 0 and 30 °C. Jounal of Food Science. 46:23.

Judge, M.D., E.S. Reeves & E.D. Aberle, 1980. Effect of electrical stimulation on thermal shrinkage of bovine muscle collagen. Proc. 26th European Meeting of Meat Research Workers, Colorado Springs, J3:74.

Locker, R.H., 1960. Degree of muscular contraction as a factor in tenderness of beef. Food Res. 25:304.

Locker, R.H. & C.J. Hagyard, 1963. A cold shortening effect in beef muscles. J. Sci. Food. Agric. 14:787.

Marsh, B.B., T.P. Ringkob, R.L. Russell, D.R. Swartz & L.A. Pagel, 1987. Effects of early post-mortem glycolytic rate on beef tenderness. Meat Science. 21:241.

Savell, J.W., T.R. Dutson, G.C. Smith & Z.L. Carpenter, 1978. Structural changes in electrically stimulated beef muscle. Journal of Food Science 43:1606.
Sleper, P.S., M.C. Hunt, D.H. Kropf, C.L. Kastner & M.E. Dikeman, 1983. Electrical stimulation effects on myoglobin properties of bovine longissimus muscle. Journal of Food Science 48:479.
Smulders, F.J.M., 1984. Electrical stimulation and the sensory quality of beef and veal. Thesis, University of Utrecht, Dissertation Abstracts International C 46:(1)114.
Smulders, F.J.M., 1986. Sensory meat quality and its assessment. Veternary Quartery 8:158–167.
Smulders, F.J.M., F. Korteknie, C.H.J. Woolthuis & G. Eikelenboom, 1984. The effect of early post mortem conditions on the sensory and bacteriological quality of electrically stimulated, hot boned beef longissimus. Proceedings 30th European Meeting of the Meat Research Workers. Bristol:75.
Smulders, F.J.M., G. Eikelenboom & J.G. van Logtestijn, 1986. The effect of electrical stimulation on the quality of three bovine muscles. Meat Science 16:91.
Smulders, F.J.M., B.B. Marsh, D.R. Swartz, R.L. Russell & M.E. Hoenecke, 1990. Beef tenderness and sarcomere length. Meat Science (in press).
West, R.L., 1982. Electrical stimulation-United States. In: K.R. Franklin & H.R. Cross (Eds.). Proceedings Meat Science and Technology Symposium, Lincoln (NE), Publ. NLMB, Chicago (IL):91.
Wu, F.Y., T.R. Dutson, C. Valin, H.R. Cross & S.B. Smith, 1985. Aging index, lysosomal enzyme activities, and meat tenderness in electrically stimulated bull and steer carcasses, Journal of Food Science 50:1025.

Integrated quality control in veal production

L.J.M.H. Senden *[1], D. Oostendorp * & J. van den Berg **

* *Research Station for Cattle, Sheep and Horse Husbandry (PR), Lelystad, NL*

** *National Inspection Service for Livestock and Meat (RVV), Rotterdam, NL*

Summary

To develop or to improve quality guarantees for veal, an integrated quality control model was set up and tested by the Dutch veal industry and the Dutch Government.

In 4 consecutive experiments during the period March 1987 to February 1989, 7064 Black and White bull calves were reared in crates on 6 farms. In total, 108 calves died prematurely and 38 calves had to be slaughtered in an emergency. The remaining 6918 healthy fattened calves were slaughtered at 2 export slaughterhouses. On the basis of aberrations in these calves, the current veterinary meat inspection (EC) was compared with a visual inspection (no palpation or incisions) (IKB) integrated with advance information supplied by the veal farmers.

Advance information from the producer and separation of calves into a risk-bearing group (B calves and emergency slaughters) and a healthy group (A calves) give better guarantees of preventing the presence of residues in veal, than the present-day inspection. The results of the IKB trial show that the veal sector in the Netherlands has an infrastructure that allows introduction of an integrated system of quality control.

Descriptors: veal; integrated quality control; meat inspection.

Introduction

The quality of veal depends on the quality of the starting product (newborn calf) and treatment during transportation, on the veal farm and in the slaughterhouse. To meet increasing requirements in quality and product safety, a field trial 'Integrated quality control (IKB) in veal calf production' was set up by the Dutch veal industry and the Dutch Government (Senden, 1991). In this project, a model developed for integrated quality care was used and tested on several veal farms and in slaughterhouses. The aim of this project was to develop or to improve quality guarantees; this involves all the links in the veal-production chain. Measuring points were fixed in the different links of the chain, where the quality of the product and the way of production could be determined. Information on preceding links and feedback were essential, to achieve quality.

The EC system of meat inspection does not incorporate requirements about information supplied by the farmer. It is based on relatively stringent legal requirements, and the examination of meat for residues of veterinary drugs may involve difficulties in testing and be costly (Logtestijn, 1984a). The efficiency and scientific basis of this inspection are currently being questioned because of developments in the veal sector, our increasing understanding and better detection of animal diseases and zoonosis, and technological developments in slaughterhouses (Logtestijn, 1984b).

Some links in the production chain in the veal sector were studied in three subprojects: the quality of the newborn calf, slaughter quality and feedback.

1 Present address: Wolff vlees Nederland B.V., P.O. Box 61, 7392 AB Twello, NL.

Quality of newborn calf

Material and methods

In the subproject 'Quality of the newborn calf', the effect of the quality of the newborn calf was described and the effect of cleaning and disinfecting livestock sheds at collection points and cattle trucks was measured. A way of systematically checking hygiene during transport was tested by the agar-offprint method with a violet-red-bile-glucose (VRBG) nutrient medium (Gerats et al., 1982). To assess the quality of the calves at the collection points, their initial weight and age were recorded and their navel status was classified.

Within the first week of arrival on the veal farms, blood samples were taken from the young calves to measure the amount of immunoglobulin G_1 (IgG_1) in the blood serum (Price et al., 1983). At the same time, haemoglobin and urea concentrations were measured (Hellemond et al., 1979). The effect of the level of these indexes on growth (daily gain) and state of health (incidence of illness) was determined for the first 6 weeks of the growing period. To do this, the calves were reweighed at the age of 6 weeks and farmers recorded in a 'logbook' the state of health and all medicine use the calves.

In 4 consecutive experiments (fattening periods) during the period March 1987 to February 1989, 7064 Black and White (HF/FH) bull calves were reared in crates on 6 farms in the region under the aegis of the Regional Animal Health Service of the Province Gelderland. The average initial liveweight was 44.5 kg. In total 108 (1.5%) calves died prematurely and 38 (0.5%) calves had to be slaughtered in an emergency. The remaining 6918 healthy fattened calves were slaughtered at 2 export slaughterhouses in Apeldoorn within the region of the National Inspection Service for Livestock and Meat (RVV) in Apeldoorn. Twenty collection points for newborn calves and 28 cattle trucks were involved in the IKB project.

Results

In spite of a favourable visual judgment the bacteriological analysis showed that the cleaning and disinfection of the collection points and cattle trucks was not optimum. There were significant differences between the collection points and the cattle trucks. Cleaning was rarely followed by disinfection. The effect of cleaning and/or disinfection could clearly be shown.

The calves were assessed at the collecting points. Nearly all met the age requirements, except for 5 that were recorded as 'too young'. None of those 5 calves died prematurely but their daily gain was lower than that of calves of the required age. Incidence of light to severe navel infection at the start was 4.2%.

The average daily gain of calves with a light aberration of the navel was 39 g lower than calves without a navel infection; calves with a severe aberration of the navel had an average daily gain 90 g less than calves without a navel aberration. The death rate of calves with a navel infection was higher than calves without a navel infection. The average IgG_1, Hb and urea concentrations were respectively, 13.5 g/l, 6.8 mmol/l and 4.2 mmol/l. Up to an average of 44.1 days after the start of the experiment, the average daily gain was 536 g. Growth was significantly influenced by farms and experiments. It seems that the IgG_1 and Hb content had a significantly positive influence on daily gain after corrections had been made for initial weight, age at weighing and urea concentration. Concentration of IgG_1 also had a significant positive effect on culling rate and on incidence of respiratory and enteric infections.

About 10% of the calves had less than 5 g/l IgG_1. The daily weight gain of these calves was 20 g less than the other calves, their death rate was 3 to 4 times higher and they had a 1.5 to 2 times as high a chance of contracting respiratory or enteric infections during the first 6 weeks of the fattening period.

Conclusions

On the basis of these findings it is recommended to improve hygiene by taking various measures, such as regular cleaning and disinfecting of sheds at collection points and cattle trucks, and regular monitoring of these measures. The agar offprint method is very useful for this. It might be advisable to add blood testing as an extra quality criterion to already existing quality criteria such as breed, sex, type (quality) and weight, or to take random blood samples as a control measure. Blood testing can be an objective help with selection and valuation of newborn calves. At dairy farms, the sale of healthy newborn calves of a suitable age should be stimulated (by selection, payment, advisory extension service, regulations, identification and registration).

Slaughter quality and feedback

Material and methods

The main purpose of the subproject 'Slaughter quality and feedback' was to develop an integrated inspection system by means of advance information and feedback between the links veal farm-transport-slaughterhouse. In accordance with regulation 86/587/EEC, the current veterinary meat inspection consists of visual inspection, supplemented with incisions and/or palpitation of organs and lymphatic glands and, if necessary laboratory investigations.

In the IKB project, a model was developed so that the meat inspection could use advance information from the veal farm in order to carry out a more appropriate inspection. This information is recorded on a quality information card by the veal farmer, thereby providing the meat inspection with relevant information that is used when inspecting for residues.

Veal farmers were asked to keep detailed records in a 'logbook' of the state of health of the calves and any medicines used. When delivered for slaughter, the animals had to be divided into those not mentioned in the logbook (A calves: no illness or medicinal treatment during the last 6 weeks of the growing period) and those recorded in the logbook (B calves). The A and B calves had to be kept seperate during transport and slaughter.

As well as comparing the current veterinary meat inspection stipulated by EC legislation a comparision was made with visual inspection (no palpation and/or incisions of organs and lymphatic glands) (IKB inspection). Any disorders and the inspectors' decisions were recorded on a specially developed form.

Besides the statutory testing for residues of antibiotics and chemotherapeutics and histological examination for hormones, extra samples were taken at random (5% of the A calves, all B calves and any calves that had to be slaughtered in an emergency) to look for residues of inhibitors of bacterial growth. Examination of kidneys for residues of antibiotics and chemotherapeutics was done by the new Netherlands kidney test (NNNT). Urine was tested for residues of chloramfenicol with the Quick-card (CAP test). Research for residues of veterinary medicines containing nitrofurane was done with meat and blood samples. In accordance with the Dutch residue-monitoring programme (VREK), samples were taken to test for environmental pollution. Blood samples were taken from all calves for serological (ELISA) testing for the occurrence of cysticercosis (Knapen et al., 1979) and toxoplasmosis (Knapen, 1984).

The farm advisors received the IKB feedback information on the farm and slaughterhouse phase. The feedback information consisted of an 'individual review of results' and a 'review of the flock with regard to the farm and IKB results'.

Results

Apart from the aberrations 'pneumonia', 'spot kidney', 'muscle bleeding', 'liver trouble', 'injection spot' and 'carcass contamination', very few aberrations, particularly with the A-calves ($n = 6392$), were found. There were differences in the incidence of pathological and anatomical findings between herds. In spite of the similarity between the IKB and the EC meat inspections, several aberrations were missed by one or both inspections. There was no statistically significant difference between the two inspections in the aberration 'pneumonia' among A calves.

More aberrations were found in the B calves ($n = 160$) than in the A calves. In the B calves, the aberrations 'spot kidney' and 'muscle injuries' were more frequently reported by the IKB inspections than by the EC inspection. By contrast, the EC inspection reported more 'liver abscess' and 'injection spots' instead. There were very few incidences of residues of medicines and contaminants. No positive NNNT samples were found in the A calves but 4 positive CAP samples were found. In the B calves, 1, 17 and 2 calves reacted positively to NNNT, CAP test and nitrofurane testing in the blood, respectively. In the calves that had to be slaughtered early, 1 calf reacted positively to the NNNT test and 3 calves to the CAP test. In one A calf tested in accordance with the VREK programme, no levels higher than the legal limits were found.

Not all the positive results of the residue testing could be confirmed by the information from the farmer. There is a problem with residues of chloramfenicol in veal calves, tested with the CAP test.

Serological testing showed that 0.6 % of the calves were positive for cysticercosis and 0.7% for toxoplasmosis. These samples appeared to be spread over time and over farms, which means that they were 'false positives' and that cysticercosis and toxoplasmosis do not occur with this form of veal production.

The IKB model indicated that the feedback of data about slaughter quality and diagnosed aberrations found by meat inspection from a data bank was practicable. It has yet to be established to what extent this can lead to improvement of farm management.

Conclusions

The results of the IKB project indicate that information obtained during fattening can be incorporated in meat inspection and the control on residues of medicines, zoonosis and contaminants. Advance information from the producer and separating calves into a risk- bearing group (treated and growth-retarded calves) and a healthy group (A calves) give better guarantees of preventing the presence of residues in veal, than does the present day-inspection. By using information collected from fattening herds the veterinary meat inspection can be simplified. The absence of cysticercosis and toxoplasmosis in veal calves can be guaranteed simply and reliably by immunological testing (ELISA). Incisions of the masticatory muscles to trace cysticercosis will then no longer be necessary.

Post mortem inspection of veal calves

Material and methods

An extra study was done to simplify the post mortem inspection of veal calves. In this study, the IKB and the EC inspection were compared with a third inspection (reference or standard) in order to obtain a better understanding of the inspection methods. At one export slaughterhouse in Apeldoorn, all normal carcasses were inspected by the three inspections for two consecutive weeks.

The IKB and the EC inspection were compared on the basis of aberrations in 8616 calves and decisions were based upon these aberrations. Moreover, the results were compared with the findings of two experienced meat inspectors and one researcher (standard). This 'standard' inspection covered about 40% of the animals.

Results and discussion

Although on average there were differences between animals the results of the EC, the IKB and the standard inspection, the total numbers of each aberration recorded by means of the three inspections were very similar. More research is necessary on estimating the risk to the consumer of veal in relation to the incidence and tracing of aberrations. Palpation and incisions seem to be necessary for the aberrations 'liver abscess' and 'injection spot'.

Conclusion

On the basis of on the results obtained it can be stated that visual inspection (IKB) is almost as efficient in tracing aberrations as regular inspection (EC).

References

Gerats, G.E. & J.M.A. Snijders, 1982. Hygienemeting in varkensslachtlijnen. VVDO rapport nr. H8202.

Hellemond, K.K. & J.E. Sprietsma, 1979. Het verloop van het hemoglobinegehalte en de hematocrietwaarde van het bloed van mestkalveren. Tijdschrift voor Diergeneeskunde 104: 501–510.

Price, C.P., K. Spence & J. Whicher, 1983. Light scattering immunoassay of specific proteins: a review. Annals of Clinical Biochemistry 20:1–14.

Senden, L.J.M.H., 1991. Integrale keten begeleiding vleeskalveren. Proefstation voor de Rundveehouderij, Schapenhouderij en Paardenhouderij, Lelystad. In press.

van Knapen, F., 1984. Immunodiagnosis of toxoplasmosis. Thesis, Amsterdam.

van Knapen, F., S. Fridas & J.H. Franchimont, 1979. The detection of circulating antigens in Taenia saginata cysticercosis by means of the enzyme-linked immunosorbent assay (ELISA). RIV rapport nr. 179/79.

van Logtestijn, J.G., 1984a. Integrale bewaking van de vleesproduktieketen (IKB). Tijdschrift voor Diergeneeskunde 109:277–278.

van Logtestijn, J.G., 1984b. De IKB en de keuring van slachtdieren. Tijdschrift voor Diergeneeskunde 109:283–285

Use of growth stimulants and residue control in North American veal production

D.M. Kinsman
The University of Connecticut, Storrs, CT, US 06269-4040

Summary

The veal industry in the United States and Canada is not large when compared with the huge cattle (beef) and swine (pork) industries, but it is nonetheless an important factor in North American meat production. Approximately 2.5 million calves in the USA and 0.7 million calves in Canada are slaughtered annually for veal. About half of these are special-fed or formula-fed calves.

Some growth stimulants are approved for use with feed-lot cattle in both the USA and Canada, but none are permitted for veal production.

Some growth stimulants including antibiotics, hormones and other pharmaceuticals are approved for therapeutic veterinary use with bovines, but not for veal. In general, the USA and Canada are in harmony in their respective programs relative to the approval, use, levels, tolerances, withholding periods and monitoring for all pharmaceuticals.

Residue monitoring for antibiotics, sulfa drugs, hormones, heavy metals and agricultural chemicals is considered for domestic as well as for imported meats, with emphasis on veal.

There is a lack of definitive information about the use of most pharmaceuticals, including growth stimulants, on the pre-ruminant veal calf. Considerable research needs to be conducted in this area to further elucidate the contributions and safety of these products in the production system and food chain.

Descriptors: growth stimulants; residue control; veal production.

Introduction

The veal industry in North America (Canada and the United States) represents a relatively small proportion of the livestock industry on this continent, especially when compared to the huge cattle (beef) and swine (pork) industries (Table 1). This fact is also reflected in the muscle food consumption patterns of these two countries (Tables 2 and 3).

Nonetheless, veal is an important portion of the livestock and meat industry of North America. In 1988, in the USA, 2.4 million calves were processed, chiefly dairy breeding, for veal: 44% as bob veal (< 150 lb or 68 kg); 42% were formula-fed veal calves (150–400 lb or 68–180 kg); 6% non-formula-fed of the same weight and 8% were heavy calves (> 400 lb or 180 kg) (Kinsman, 1989; Slack & Warner, 1965).

Use of growth stimulants for veal production

There is a definite trend in the USA and Canada to produce more lean meat (muscle) and less fat (adipose tissue) without sacrificing quality of product. By growth management, it is possible to achieve this goal by the proper control of genetics, nutrition and growth regulation (Table 4). Genetically, the patterns, limits and types of growth can be identified and selected for maximum growth and development. Nutritionally, the rate of deposition of protein and fat may be controlled.

Table 1. Numbers of livestock in the USA and Canada in 1989 (in thousands). Data from Birchfield (1990) and US Department of Agriculture (1989a).

Species	USA	Canada	Total
Poultry	5 000 000	150 000	5 150 000
Cattle	98 994	11 004	109 998
Swine	53 795	10 640	64 435
Sheep	10 774	730	11 504
Veal	2 500	700	3 200

Table 2. Consumption of muscle food per capita in 1988 (carcass basis). Data from American Meat Industries (1989), Birchfield (1990) and US Department of Agriculture (1989b).

Species	USA		Canada	
	(lb)	(kg)	(lb)	(kg)
Beef	102.0	46.0	90.0	41.0
Poultry	85.0	39.0	90.0	41.0
Pork	63.0	29.0	65.0	30.0
Fish	20.0	9.0	21.0	9.5
Veal	1.7	0.8	3.4	1.5
Lamb/Mutton	1.6	0.7	2.0	0.9
Total	273.0	125.0	271.0	123.0

Table 3. Consumption of muscle food per capita in 1988 (retail basis). Data from American Meat Industry (1989) and Birchfield (1990).

Species	USA		Canada	
	lbs.	kgs.	lbs.	kgs.
Beef	72.0	33.0	63.0	29.0
Poultry	80.0	30.0	63.0	29.0
Pork	49.0	22.0	50.0	23.0
Fish	15.0	7.0	16.0	7.0
Veal	1.4	0.6	2.3	1.0
Lamb/Mutton	1.4	0.6	1.5	0.7
Total	219.0	93.0	196.0	90.0

Through the mechanism of growth regulation capitalizing on growth hormones, growth-hormone-releasing factors, beta-adrenergic agonists and immunization strategies, it is possible to induce more rapid weight gain, chiefly muscle, at early growth stages and produce a dramatic increase in mass of skeletal muscle while greatly reducing the fat content of the body.

Table 4. Growth management. After National Research Council (1988).

Genetics	Nutrition	Growth regulation
– Patterns of growth – Limits of growth – Types of growth	– Rate of deposition of protein and fat	– Growth hormones – Growth-hormone-releasing factors – Beta-adrenergic agonists – Immunization strategies

Table 5. Hormones that affect growth. From Mitchell (1989).

1. Growth hormone
2. Insulin
3. Thyroid hormones
4. Glucocorticoids
5. Prolactin
6. Gonadal steroids (androgens and oestrogens)

Table 6. Drugs not permitted. From Honstead (1990).

1. Diethylstilbesterol (DES)
2. Chloramphenicol
3. Dimetridazole
4. Ipronidazole

Table 7. Beta-adrenergic agonists (beta-agonists) and somatotropin. From Yang & McElligot (1989).

Analogues of epinephrine and nor-epinephrine
- stimulate protein synthesis
- decrease protein degradation
- stimulate nuclear accretion

Somatropin
- increases muscle
- decreases fat content
- improves feed efficiency

The hormones that affect growth (Table 5) have been employed to various degrees with several species of livestock with varying degrees of success. Some are approved for use for beef but not for veal. Some are not permitted in any species (Table 6). Thus, the use of growth

Table 8. Immunization strategies and other mechanisms. From National Research Council (1988).

1. Anabolic implants: growth promotors, shifting nutrients from fat to protein and enhancing growth rates.
2. Synchronize nutrition with animal's needs for protein growth.
3. Continuous delivery of repartitioning agents in all phases of growth.
4. Use of intact males: bull calves (veal production).

Other mechanisms for enhancing desirable growth patterns.

Bio-assays and cell structure

1. Differentiation inhibitor: occurs during embryonic myoblast growth.
2. Transferrin (iron-saturated): stimulates proliferation and differentiation of myotubes.
3. Fibroblast growth factor: stimulates proliferation of myogenic cells.

Table 9. New animal drug application (NADA) approval process. From Mitchell (1989).

1. Safety in animal and resultant food.
2. Effectiveness in research and field studies.
3. Manufacturing procedures and facilities before release.
4. Environmental impact analysis
5. Analytical method for residues established.
6. Label information and directions for use approved.

stimulants is still largely in the experimental stage, although some are in use with beef and pork production in particular. The immunization strategies are receiving the greatest emphasis today along with the use of beta-agonists and somatotropin (Table 7). However none of these growth stimulants are approved for veal calf production in North America. The pre-ruminant (veal) calf requires a great deal more research relative to the use of growth stimulants, as opposed to its more mature ruminating brethren destined for beef. This is also true of withdrawal periods for medication and other treatments of preruminant (veal) calves.

There are some new techniques being developed by biotechnology that will merit consideration in the future but none of these are yet in use on a practical basis (Table 8). The introduction of these or any new animal drugs must, of course, go through a rigorous approval process (Table 9). It should be emphasized, however, that no growth stimulants are approved for use with veal in North America (Desgagne et al., 1989; Guide for the care and use of agricultural animals in agricultural research and teaching, 1988).

Monitoring residues in veal tissues

All non-naturally occurring substances found in our meat supply are either approved by the Food and Drug Administration (FDA) or are banned, and therefore the product is condemned when such substances are found. The Food Safety and Inspection Services (FSIS) of the United

Table 10. Survey of 130 veal farms by FDA in 1983. Data from Hays & Black (1989).

	Veal farms (%)	Antibiotic
Medicated formula feeds:	38	with oxytetracycline and neomycin
	35	with chlortetracycline
	27	with oxytetracycline
Oral dosage forms:	33	nitrofurazone
	26	chlortetracycline
	15	sulfamethazine
	12	oxytetracycline
	11	neomycin
	11	chlorotetracycline and sulfamethazine
Single animal treatments:	51	penicillin
	50	tylosin
	35	oxytetracycline
	16	penicillin-streptomycin
	15	erythromycin
	15	selenium

States Department of Agriculture (USDA) and corresponding agencies in Agriculture Canada, monitor this program.

In general, the residue-testing program operates at three levels:

1. monitoring, wherein random samples are taken periodically;
2. surveillance, when problems are found and a concentrated sampling is warranted; and
3. exploratory, when new direction or studies are indicated.

FSIS is aware of approximately 400 potential residues from all sources in animal tissues and it regularly monitors about 100 of these. Fewest residues are found in our feed-lot (grain-fed) cattle, whereas the greater incidence of residues is detected in cull cows, hogs and veal calves. A Washington State University study determined that 85% of the antibiotics administered to veal calves was to 20% of the calves, namely the problem calves.

A 1983 FDA survey of 130 veal farms (Table 10) identified the mode of administration of antibiotics and the prevalence of these antibiotics by usage in veal production. Oxytetracycline and chlortetracycline, alone or in combination with others, were the predominant antibiotics used by all systems, although penicillin was the principal one for injections with illness.

The 1989 Canadian annual report on chemical and biological testing showed some interesting data relative to the decreased use or prevalence of residues in meat during the past five years. This includes antibiotics, heavy metals, steroids, hormonal drugs and pesticides or herbicides. Their diminished use and presence in animal tissue is very encouraging.

Furthermore, they sampled all imported meats, thus monitoring for residues from those sources as well (Agriculture Canada, 1989).

A 1988 USDA study (Table 11) identified 48 violations among formula-fed veal calves, citing the number and percentage of these violations by antibiotic. These violations reflect dosage as well as withdrawal times as possible causes of non-compliance. In general, 10 days is a minimum period of withdrawal, with some substances requiring up to 90 days or 13 weeks.

The concerns by the consumer about to freedom from drug or pesticide residues of animal-derived food products are also of vital interest and importance to all of us in the food chain: producers, processors and retailers. These concerns are reflected in the consumers' selections

Table 11. USDA violative residue profile for formula-fed calves (1988) (n = 48). Data from Geleta et al. (1988).

Drug	Violations number	%	Range of violations (ppm)	Tolerance (ppm)
Tetracycline	13	27	0.20 – 3.75	0.25
Gentamicin	10	21	0.01 – 14.08	0.0
Neomycin	7	15	3.24 – 132.00	0.25
Oxytetracycline	7	15	0.41 – 2.48	0.1
Streptomycin	6	13	2.13 – 4.87	2.0
Penicillin	3	6	0.07 – 0.16	0.05
Sulfamethazine	2	4	0.39 – 1.36	0.1

of food. Consumers today are both nutrition-conscious and health-conscious, and they demand and deserve proper and responsible use of any substance incorporated into their food supply, be it animal drugs or food additives.

This is a most important issue for all of us to address who are involved in animal agriculture. A veal quality assurance program has been developed to insure compliance with FDA, USDA and EPA specifications.

References

Agriculture Canada, 1989. Annual report on chemical and biological testing for the fiscal year 1988-89. Ottawa, Ontario K1A 1B7.

American Meat Institute, 1989. Meat facts. Washington, DC 20007.

Birchfield, G., 1990. Livestock inventories and meat consumption. Statistics Canada; Agriculture Canada, Ottawa, Ontario K1A 1B7.

Desgagne, M., J. LeCavalier & I. Malyniwsky, 1989. Anabolic steroids: use and abuse profile in Canada. Bureau of Dangerous Drugs. Ottawa, Ontario, K1A 1B7.

Geleta, J.N., C.J. Barnes, G.L. Morris, J. Matusik, G.C. Guyer, L. Frank, & D.J. Weber, 1988. Chemotherapeutic agents and their disposition in food producing animals. I. Disposition of sulfamethazine in tissues and plasma of bob veal, fancy veal and replacement calves after oral administration. USDA, Division of Veterinary Medical Research Report 290.1-290.01 (INAD No 4265), Beltsville, MD 20705.

Guide for the care and use of agricultural animals in agricultural research and teaching, 1988. First Edition, Chapter 11: Guidelines for veal calf husbandry. Champaign, IL 61820.

Hays, V.W. & C.A. Black, 1989. Antibiotics for animals: the antibiotic resistance issue. Council for Agricultural Science and Technology, 1989-2, Ames, Iowa.

Honstead, J., 1990. Personal communication regrowth promotants and residue evaluation, USDA, FSIS, Washington, DC 20205.

Kinsman, D.M. 1989. Veal: meat for modern menus, 50 pp. National Live Stock and Meat Board, Chicago, IL 60611.

Mitchell, G.A., 1989. Proper drug use in veal calves. Presented to American Veal Association Veal Health Seminar, Lancaster, PA.

National Research Council, 1988. Designing foods: animal product options in the market-place. National Academy Press, Washington, DC. p. 135–193.

Slack, S.T. & R.G. Warner, 1965. Veal production. Cornell University Animal Husbandry Mimeo Series No 4, Ithaca, NY.

US Department of Agriculture, 1989a. Livestock and meat statistics. Economic Research Service. Statistical Bulletin Number 784, Washington DC 2005-4788.
US Department of Agriculture, 1989b. Livestock and poultry situation and outlook yearbook. Economic Research Services, LPS-38, Washington, DC 20005-4788.
Yang, Y.T. & M.A. McElligott, 1989. Multiple actions of beta-adrenergic agonists on skeletal muscle and adipose tissue. Biochemical Journal 261:1–10.

Multiresidue test for β-agonists in a variety of matrices

L.A. van Ginkel, R.W. Stephany, H.J. van Rossum & J. Farla
Laboratory for Residue Analysis, National Institute for Public Health and Environmental Protection, P.O. Box 1, 3720 BA Bilthoven, NL

Summary

A multiresidue method for the detection and identification of β-agonists was developed. The method is based on sample clean-up with multi-immuno-affinity chromatography and detection and identification by gas chromatography-mass spectrometry. For salbutamol in urine of veal calves, a half-life was found of bout 2 days.
Descriptors: immuno-affinity; analysis; salbutamol.

Introduction

The use of β-agonists as growth-promoting agents has confronted analytical chemists with a new challenge, the development of reliable and sensitive methods suitable for the detection and identification of a number of analytes in a variety of matrices. Previous work done in our laboratory (Ginkel et al., 1989) showed that the use of antibodies (immuno-affinity) for isolating analytes from a complex mixture is a very suitable method for sample clean-up. When immuno-affinity chromatography (IAC) is combined with gas chromatography-mass spectrometry (GC-MS), a potentially very adequate analytical strategy is available. On the strategy, we developed a method for detection and identification of β-agonists containing an *N*- tertbutyl group (Figure 1).

Material and methods

Standard clenbuterol and carazolol was kindly donated by Boehringer Ingelheim (Ingelheim am Rhein, FRG) and standards salbutamol, terbutaline, carbuterol, pirbuterol and sotalol by the National Institute for Quality Control of Drugs (Leiden, the Netherlands).

GC-MSD system

The GC-MSD system (Model 5970 mass-selective detector combined with a Model 5890 gas chromatograph; Hewlett-Packard, Avondale, PA, USA) was used in the selected-ion mode, using a holding time of 100 ms. The gas chromatograph was equipped with a SE52 capillary column (Machery Nagel, Dren, Swiss) (inner diam. 25 m × 0.25 mm) using helium as the carrier gas at a column head pressure of 70 kPa. The oven was kept at 100 °C for 3 min, then the temperature was raised to 280 °C at 20 °C/min.

Preparation of the IAC material

The immunogen used was prepared as described earlier by coupling diazonium-clenbuterol to bovine serum albumin (BSA, Sigma)(Yamamoto et al., 1982). The antiserum used to prepare the immuno-affinity matrix was obtained by immunizing a rabbit four times over a period of 5 months, 2 mg of the immunogen each time. The isolation of the antiserum immunoglobulin

G (IgG) fraction and coupling of this fraction to the activated matrix (Tresyl activated Sepharose; Pharmacia) was performed as described previously (Ginkel et al., 1989).

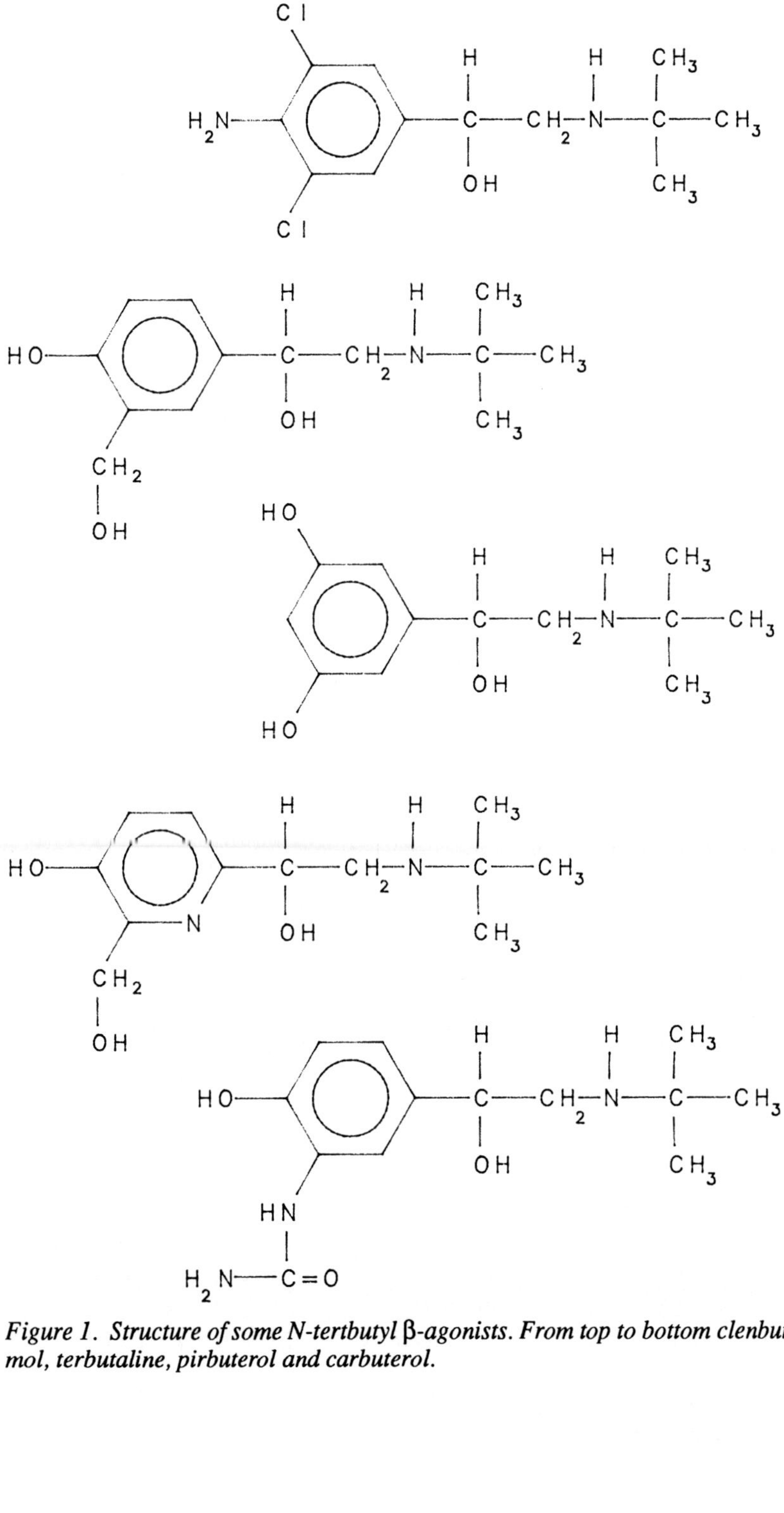

Figure 1. Structure of some N-tertbutyl β-agonists. From top to bottom clenbuterol, salbutamol, terbutaline, pirbuterol and carbuterol.

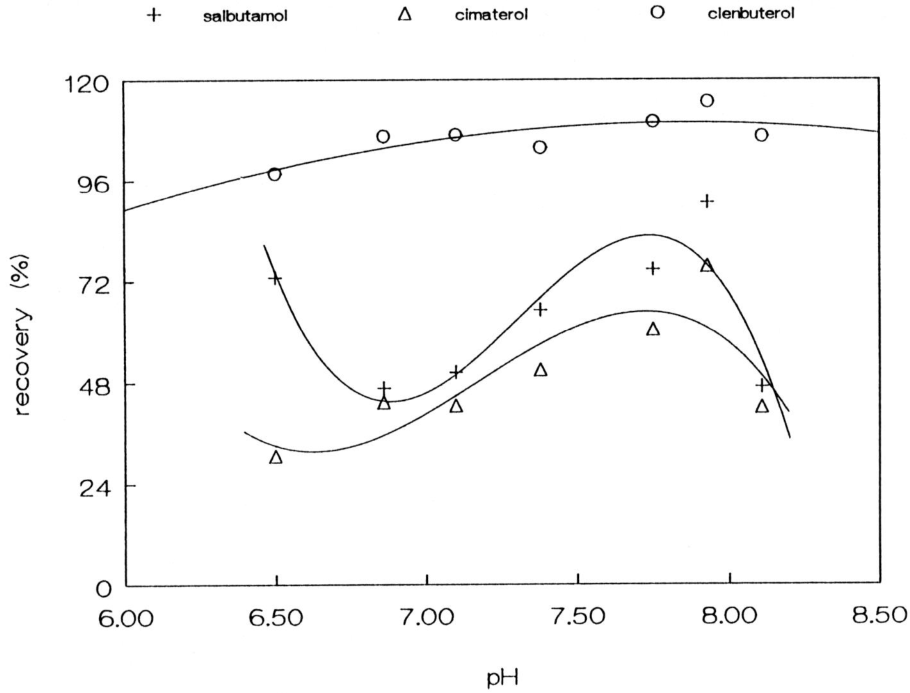

Figure 2. Optimization of ion-pair extraction for some β-agonists.

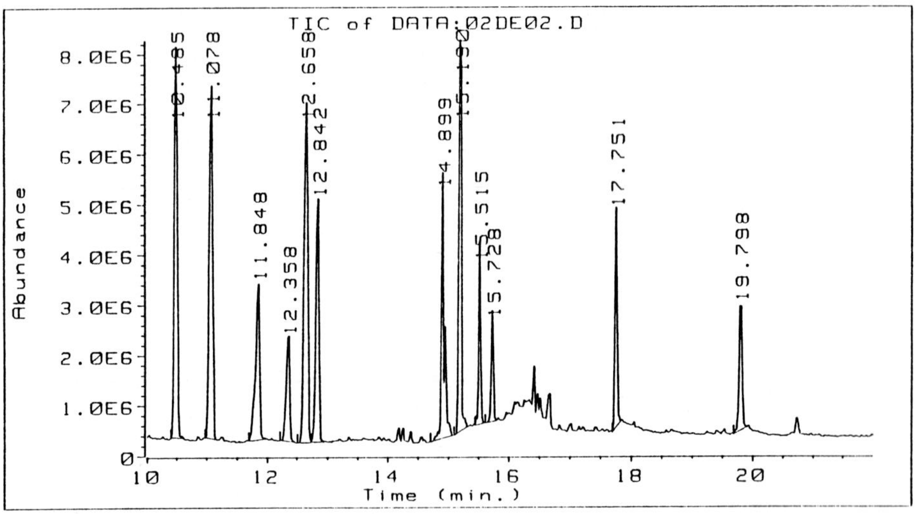

Figure 3. GC-separation of β-agonists and related compounds (10.49:isoprenaline-tri-TMS; 11.08: terbutaline-tri-TMS; 11.78: carnuterol-di-TMS; 11.85: clenbuterol-TMS; 12.36: cimaterol-TMS; 12.66: salbutamol-tri-TMS; 15.52: sotalol-di-TMS; 15.73: sotalol-TMS; 19.80: carazolol-TMS; 20.72: carazolol-di-TMS).

Analytical procedure for urine

Extraction

Usually clenbuterol is extracted under basic conditions. However since most of the agonists tested are phenolic compounds, these conditions cannot be used for multi-analyte extraction. Therefore we tried ion-pair extraction, previously used by, for instance, Brand et al. (1982) for the extraction of terbutaline. We used sodium dodecyl sulphate, since this counterion proved to be suitable in the HPLC analysis of clenbuterol and cimaterol (Courtheyn et al., 1988). Figure 2 shows the results of a pH optimization experiment from which it was concluded that the best results are obtained at pH 7.8. The use of bis(-2-ethyl)hexyl resulted in slightly lower recoveries whereas no significant differences were observed between solutions buffered with phosphate or Tris.

IAC

After evaporation of the solvent the residue is dissolved in water and applied to the IAC column. After washing the column with water, the β-agonists are eluted with acetate buffer (0.1 mol/l), saline, pH 4.8/ethanol (1:1, v/v). The eluate is evaporated, desalted and derivatized with BSTFA/1% TMCS (Pierce) to form TMS-derivatives.

GC-MS

Figure 3 shows the GC-separation after derivatization of the five β-agonists tested and a few additional test compounds. Base-line separation, using GC-MS (TIC) detection, was achieved between all compounds with the exception of carbuterol and clenbuterol, which were impossible to separate on the column used. Sotalol and carazolol formed two different derivatives. Based on the mass spectra obtained, a fragmentation pattern as shown in Figure 4 is suggested. For clenbuterol and salbutamol the fragments have the following nominal masses:

	M_o	M_d	[A]	[B]	[C]	[D]	[E]	[F]
Clenbuterol-TMS	276	348	86	262	243	333	277	187
Salbutamol-triTMS	239	455	86	369	350	440	384	294

M_o= molecular weight
M_d= molecular weight after derivatization

Results and discussion

The procedure was first tested by analysing blank urine samples and samples were spiked with one or more of the analytes. The extraction procedure, performed in duplo for each sample, resulted in a recovery of 90% or more. Up to a level of 100 ng, no losses were observed during the IAC clean-up. Samples from animals treated with salbutamol were analysed using clenbuterol as an internal standard. The validity of this approach was evaluated by analysing different volumes (1 to 4 ml) of a sample. The resulting average value (± SD) was 18.0 ± 2 µg/l indicating parallellism between volume analysed and amount of salbutamol detected. Figure 5 shows the salbutamol concentration in urine of veal calves after treatment with salbutamol (salbutamol hemisulphate 32 mg, twice a day (Ventolin 4 ex. Glaxo) for 5 days). From Figure 5, it can be concluded that the salbutamol concentration in urine declines rapidly, *T* 2.6 days (Animal 1) and 1.7 days (Animal 2), after withdrawal of salbutamol from the feed.

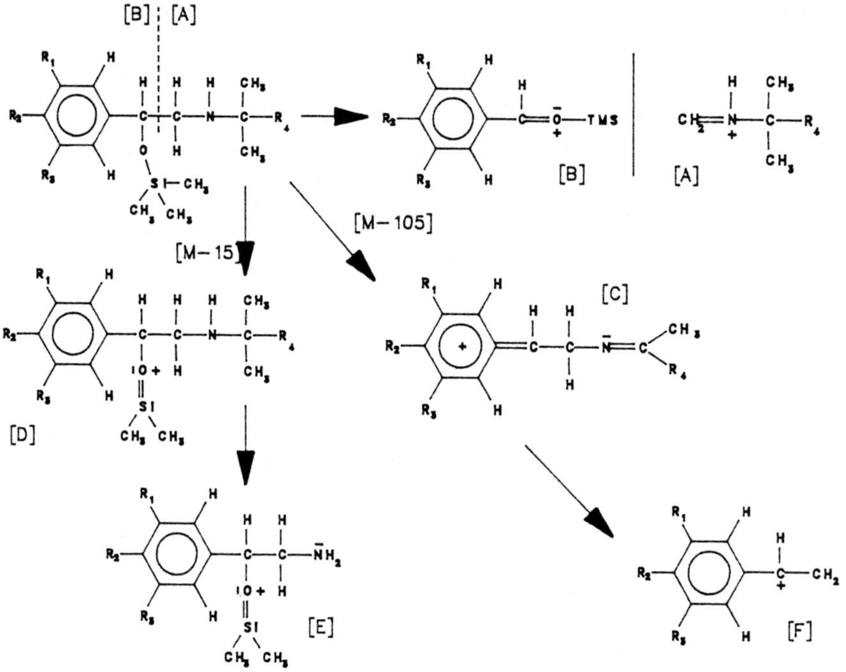

Figure 4. General EI-fragmentation pattern of β-agonists.

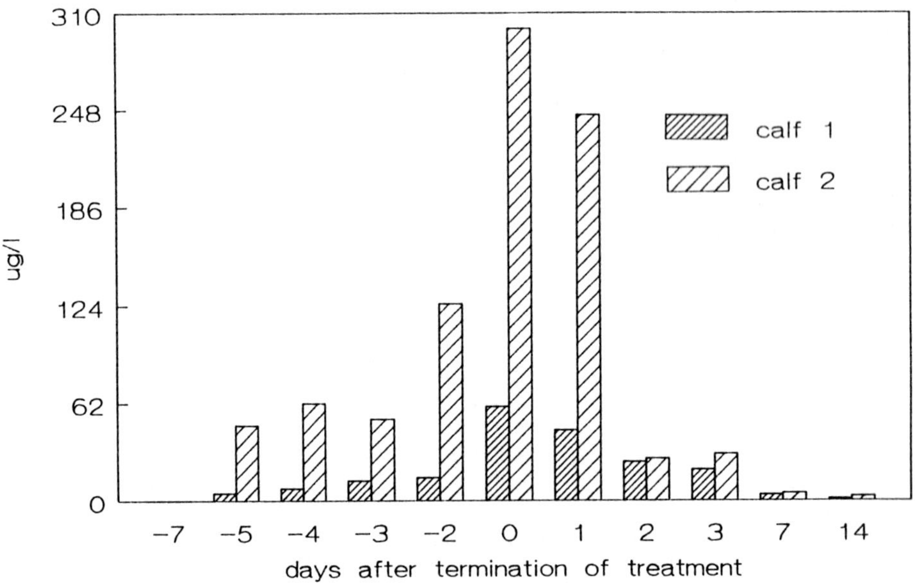

Figure 5. Salbutamol in urine (miac/gc-msd) during and after treatment with salbutamol.

Two weeks after treatment, salbutamol concentrations were in the range of the limit of detection of the method. However the limit of detection did not represent the limit of identification. Within the EC, there were criteria for identification of residues of hormones and related compounds which had to be fulfilled to allow positive identification with GC-MS. Due to the severe fragmentation with electron-impact ionization, it was not possible to identify β-agonists in accordance with these criteria (4-ions with correct relative abundance) at the low ppb level. With other ionization techniques or combinations or techniques and derivatives, it is possible to circumvent these problems, if necessary (Ginkel et al., 1990). The procedure described was also used (with appropriate modifications for the extraction) for feed, stomach contents and liver and has proved to be a reliable and very flexible multiresidue method.

References

Brandts, P.M., R.A.A. Maas, J.F. Leferink, C.L. de Ligny & G.H.E. Nieuwdorp, 1982. Ion-pair extraction of some sympathomimetics. Description of an extraction model for terbutaline and investigation of some factors influencing the recovery of sympathomimetics. Analytica Chimica Acta 135:82.

Courtheyn R., R. Verne, R. Schilt, H. Hooijerink, C. Desaever & V. Bakeroot, 1988. Detectie en identificatie van clenbuterol en cimaterol met HPLC. Benelux Working Group "Hormones and anti-Hormones", Brussels, document SP/LAB/h (88) 43.

van Ginkel L.A., H. van Blitterswijk, P.W. Zoontjes, D. van den Bosch & R.W. Stephany, 1988. Assay for trenbolone and its reduced metabolite 17α-trenbolone in bovine urine based on immunoaffinity chromatographic clean-up and off-line high-performance liquid chromatography-thin layer chromatography, Journal of Chromatography 445:385–392.

van Ginkel L.A., R.W. Stephany, H.J. van Rossum, H.M. Steinbuch, G. Zomer, E. van der Heeft & A.P.J.M. de Jong, 1989. Multi-immunoaffinity chromatography: a simple and highly selective clean-up method for multi-anabolic residue analysis of meat. Journal of Chromatography 489:111–120.

van Ginkel L.A., R.W. Stephany et al., 1990. In preparation.

Yamamoto, I & K. Iwata, 1982. Enzyme immunoassay for clenbuterol, a beta-2-adrenergic stimulant. Journal of Immunoassay 3:15.

8. Nutrition: protein sources

Introductory remarks: some aspects of skim-milk replacement by other protein sources in veal-calf diets

G.J.M. van Kempen & J. Huisman
ILOB-TNO Institute for Animal Nutrition and Physiology, P.O. Box 15, 6700 AA Wageningen, NL

Summary

Different limitations for replacement of skim milk by other protein sources are discussed. The majority of alternative protein sources have a lower protein digestibility than skim-milk-powder. Most of these proteins do not clot in the abomasum.

The abomasal passage rate of non-clotting whey and soya proteins is much faster than that of skim-milk protein. Curd formation may be essential for optimum protein digestibility. Fat digestibility was decreased when skim milk is replaced by other protein sources. The possible relation of lower fat digestibility with non-curd formation, emulsifiers, carbohydrates and some minerals is discussed. Other subjects mentioned are content of available iron and amino acid requirement.

Descriptors: veal calves; digestibility; protein sources; passage rate.

Introduction

The veal calf industry in Europe was established in the late 1950s and early 1960s as a consequence of the increase of milk production in Europe. Since that time, veal calf production has always played an important role in the economics of dairy industry and meat production.

The success of the veal calf industry was based on the fact that industries succeeded in composing a milk replacer in which the butterfat was replaced by other fats. In the 1950s and 1960s diets were based mainly on skim-milk powder, some whey powder and non-butterfats. In this period, research into fat quality and emulsifiers were important areas (Huisman & van Weerden, 1984; Toullec et al., 1980). Later, when milk proteins increased strongly in prices, research into the possibilities to replace substantial parts of skim-milk powder by other protein sources was initiated. However, it soon became clear that there were serious limitations for these replacements. In the following, some aspects of these limitations are discussed and suggestions for research to reduce these limitations are given.

Digestibility of protein and fats in relation to abomasal passage, contents of antinutritional factors and antigenicity

The protein digestibility of the majority of alternative protein sources is lower than for skim-milk powder (Table 1). An exception is wheat gluten. It is not clear what the reason is for this difference in digestibility. But, it is striking that skim-milk powder curdles in the abomasum and also wheat gluten has some clotting ability. Other proteins remain unclotted in the abomasum. This suggests that the structure of protein may be an important factor. The following topics were studied:

Table 1. Faecal digestibility of various protein sources with scp as single-cell proteins (scp).

Protein source	Mean protein digestibility
Skim-milk powder	95
Wheat gluten	96
Soya concentrate and flour	80 – 85
Fish protein concentrate	80
Yeast scp	80 – 85
Bacterial scp	92 – 94
Potato protein	80

- duodenal passage of protein and fat in veal calves fed on different protein sources;
- digestibility of protein and fat in diets in which a substantial part of the skim-milk powder was replaced by other protein sources;
- influence of antigenicity of proteins on performance and digestion.

Duodenal passage was measured with diets containing the following sole protein sources: skim milk protein, whey protein (mix of whey powder and whey concentrates), soya concentrate and single cell proteins. Each diet contained about 25% of one of these protein sources as the sole protein source. The dietary fat content was 18–20%. The fat mixture consisted of 35% beef tallow, 35% lard, 25% coconut or palm kernel oil and about 5% emulsifiers. Criteria were the passage of protein, fats, carbohydrates, pH course and buffer capacity. As an example in Figures 1 and 2, the duodenal passage of protein and fat obtained with skim-milk powder, whey powder and soya concentrate are given.

The skim-milk protein passed the duodenum at a relatively constant rate during the whole day, whereas whey and soya proteins passed at a much faster rate. Within about six hours, the main part of whey and soya proteins passed the duodenum. The same pattern was observed for crude fat. A relatively constant fat-passage rate with skim-milk protein-based diet was observed and a fast fat-passage rate with the whey and soya proteins based diet. These differences in passage rate of protein and crude fat may be attributed to the fact that casein has curdling capabilities while the other proteins do not have this capacity. With curd formation, fat is included in the clot, but whey is squeezed out. Because of this, casein and fat are retained for a longer time in the abomasum while other constituents pass the abomasum more rapidly. This information suggests that curd formation may be essential for digestibility for the following reasons:

- due to curd formation, the protein remains longer in the abomasum. This implies that protein and fat are exposed to pepsin and HCl for a longer time. This first pre-digestion step may be of importance for the further digestion process;
- due to the slower passage rate of skim-milk powder based diets, the small intestine is more regularly filled with protein and fat than when whey and soya-based diets are given;
- the fat in the skim-milk diet is more included in protein clots. With other proteins, the fat remains separated from the protein. When non-clotting diets are given, the fat passes rapidly together with the liquid phase. The structure of the protein and fat in the chyme is thus different from that of milk;
- the structure of clotted protein is different from non-clotting proteins. The structure of protein may be important for optimum digestion.

As indicated in Table 1 most alternative proteins have a lower digestibility. These proteins have one property in common, namely that they do not clot in the abomasum. The results of

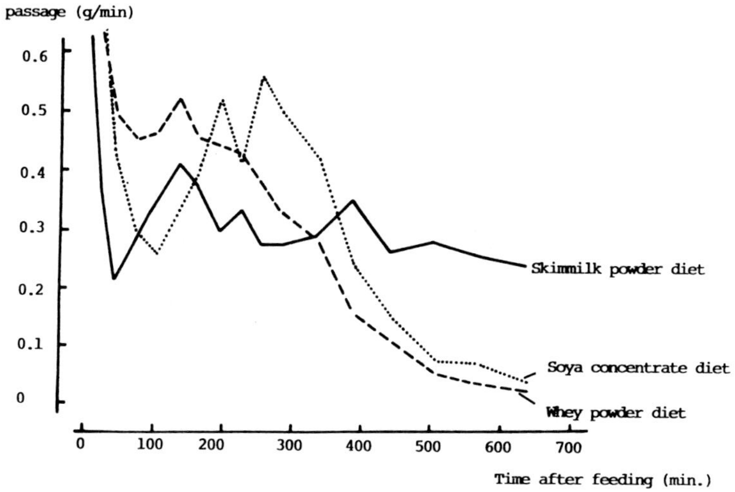

Figure 1. Duodenal passage of crude protein.

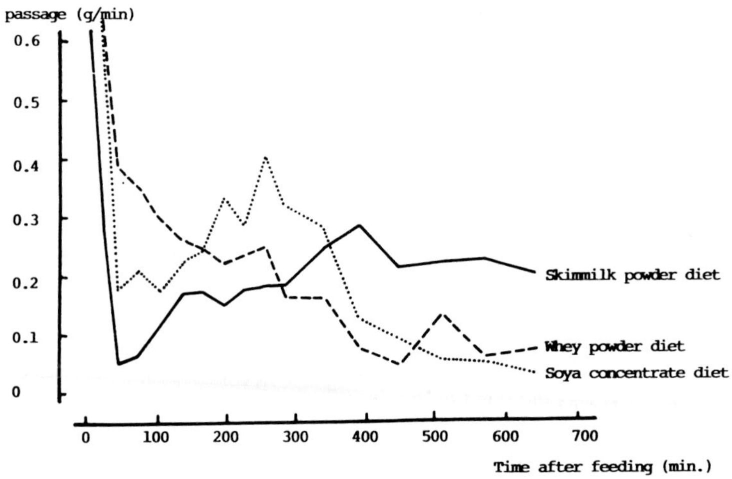

Figure 2. Duodenal passage of fat.

duodenal passage indicate that this observation may be important. For plant proteins also other factors may play a role, such as trypsin inhibitors, lectins and antigenicity of protein. Trypsin inhibitors can cause reduced (chymo)trypsin activity and losses of endogenous protein by hyper stimulation of the pancreas. Lectins and antigenic proteins can cause gutwall damage resulting in reduced intestinal absorption of nutrients, reduced secretion of brushborder enzymes and enhanced losses of endogenous protein. Moreover intestinal and humoral immunological reactions may occur. Some aspects of lectins, trypsin inhibitors and antigenic proteins will be presented in this meeting. Inclusion of alternative proteins may effect fat

Table 2. Effect of non-milk protein sources on fat digestion.

Amount of product in the test diet	Digestibility of fat control diet		test diet		difference	Number of calves
25% soya concentrate	98.0	(0.2)	94.2	(2.5)	– 3.8	4
20% soya flour	94.4	(1.3)	88.7*	(1.4)	– 5.7	4
35% soya flour	96.0	(0.6)	86.7***	(1.3)	– 9.3	4
20% bacterial scp	96.2	(0.6)	96.9	(0.3)	+ 0.7	10
25% low fat yeast scp	97.2	(0.8)	94.2*	(0.9)	– 3.0	15
20% fish concentrate A	97.4	(0.4)	93.9**	(0.6)	– 3.5	6
11% fish concentrate B	93.5	(0.6)	85.2***	(1.3)	– 8.3	4
22% fish concentrate B	93.5	(0.6)	81.4**	(3.0)	–12.1	4

* Significantly different from control ($P < 0.05$)
** Significantly different from control ($P < 0.05$)
*** Significantly different from control ($P < 0.001$)
Between brackets the Standard Error (SE) is given

digestibility negatively. In various digestibility experiments a substantial part of skim-milk powder was replaced by other proteins. In Table 2 results of the fat digestibility are summarized. These data were obtained in experiments with calves of 6 to 10 weeks of age. The fat mixture in these diets consisted in average of: 35% beef tallow, 35% lard, 25% coconut or palm kernel oil and about 5% emulsifiers.

With the exception of bacterial scp (single cell protein) the fat digestibilities were lower when a substantial part of the skim-milk powder was replaced by alternative proteins. Different reasons may have played a role.

- As already discussed, the fact that the tested alternative proteins do not clot in the abomasum may effect the digestion process, fat digestion included. This hypothesis was tested in an experiment in which the non-clotting whey protein was compared with skim milk protein. A diet was formulated consisting of delactosed whey powder as the sole protein source. The fat mixture consisted of 30% beef tallow, 30% lard, 11% coconut oil, 21% palm kernel oil and 8% soya lecithin and was included in the diet at a level of 19%. The control diet comprised 70% skim milk and the same fat mixture. The diets were further balanced for protein content (25%) and minerals. The digestibility coefficients for the skim-milk and the whey diets were for protein 96 and 91, respectively, and for fat 96 and 89, respectively. These results underline the hypothesis that clotting in the abomasum may be of importance for the digestion process.
- The emulsifiers used in veal calf diets are mainly developed for diets containing milk protein as the major protein source. With alternative proteins other qualities and properties of emulsifiers may be necessary.
- Soya and scp's contain carbohydrates which are digested in the large intestine because of a lack of appropiate enzymes in the calf's gastro-intestinal tract. These carbohydrates are digested in the large intestine by bacterial fermentation. Due to this fermentation the amount of bacteria excreted with faeces, increased. Bacteria contain fat, thus also more fat is excreted with faeces.

- Fish protein contains relatively high levels of Ca and other minerals. Thomke (1963) demonstrated that fat digestibility may be negatively affected when high levels of Ca are present in diets for calves.

Content of available iron

For the production of white veal it is important that the content of available iron in the diets is low. Iron content can limit the possibilities for use of protein sources for replacement of skim-milk powder. The availibility of iron can differ markedly between protein sources. Van Weerden et al. (1977, 1978) demonstrated that the availability of iron in fish is high and in soya concentrate low. There are possibilities to reduce Fe-availability with special agents, such as EDTA preparations (van Weerden, 1973, 1974).

Amino acid requirement

Determination of the amino acid requirement becomes more and more important because of the tendencies to lower the protein content in the diets and to replace the skim-milk protein by other protein sources. The alternative proteins have in general a different amino acid profile and a lower protein and amino acid digestibility. The differences in amino acid profile and digestibilities can be satisfied by enrichment of the diets with synthetic amino acids. However, to achieve optimal diets it is important to know the requirements. On behalf of the Dutch veal calf milk-replacer industry experiments are running for the determination of the amino acid requirement of the milk fed veal calf. A part of these results are published (van Weerden & Huisman, 1985).

References

Huisman, J. & E.J. van Weerden, 1984. Dietary fat for veal calves. 35th Annual meeting of the European Association for Animal Production, The Hague, the Netherlands, 17 pages.

Thomke, S., 1963. Die Verdaulichlichkeit vonn Vollmilch und Milchaustauschfuttermitteln mit Zusatz eines Rindertalg-Schmalz-Gemisches bzw. von Knochenfett. Züchtungskunde, 35:214–231.

Toullec, R., M. Theriez & P. Thivend, 1980. Milk replacers for calves and lambs. World Animal Review, 33:32–42.

van Weerden, E.J., 1973. Het verlagen van de beschikbaarheid van het ijzer in het rantsoen van mestkalveren. Landbouwkundig Tijdschrift, 85:388–394.

van Weerden, E.J., 1974. Low fat fish meal in milk replacers. 3rd European Symposium on the use of Fish Meal in Animal Feeding, Amsterdam, NL.

van Weerden, E.J., J. Huisman & J.E. Sprietsma, 1977. De beschikbaarheid van het ijzer in enkele eiwitrijke voedercomponenten voor mestkalveren. Landbouwkundig Tijdschrift, 89:191–196.

van Weerden, E.J., J. Huisman & E.J. Sprietsma, 1978. Availability of some iron sources in veal calves. Zeitschrift für Tierphysiologie, Tierernährung und Futtermittelkunde, 40(4): 209–221.

van Weerden, E.J. & J. Huisman, 1985. Amino acid requirement of the young veal calf. Zeitschrift für Tierphysiologie, Tierernährung und Futtermittelkunde, 53:232–244.

The surplus of milk proteins for nutrition of veal calves

E. Hetzner
Milchindustrie-Verband, Schedestr. 11, 5300 Bonn 1, FRG

Summary

The milk quota introduced by the EC in 1984, and the hormone scandals of 1988 and 1989 have led to less skim-milk powder being used in veal calf nutrition. Unexpectedly the resulting fall in prices has not led to more spray skim-milk powder being incorporated in milk replacer. The consumption of skim-milk powder can be expected to worsen, as a result of developments in the Eastern Block.

The EC Commission should re-introduce the obligatory incorporation of skim-milk powder in animal feed and aid should be increased to save the dairy industry from a considerable increase of stocks. The dairy industry is very interested in using skim-milk powder in veal calf nutrition.
Descriptors: milk proteins; EC policy; skim-milk powder; veal calf nutrition.

Introduction

This seminar has so far dealt only with the aspects of the physiology of nutrition in veal calf nutrition. I intend to discuss the market situation resulting from the management of the European milk market. The general situation is influenced on the one hand by the amount of milk production in general, and on the other hand by the consumption and export of all dairy products that cannot be reckoned as so-called intervention products (Figure 1). The production of butter and skim-milk powder is a residual factor. Here I will concentrate on the proteins.

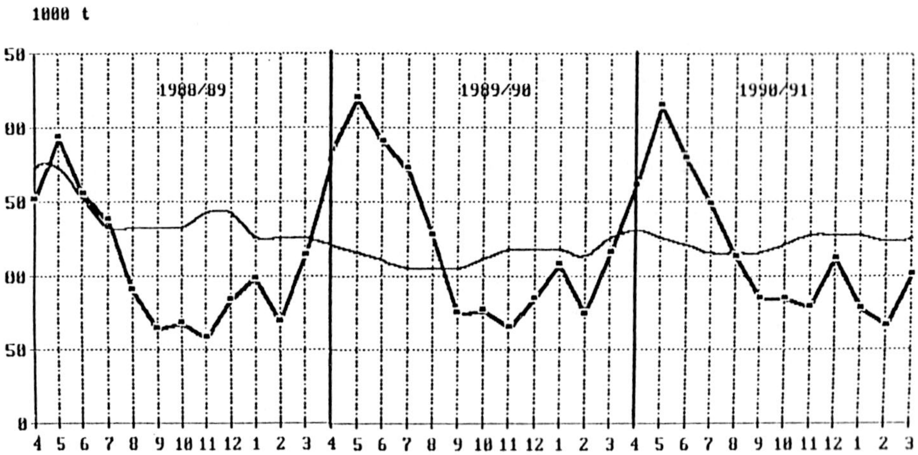

Figure 1. EC skim-milk powder production and demand in 1000 t with —■— as production and —— as demand including exports.

Availability of skim-milk powder in recent years

The availability of skim-milk powder in the EC was drastically reduced by the introduction of the quota system in 1984. The efficacy of the system, which was unsatisfactory at the beginning, was intensified in 1986. Since then the production of skim-milk powder has been reduced by more than 700 000 t. In 1987 and 1988, this reduction could be compensated for by a massive reduction of intervention stocks. In 1989, these amounts were no longer available. The EC Commission suspended the obligation to incorporate skim-milk powder in milk replacers as part of a cut in subsidies (from 80 to 60 ECU/100 kg) before the end of the year 1988. This procedure alone has resulted in nearly 400 000 t less skim-milk powder being used in veal nutrition today.

If the Federal Republic is taken as an example, we see that the quantity of skim-milk powder used in the mixtures for veal nutrition has fallen dramatically (Figure 2). Unfortunately we do not have comparable data (by months) for the different member states (Figure 3).

Development of production dependent on the raw material

The information on intervention measures in the skim-milk powder sector recently published by the EC Commission shows that from 1987 to 1989 the use of skim-milk powder declined dramatically. The most important producer countries France, Italy, the Netherlands and Federal Republic of Germany made great efforts to use less skim-milk powder. But it would be incorrect to say that only the percentage was reduced. The absolute consumption of milk replacers was reduced in a very clear way too, for three main reasons:

- The quota system, which meant that veal calves were given the surplus milk in liquid form, especially at the end of the quota period to avoid having to pay a special tax. Thus the use of milk replacers was reduced.
- Fewer calves because there are fewer cows.
- The hormone scandals of 1988 and 1989, which affected the consumption of veal especially in France.

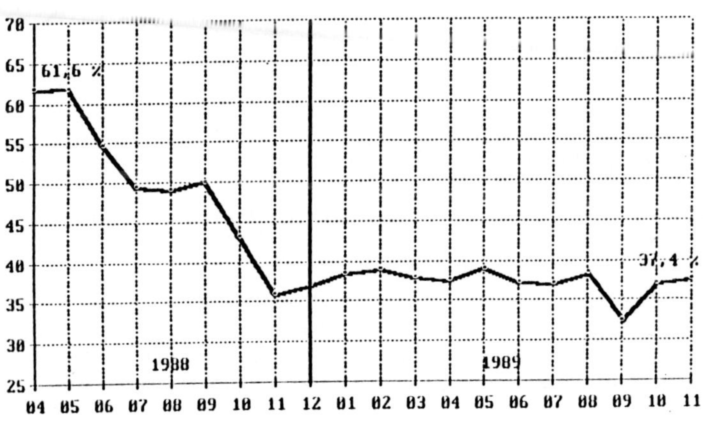

Figure 2. Incorporation of skim-milk powder in milk replacers in West Germany in per cent.

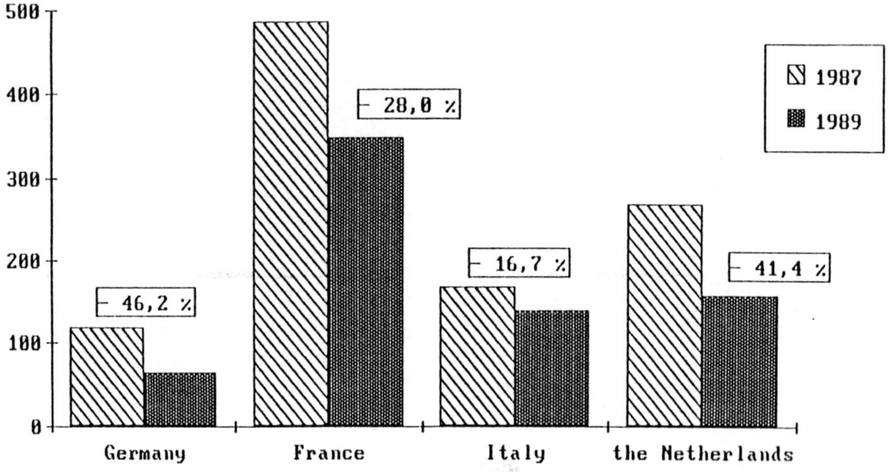

Figure 3. Incorporation of skim-milk powder in milk replacers in member states of the European Community in 1000 t.

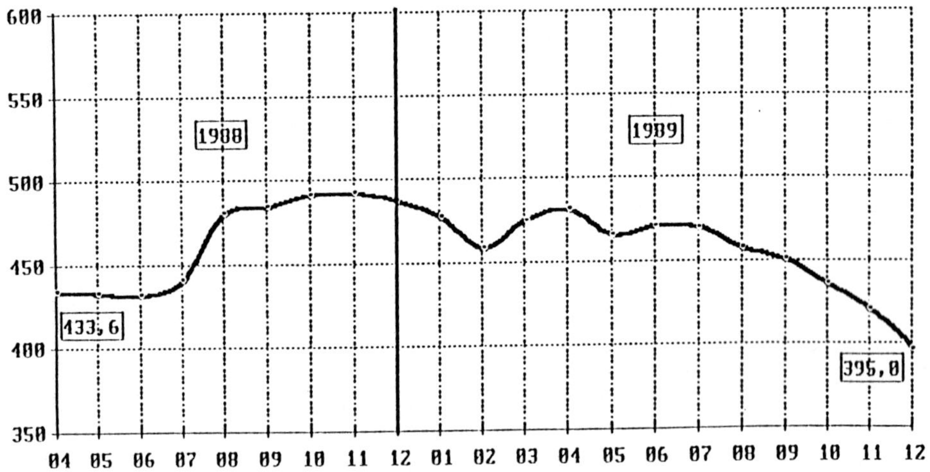

Figure 4. Average prices for spray skim-milk powder in DM per 100 kg.

The hopes of the producers of milk powder in the EC that the important reduction of prices for spray skim-milk powder (Figure 4) would lead to more skim-milk powder being incorporated in milk replacers have so far not materialized. Therefore the EC Commission is considering reintroducing the obligation to incorporate a minimum percentage of subsidized skim-milk powder in milk replacers.

Those in favour of this procedure believe that consumption could be boosted to 100 000–150 000 t, particularly if aid (subsidies) is increased simultaneously.

Opponents of this strategy doubt whether consumption will increase. They expect that the price of milk replacers will rise unless aid is raised correspondingly. Moreover, higher prices for veal calves and veal would affect consumption (Figure 5).

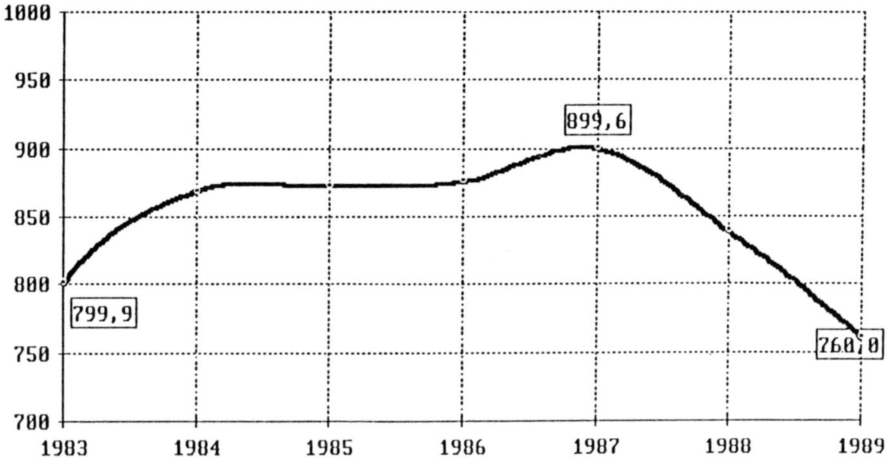

Figure 5. Consumption of veal in the EC (Community of the ten) in 1000 t.

Table 1. Anticipated trends in skim-milk powder balance in the EC (Community of the ten) in 1000 t.

	1984	1985	1986	1987	1988	1989	1990 opti.	1990 pessim.
Initial stocks	1 100	750	655	858	654	103	107	107
Production	2 060	1 915	2 120	1 604	1 297	1 399	1 440	1 470
Imports	-	-	-	2	5	55	30	50
Available	3 160	2 665	2 775	2 464	1 956	1 557	1 577	1 627
EC consumption								
total	2 090	1 700	1 650	1 420	1 234	1 030	1 100	980
without subsidies	220	200	200	270	280	278	300	280
feeding to calves	1 260	1 140	1 100	1 140	954	752	800	700
special sales for pigs	610	360	350	10	0	0	0	0
Exports	320	310	267	390	619	420	400	350
at world market prices	150	200	157	280	505	336	300	250
food aid programme	170	110	110	110	114	84	100	100
EC stocks	750	655	858	654	103	107	77	297
public	617	520	772	473	0	5	50	130
private	133	135	86	181	103	102	27	167
Changes in stocks	–350	–95	203	–204	–551	4	–30	190

Exports at world market prices:

trends in the consumption of skim-milk powder

The difficult situation facing the EC can be seen once more in the skim-milk powder balance. Table 1 shows optimistic and pessimistic trends for 1990.

Concerning production in the pessimistic version, it must be borne in mind that cheap casein from the Eastern Block will affect the EC market. Therefore EC producers will be unable to use the equivalent quantity of milk in the casein sector and therefore more skim milk will be made into skim-milk powder The same holds for the opening of Eastern European markets for the product itself.

If the incorporation of skim-milk powder into milk replacer is not reintroduced, we can expect a further decrease of consumption in the veal sector. This decrease might be less in countries which do have an incorporation of 30 to 35 per cent; the decrease will be greater in countries which use an average of 45 to 50 per cent.

Conclusions

The decreasing consumption and exports of skim milk mean that the European dairy industry is faced with a considerable increase in stocks. The EC Commission is obliged to take over only 100 000 t of skim-milk powder by intervention. From now on, skim-milk powder will be threatened by a free fall of prices without any further national aid. This cannot be accepted by the dairy sector as a whole.

You will understand that the dairy industry is deeply interested in using skim-milk powder in veal nutrition and that this interest should be shared by EC Commission. The experiences of the past have taught us that most of the stocks are used in the pig sector accompanied by high subsidies. This is undoubtedly the most expensive of all options.

Alternative protein use in calf milk replacers

C. W. Kolar & T. J. Wagner
Protein Technologies International, St. Louis, Missouri, US

Summary

Traditionally, milk protein has been the standard for feeding veal calves. Recent change in the cost of milk protein has created a need for alternative proteins for veal rations. A number of protein sources have been considered for veal rations, including proteins from soya bean, wheat, potato, pea, fish, and yeast. Important factors for assessing alternative proteins include chemical composition, antinutritional factors, digestibility, amino acid content, physical and microbiological properties, economics, and commercial availability. Specially treated soya protein isolate and alcohol-treated soya concentrate appear to be the best alternative proteins for veal rations. Of the two, specially processed soya protein isolate appears to provide the highest digestibility (88%), with the lowest level of antinutritional factors, and has the best physical (functional) properties.
Descriptors: proteins; calf; milk replacer; amino acids; digestibility; antinutritional factors.

Introduction

Traditionally, milk in the form of skim-milk powder and dried cheese whey has been used for the feeding of veal calves. Other protein products from milk such as casein and whey protein concentrate, are also used. These protein ingredients have established the standard of performance for protein used in the raising of veal calves. However the increasing cost of milk proteins, which impacts the cost of growing veal calves, has put increased emphasis on the need to find alternative proteins. The purpose of this review is to examine alternative proteins to skim milk evaluated by animal nutritionists over the past several years.

Alternative proteins and factors important to their use

Alternative proteins evaluated for feeding calves include protein from soya bean, wheat, potato, fish, pea and yeast. In most cases, the economics of these protein sources are quite good in comparison to milk proteins. Potential drawbacks can include poor digestibility and weight gain, thought to stem from antinutritional factors, and limited commercial supply of the protein ingredient. Also, the protein content of these products can be relatively low, and other components such as starch or polysaccharide may not be fully utilized.

Several factors should be considered in the evaluation of alternative protein ingredients for feeding veal calves. The most important of these include nutrient composition, antinutritional factors, digestibility, physical (functional) properties, microbiological profile, economics with regard to the cost of weight gain, and commercial availability.

A protein ingredient that is high in protein (90%) with a minimum of other non-protein components is preferred to allow maximum flexibility of formulation. Lower-protein ingredients could be used provided the other components present contribute to energy and are easily utilized by the calf. The iron content of the ingredient and bioavailability of the iron are critical and must not be greater than can be effectively managed by the veal grower.

Several antinutritional factors have been identified that contribute to the reduced performance of non-milk proteins. The mechanisms involved and economic methods to overcome these limitations require further exploration. Some identified antinutritional factors such as

Table 1. Summary of assessment of proteins for calf milk replacers.

Protein source	Protein content	Anti-nutr. factors	Digestibility	Calf performance	Functional property	Supply
Milk (SMP)	35%	none	95%[a]	excellent	excellent	commercial
Soya protein isolate	87%	low	87–88%[bc]	excellent	excellent	commercial
Soya conc.	65%	low	80–88%[bd]	good to excellent	fair suspension	commercial
Soya flour	50%	high	63–74%[b]	poor	poor	commercial
Wheat, soluble	85%	unknown	unknown	unknown	fair	commercial
Potato	75–85%	low	74–82%[b]	good	poor suspension	by-product
Fish	69–92%	low	80–92%[be]	good to excellent	variable	commercial
Pea	variable	high	50–79%[bf]	poor	unknown	development
Yeast	variable	low	75–84%[b]	good to excellent	variable	development

[a] Toullec, 1989. [d] Guilloteau, 1977.
[b] Joussellin, 1988. [e] Paruelle, 1974.
[c] Tolman, 1988. [f] Bell, 1974.

enzyme inhibitors (e.g. trypsin and chymotrypsin), haemagglutinins (lectins), and antigenic proteins are present in many plant protein ingredients. The presence of enzyme inhibitors not only reduces digestibility, but the binding of proteases causes the pancreas to produce more of these enzymes, which is a waste of feed-derived amino acids. Haemagglutinins (lectins) have been shown to react with the villi in the intestine, resulting in a disruption of the cell structure. This may lead to poor nutrient absorption and, possibly, the leakage of large protein molecules into the circulatory system, causing antibody formation, which may result in allergic reactions in the gut and may result in poor feed and nutrient utilization. Thus, it is essential to have a protein source that is low in haemagglutinins and has relatively low antigenic properties.

High digestibility is desirable as it allows for maximum utilization of amino acids present in the protein. A digestibility value less than of milk is not an antinutritional factor, but reduces the availability of amino acids, and thus reduces the value of the protein to some degree. It is desirable that the amino acid composition of the protein ingredient meet the requirements of the calf. If not, it should be easily and cost-effectively fortified.

Physical or functional properties of a protein ingredient can be important for the preparation of a uniform fluid milk replacer. It is desirable to have a protein ingredient that can be easily combined with the other ingredients used in the ration and, when reconstituted in water, the milk replacer should give excellent solubility, and/or suspension properties similar to that of skim-milk powder or milk-based ingredients. The ingredient should not settle out or cause other ingredients to settle out during preparation or feeding of the milk replacer, as this could result in the calf not receiving all of the nutrients in the ration.

The microbiological quality of the ingredients used in milk replacers for young animals must be of food-grade quality. They should have a low total plate count and be free from salmonella and other pathogens, because these micro-organisms can contribute to digestive disturbances, sickness, and poor performance.

Obviously, a low-cost protein ingredient is desirable in relation to milk proteins on a protein basis, as the cost for growing the calf is critical. In addition to cost, a high-quality carcass must be produced and marketed.

Finally, the protein should be available commercially, as large amount are required for growing veal calves. Significant time may be required to develop new products that are acceptable for veal feeding from existing raw materials. Capital may be required to construct plants for manufacturing new protein ingredients. Thus, several years may pass before commercial amounts are available for use by the veal industry.

The factors outlined as important in the selection of protein ingredients for growing veal calves are summarized in Table 1. When one considers all the factors involved in selecting an alternative protein to milk protein for growing veal calves, soya protein isolate and soya concentrate are protein ingredients that rank highest. These products are manufactured and marketed in large amounts.

In particular, soya protein isolate is most similar to milk protein, except that its digestibility is somewhat lower for calves. Nevertheless, the protein content of soya protein isolate is high with the absence of insoluble and soluble carbohydrates. Compared with soya protein isolate the soya concentrate has lower protein content, lower digestibility, and only fair suspendability.

Soluble wheat protein may have some promise as an alternative protein to skim milk, as it is reported to have high digestibility, but little is known about its digestibility, antinutritional factor content, and performance in calves. Potato protein is another potential ingredient, but it is a by-product of the potato-processing industry and consequently, there is a limited supply of this ingredient. It is probably limited to a low use level in veal rations because of poor suspendability, supply, and digestibility.

Several studies on the value of fish protein for use in veal rations have been reported. The quality of fish protein, from the standpoint of amino acids and digestibility, is good to excellent. However fish protein has distracting factors, such as colour, odour, and variable physical properties, which limit its use in calf milk replacers.

Soya flour has been evaluated extensively. It is a low-cost ingredient, has an intermediate protein content, and is available in large amounts. However it is severely limited by its high level of antinutritional factors and by poor calf performance. Thus it is not a viable alternative to milk proteins.

Pea protein products are variable in protein content and experimental products have had poor suspension properties. The antinutritional factors in the pea are significant and processes have not been developed on a commercial scale to eliminate them. So acceptable pea protein ingredients are not available with a high protein content and with low antinutritional factors.

Research has been conducted on the use of yeast and yeast protein as a protein source for veal. Digestibility studies indicate high digestibility and excellent calf performance. However commercial supply and functional properties limit its use.

After assessing the various protein ingredients, the major commercial supplies of alternative proteins to milk protein for the veal industry are soya protein isolate, soya concentrate, and possibly a soluble wheat protein product.

The composition of soya protein isolate, soya concentrate, and soluble wheat protein is shown in Table 2. Soya protein isolate is primarily pure protein. It is virtually free from crude fibre, insoluble carbohydrates, or soluble carbohydrates. Soya concentrate is about 65% protein and the remaining portion of the soya concentrate is primarily insoluble carbohydrates, a portion of which is crude fibre. Soluble wheat protein is slightly higher in protein than soya concentrate, has more lipids than the soya ingredients and contains about 5% carbohydrate.

Table 2. Typical composition of soya protein isolate, soya concentrate and soluble wheat protein.

Assay	Soya protein isolate (%)	Soya concentrate (%)	Wheat protein (%)
Moisture	5.0	6.0	7.0
Protein ($N \times 6.25$)[1]	87.4	64.9	79.1
Fat	0.5	0.7	7.5
Crude fibre	< 0.2	4.0	0.4
Ash	4.5	6.0	1.1
Total carbohydrates[2]	2.5	18.4	4.9

1 As basis.
2 Includes crude fibre.

Table 3. Typical amino acid composition of various protein products (g/100 g protein).

	Soya protein isolate[a]	Soya concentrate[b]	Wheat protein[c]	Skim milk powder[d]
Alanine	4.3	4.2	2.5	3.5
Arginine	7.6	7.2	3.0	3.7
Aspartic acid	11.6	11.6	3.1	7.7
Cysteine	1.3	1.5	2.1	0.9
Glutamic acid	19.1	19.5	37.5	21.4
Glycine	4.2	4.1	3.2	2.2
Histidine	2.6	2.5	2.2	2.8
Isoleucine	4.9	5.0	4.1	6.2
Leucine	8.2	7.7	6.9	10.0
Lysine	6.3	6.2	1.4	8.1
Methionine	1.3	1.3	1.6	2.6
Phenylalanine	5.2	5.0	5.2	4.9
Proline	5.1	5.2	13.2	9.9
Serine	5.2	5.5	5.0	5.6
Threonine	3.8	4.0	2.5	4.6
Tryptophan	1.3	1.5	1.0	1.4
Tyrosine	3.8	3.7	3.7	4.9
Valine	5.0	5.0	4.3	6.8
Total sulphur AA	2.6	2.8	3.7	3.5
Total aromatic AA	9.0	8.7	8.8	9.9

a Kolar, 1990.
b Loders Croklaan B.V., NL.
c Nutr. Div., FAO, 1972.
d Posati & Orr, 1976.

The typical amino acid composition for soya protein isolate, soya concentrate, wheat protein, and skim milk powder are given in Table 3.

Table 4. Antinutritional factors found in soya isolates and soya concentrate.

Analysis	Soya protein isolate 'A'	Soya protein isolate 'B'	Soya concentrate
Trypsin inhibitor (mg/g)[a]	1.22	3.95	2.87
Haemagglutinins (HA)[1]	< 0.05	0.05	0.2
Antigenicity (LPDB)[2b] (calf)	< 0.4	5.1	5.1

a Tolman, 1988.
b Grev, 1989.
1 Haemagglutinating activity (HA) is expressed as the highest dilution of the purified lectin extract producing visible agglutination divided by 1000.
2 Log protein dilution basis; antibodies to soya flour.

The soya proteins are relatively good sources of amino acids for the veal calf; they are normally supplemented with certain amino acids when combined with skim milk or whey proteins to meet the amino acid requirements of the calf. Methionine is used when soya protein is included in the formula. Wheat protein has lower levels (70% or less than that of skim-milk protein) of isoleucine, leucine, lysine, threonine and valine and is also likely to require additional supplementation.

The antinutritional factors present in soya protein isolate can vary from one soya protein isolate to another. The same is true for soya concentrates. Information in Table 4 illustrates the difference between two soya protein isolates and a soya concentrate reported to be antigen-free. These proteins were evaluated for antigenicity using the ELISA technique with antibodies to soya flour.

The soya protein isolate 'A' with the lowest trypsin inhibitor, haemagglutinin (lectins levels, and antigenicity is recommended for use in veal rations, while soya protein isolate 'B' would not be recommended even though it is a high-quality, food grade product. The soya concentrate is an alcohol processed product that is recommended for veal feeding. It has slightly higher levels of trypsin inhibitor, haemagglutinins, and antigenicity than the soya protein isolate recommended for veal rations.

In summary, a variety of alternative proteins could theoretically be used to replace milk protein in raising veal calves. Soya protein, soluble wheat protein, fish and potato proteins are used in veal rations. Specially processed soya protein isolate and alcohol-treated soya concentrate are the primary alternative proteins used.

References

Bell, J.M., G.F. Royan & C.G. Youngs, 1974. Digestibility of pea protein concentrate and enzyme-treated pea flour in milk replacers for calves. Canadian Journal of Animal Science 54:355.

Food Policy and Food Science Serv. Nutr. Div., FAO, 1972. Amino acid content of foods and biological data on proteins. p. 42. Food & Agriculture Organization, Rome, Italy.

Grev, A.H., 1989. Protein Technologies International. Unpublished data.

Guilloteau, P., R. Toullec, J. Culioli & D. Le Douaron, 1977. Utilization of proteins by the preruminant veal calf. Annales de Zootechnie 26(1):15–28.

Joussellin, W., 1988. Quel veau de produire demain? Association Francaise des Techniciens De l'Alimentation Animale, Rennes Cedex, France. p. 9.

Kolar, C.W., 1990. Protein Technologies International. Unpublished data.
Paruelle, J.L., R. Toullec, P. Patureau-Mirand & C.M. Mathieu, 1974. Utilization of protein by fattening preruminant calves. Annales de Zootechnie 23(4):519–535.
Posati, L.P. & M.L. Orr, 1976. Composition of foods. Agric. Handbook No. 81. Agricultural Research Service, US Department of Agricultyre, Washington, DC.
Tolman, G.H. & J. Huisman, 1988. ILOB-TNO Institute of Animal Nutrition and Physiology, Wageningen. Unpublished data.
Toullec, R., 1989. Veal calves. In: Ruminant nutrition: recommended allowances and feed tables. John Libbey Eurotext, Montrouge, France.

Varying protein levels in veal calf finisher formulations

P. Schneider, D. Bakke & D. Bearda
Real Veal Inc., Ixonia, Wisconsin 53036, US

Summary

The price of bob calves and dairy protein ingredients have been steadily increasing, a trend which is expected to continue. This has limited the margin available to the veal grower. This study was designed to determine responses of calves to varying levels of protein in their diets and provide the grower with alternatives in feeding management. Five hundred and four calves were fed on finisher formulations with protein contents ranging from 12.5 to 18.5% from weeks 7 to 16. Body weight gains, feed conversions and carcass grades were highest at the higher protein levels, but margins were highest at the low protein levels. Finisher formulations with protein levels of 12.5% will produce veal calves that yield higher margins to the grower.
Descriptors: veal; protein levels; feed.

Introduction

The profitability of raising calves for veal in the United States is hampered by market forces outside the veal industry. The means by which veal growers can increase their bottom line is thus limited. Management, especially feeding management, is one of the factors that the veal grower can actively control.

Due to declining numbers of dairy cattle, the price of calves has steadily increased. This trend is likely to continue into the future and to be accelerated by new technologies such as bovine somatotropin (BST), genetic engineering, and embryo transfer. The increasing demand for calves by both the veal industry and the dairy beef industry will keep pushing up the price of quality bob calves. Buying low-quality calves can jeopardize a veal operation.

Skim milk sold by the US government had been the major protein source for the veal industry until 1988. Since 1988, the government has not released skim to the veal industry. While available, skim prices had fluctuated with costs well over $1.00 per kg. The price for fresh skim on the open market has averaged slightly higher with a moderate increase in late 1988 and a huge increase in 1989. Prices have been lower in January and February of 1990. Nevertheless, skim as a source of protein is expensive and becoming scarcer.

Whey-protein concentrate has been the focus of attention at Real Veal, Inc. as a protein replacement for skim milk. The price per kilogram averaged $1.25 in 1987 and $1.60 in 1988. In 1989, the range was from $1.25 to almost $1.90.

In February 1990, the fat substitute Simplesse was cleared by FDA for human consumption. Dairy proteins are a major component of Simplesse. This is certain to increase the pressure on the veal industry for dairy proteins.

Historically, the market price for finished calves has been high enough relative to the price of bob calves for profitability to be fairly easy. Beginning in 1987, however, the price of bob calves began to climb (Figure 1). Although packer prices increased in response, the amount consumers are willing to pay creates a ceiling on what meat packers will pay for finished calves.

Protein is the most expensive nutrient in milk replacers. Published protein requirements (NRC, 1988; Roy, 1980) indicate that protein levels should be at least 18%. We designed the following study to determine the response of veal calves to varying levels of protein during the finishing phase. The objectives were to provide the grower with alternatives in feeding management in order to optimize his profit potential.

Table 1. Ingredients of low and high experimental formulations.

	12.5%	18.5%
Whey	63.7	37.0
Skim	12.1	38.1
7/60 Fat concentrate	8.6	8.6
Coconut oil	5.4	5.4
Tallow/Lard	7.7	7.7
Corn Flour	1.2	1.2
Premix	1.33	1.33

Table 2. Composition of low and high experimental formulations.

Component	Protein (%)	
	12.5	18.5
Fat (%)	19.40	19.4
Lactose (%)	52	47
Ca (%)	0.55	0.71
P (%)	0.63	0.69
Mg (%)	0.14	0.12
K (%)	1.6	1.4
Na (%)	0.55	0.46
Zn (mg/kg)	26	36
Fe (mg/kg)	6	6
Lysine (%)	0.96	1.4
Methionine (%)	0.35	0.47

Table 3. Margins based on feed intake and gains from week 7 to 16. Haemoglobin values are for week 16. Conversions are feed intake divided by gain. Carcass grades are give by meat packer: 1, best; 3, worst.

Item	12.5	13.5	14.5	16.0	16.5	17.6	18.5
Gain (kg)	86.9^{a}	89.7^{ab}	91.8^{bd}	96.6^{ce}	94.4^{de}	95.6^{ce}	98.4^{c}
Intake (kg)	171^{a}	171	170	171	169^{b}	170	170
Margin ($)	112^{a}	106^{a}	98^{b}	95^{b}	93^{b}	82^{c}	78^{c}
Hgb (g/dl)	8.89^{a}	8.34^{b}	8.21^{bd}	7.77^{c}	7.92^{cd}	7.88^{cd}	7.82^{cd}
Conversion	1.97	1.93^{a}	1.86	1.78	1.79	1.80	1.73
Carcass	1.67	1.57^{ab}	1.60	1.47	1.37	1.43	1.23

Expt. $P < 0.05$, treat × expt. $P > 0.05$, different superscripts in row: $P < 0.05$.

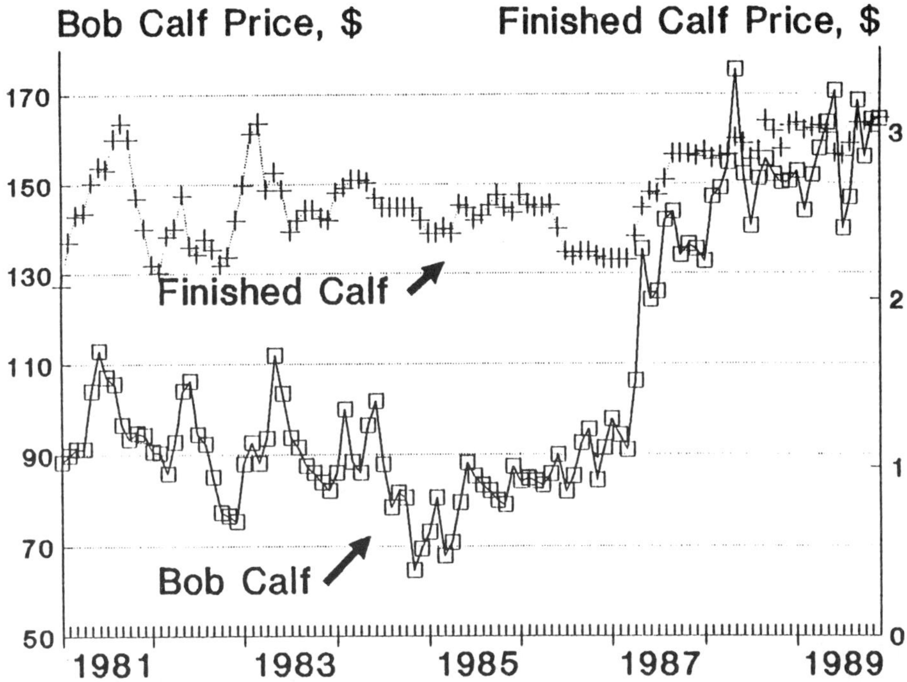

Figure 1. Prices for bob and finished calves. Bob calf price based on 45 kg average weight. Finished calf price in US dollars per kg liveweight.

Material and methods

Three experiments were conducted at the Real Veal research farm. Each experiment had 7 treatments with 24 calves per treatment. Calves were adapted to their respective formulation from the starter formulation over 7 days during week 6. Each calf remained on its dietary treatment for the entire experiment, which lasted from week 7 of the fattening phase until week 16 when the calves were slaughtered at a commercial meat packer.

Six experimental treatments were formulated to be as similar as possible except for protein (Tables 1 and 2). The protein levels were 12.5, 13.5, 14.5, 16.5, 17.5 and 18.5% on an as-is basis. Our commercial 16% product, Finmelk +, was included as an additional treatment. Ingredient composition of Finmelk + was the same as the experimental treatments. Calves were fed twice a day, and weighbacks were measured daily. Every 2 weeks, calves were weighed and blood was collected for haemoglobin determination. Iron intake and supplementation were the same across all treatments and experiments.

Results and discussion

There were no differences in feed intake except between diets with 12.5 and 16.5% protein (Table 3). Intake tended to be lower with 16.5%, because of lower intake in Experiment 3 at this protein level then in Experiments 1 and 2. Feed intakes were higher across all protein levels for Experiment 2 than in 1 and 3.

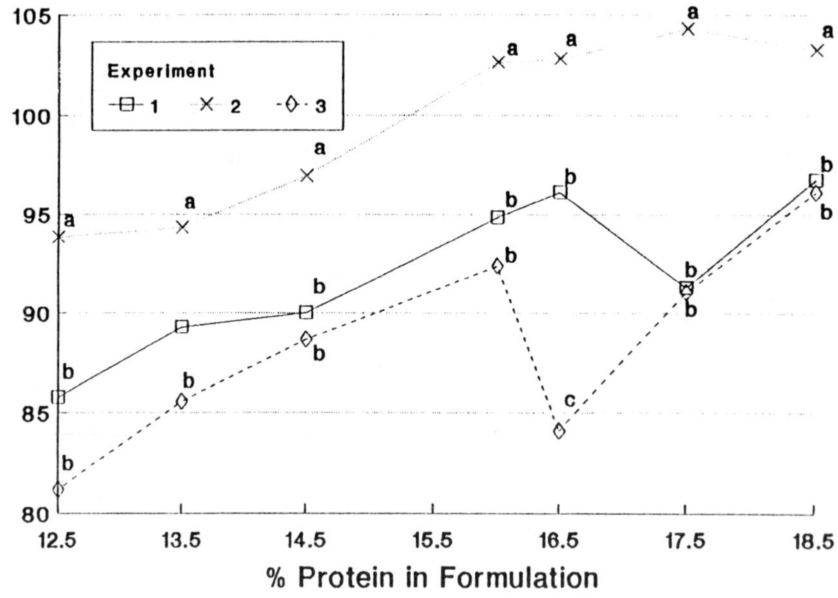

Figure 2. Effects of varying protein levels on weight gain in kg for Experiments 1, 2 and 3. Supercripts denote differences (P < 0.05) between experiments for protein levels only.

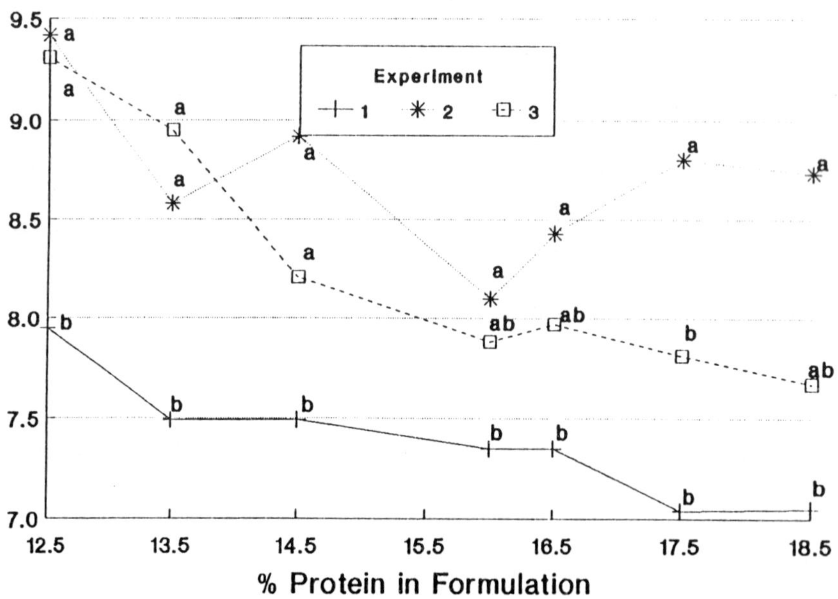

Figure 3. Effects of varying protein levels on haemoglobin values (g/dl) at week 16 for Experiments 1, 2 and 3. Superscripts denote differences (P < 0.05) between experiments for protein levels only.

Weight gain increased with increasing protein intakes. There was a decrease in gain with 16.5% protein parallel to the decrease in feed intake. This decrease occurred only in Experiment 3, where gain was significantly lower with 16.5% than in either Experiment 1 or 2 (Figure 2). Gains were consistently higher for calves in Experiment 2, as were feed intakes. Incoming body weights were higher for calves in Experiment 2 than Experiments 1 and 3. Environment may also have been a factor, in that Experiment 2 was conducted during the summer and Experiments 1 and 3 were conducted during late fall and winter. Feed conversion (feed intake/gain) tended to improve with higher protein levels.

Haemoglobin values were lower at the higher protein levels. The decrease in haemoglobin as gain increased probably reflects the higher nutrient requirements of the faster growing animal. Erythropoiesis apparently cannot keep pace with the higher growth rates (Wagner, 1990). Lower haemoglobin values also reflected the higher packer grades for animals on the higher protein levels. In the United States, light-coloured carcasses, as indicated by lower haemoglobin values, receive the highest grades. Haemoglobin values for Experiment 2 also tended to be higher than values for Experiments 1 and 3 (Figure 3).

Gross margins were determined for the fattening phase using feed intake, feed costs, weight gains and an average liveweight packer price of $2.87/kg. Margins decreased from an average of $112 per calf with 12.5% protein to $78 per calf with 18.5% protein. Thus, despite lower weight gains, packer scores and feed conversions with 12.5% than with 18.5% protein, there was a potential for higher profits.

Conclusions

Finisher feed with 12.5% protein provides the veal grower with an opportunity to exert some control over his feed costs. Body weight gains, feed conversions and carcass grades from the meat packer were all better at the higher protein levels. Nevertheless, gains, carcass scores and conversions at lower protein levels were not affected to the extent that margins were compromised. Finisher formulations with protein levels of 12.5% will produce veal calves which yield higher margins to the grower.

References

National Research Council, 1988. Nutrient requirements of dairy cattle. National Academy of Science, Washington, DC.

Roy, J.H.B., 1980. The calf. Butterworths, Inc., Boston, MA 01801.

Wagner, J.M., 1990. Personal Communication. Ixonia, WI 53036.

Soluble wheat protein in milk replacer for veal calves

A. De Laporte & M. Demeersman
Amylum n.v., Burchtstraat 10, B-9300 Aalst, BE

Summary

Over the last few years, a lot of research has been devoted to finding alternative proteins to skim-milk powder for veal-calf production.

Recently, soluble wheat protein has been introduced as an alternative for high-quality proteins. Soluble wheat protein is a creamy-coloured powder with a neutral taste and odour. The product has very high solubility, and has a high emulsifying capacity and emulsion stability.

Product consistency and a high protein content of 82% are key advantages. Concerning the essential amino acids, the complementarity with whey powder is remarkable. In combination with whey powder, an ideal pattern of essential amino acids can be obtained if synthetic lysine is added. The clotting of the protein is an important feature, and is a possible explanation for the high digestibility of soluble wheat protein (95%).

Low coloured veal can be produced with soluble wheat protein. The product contains 30 ppm iron. Due to the complex-forming properties, the digestibility of the total iron present in the milk replacer is decreased. No antinutritional components or allergic factors are known.
Descriptors : soluble wheat protein; milk replacer; veal; caseinate.

Introduction

Over the last few years, a lot of research has been devoted to finding alternatives for skim-milk powder as an ingredient of milk replacers for veal-calf production.

From the technical point of view, caseinates are an excellent alternative but they have become expensive, even if costed on a protein basis.

As a result of one of these new developments, soluble wheat protein is introduced as an alternative to milk protein in calf milk replacers. The production process, composition, and physical and nutritional characteristics are discussed. The results of digestibility tests and growth trials are summarized by Tolman & Demeersman (this book).

Production and composition of soluble wheat protein

Figure 1 gives a global overview of the production of soluble wheat protein. After cleaning and conditioning, the wheat is ground into flour with a specific milling system. The wheat flour is then mixed with water and homogenized to a dough. The primary separation results in a starch and a protein fraction. Transformation and subsequent drying of the protein fraction results in the production of a creamy-coloured powder, soluble wheat protein (De Laporte et al., 1989).

The typical composition of soluble wheat protein obtained by this proprietary transformation process is given in Table 1. Product consistency is a key advantage for application in milk replacers.

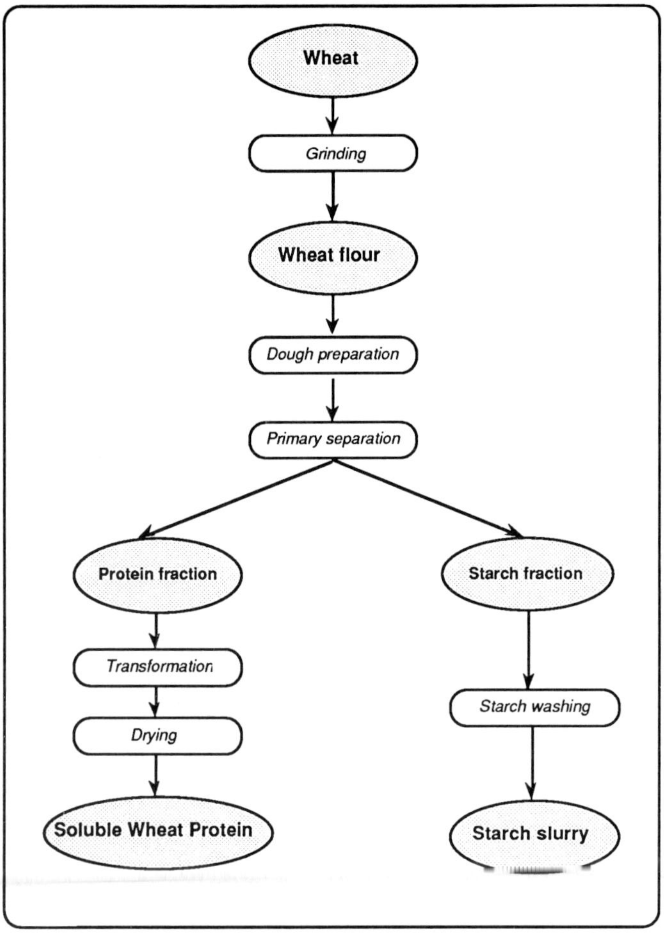

Figure 1. Production scheme for soluble wheat protein.

Table 1. Average composition of soluble wheat protein.

Component	Content
Water (%)	5
Protein (N $\times$ 6.25) (%)	82
Crude fat (%)	8
Ash (%)	4
Carbohydrates (%)	1
pH	7
Iron (ppm)	30

Table 2. Solubility and emulsifying properties of soluble wheat protein, caseinate and soya isolate (according to De Laporte et al., 1989).

Parameter	pH	Soluble wheat protein	Caseinate	Soya isolate
Solubility (nitrogen solubility index, %)	7	90	90	50
	6	90	85	20
	4	10	5	10
Emulsifying capacity (g oil/g product)	7	620	670	290
	6	570	530	220
	5	470	160	110
Emulsifying activity (%)	7	53	52	56
	6	52	51	51
	5	53	48	49
Emulsifying stability (%)	7	100	94	100
	6	100	100	100
	5	100	100	95

Table 3. Essential amino acid composition (as % of protein) of soluble wheat protein, veal, skimmilk powder and whey powder.

	Veal	Skim-milk powder	Whey powder	Soluble wheat protein
Lysine	8.4	8.2	6.4	1.6
Methionine	2.3	2.6	1.4	1.8
Methionine + cystine	3.5	3.5	3.5	4.3
Threonine	4.3	4.6	5.9	2.6
Tryptophan	1.3	1.3	1.4	1.0
Isoleucine	5.2	5.6	5.8	4.4
Leucine	7.3	9.8	8.8	7.6
Valine	5.1	6.9	5.5	4.4

Physical characteristics

Soluble wheat protein is a white creamy powder. It has a neutral taste and odour. The solubility and emulsifying properties of soluble wheat protein were compared to caseinate and soya isolate by De Laporte et al. (1989). These functional characteristics of soluble wheat protein are comparable to caseinate. Solubility and emulsifying capacity of soluble wheat protein are higher than of soya isolate (Table 2). A 20% solution of this wheat protein in water has a Brookfield viscosity of 4000–8000 mPa.s. The physical characteristics of soluble wheat protein are thus comparable to low viscosity caseinate.

Nutritional characteristics

Quality of the protein

Soluble wheat protein has a high and constant protein content of 82% on average. This wheat protein is relatively rich in arginine, histidine, sulfur amino acids and aromatic amino acids. After supplementation with synthetic lysine, an ideal essential amino acid pattern can be obtained in combination with whey powder (Table 3).

Digestibility of dry matter, organic matter and crude protein are found to be more than 95% (Tolman & Demeersman, this book). A possible explanation for this high digestibility of soluble wheat protein is the flocculation at pH = 4.5 (De Laporte et al., 1989).

Iron content and availability

Soluble wheat protein has a low iron content of only 30 ppm on average. Digestibility tests and growth trials indicate a very low digestibility of iron (Tolman & Demeersman, this book).

Antinutritional and allergic factors

Consistency of faeces, good health of the calves and the absence of diarrhoea in feeding trials with diets containing even up to 20% of soluble wheat protein are strong indications for the absence of antinutritional components and allergic factors in soluble wheat protein. Liener & Kakade (1980) indeed mentioned the absence of, for instance, flatulence factors and lectins in wheat.

Energy value

The metabolizable energy content of soluble wheat protein is 17 327 kJ/kg. This value can be calculated as follows:

$$ME = 0.82 \times 18\,800 \times 0.95 + 0.08 \times 37\,250 \times 0.9 = 17\,327 \text{ kJ/kg}.$$

Conclusion

By a proprietary process, soluble wheat protein can be produced from wheat. This protein is a creamy-coloured powder with a bland taste. The two main characteristics of this product are a high protein content (82%) and a very high solubility. Soluble wheat protein is a good emulsifier. A milk that remains stable in emulsion can be produced.

By adjusting with about 7% synthetic lysine-HCl, a good pattern of essential amino acids can be obtained because of the complementarity with other protein sources. In combination with whey powder and synthetic lysine-HCl, the essential amino acid requirements for veal calves can be met. As a rule of thumb, 3 parts of skim-milk powder can be replaced by 1 part of soluble wheat protein and 2 parts of whey powder (plus lysine).

Soluble wheat protein flocculates at acidic pH. This may explain the very high digestibility (95%) of this wheat protein. It can also explain why the fat digestibility in the milk replacer remains high, contrary to what is observed with other non-milk proteins (Troccon & Toullec, 1989). The low content and availability of iron are remarkable. Finally, unlike most other plant proteins, no antinutritional components are known.

As a result of the above mentioned properties, soluble wheat protein is an alternative to high-quality proteins, such as skim-milk powder, caseinate, whey protein concentrate, soya isolate etc. in calf milk replacers. From its physical properties, it can be concluded that soluble wheat protein could be described as 'vegetable caseinate'.

References

De Laporte, A., Ph. van Tornout, V. Anteunis & J. de Meester, 1989. Oplosbaar tarwe-eiwit: functionaliteit en toepassingsmogelijkheden. Voedingsmiddelentechnologie 22(18)22–24

Liener I.E. & M.L. Kakade, 1980. In: Toxic constituents of plant foodstuffs. I.E. Liener (Ed.). Academic Press Inc., New York. p.7.

Troccon J.L. & R. Toullec, 1989. Aliments d'allaitement pour veaux d'élevage. INRA Production Animale. 2(2)117–128.

Digestibility and growth performance of soluble wheat protein for veal calves

G.H. Tolman * & M. Demeersman **
* *ILOB-TNO, P.O. Box 15, 6700 AA Wageningen, NL*
** *Amylum n.v., Burchtstraat 10, B-9300 Aalst, BE*

Summary

The suitability of soluble wheat protein in milk replacers for veal calves was investigated. Milk replacers containing up to 20% of soluble wheat protein remained stable in solution and had no off-odours. Feed intake by calves was also good and faeces of the animals were normal. The digestibility coefficients of soluble wheat protein for dry matter, organic matter and crude protein were estimated at 95% each. They were independent of the amount of the wheat protein incorporated (10% and 20% of the diet). The performance of calves fed on a control diet with 20% of soluble wheat protein and whey powder was compared in a growth trial with that of calves fed on a diet containing 66% of skim-milk powder. Under these very extreme conditions, average liveweight, average weight gain and average feed conversion were about 4% better for the group fed on the diet with 66% skim-milk powder than for the group fed on the diet with 20% of soluble wheat protein. Meatiness, meat colour and fat covering of the carcasses for both groups were comparable. Given the exceptional performances, partial replacement of milk proteins by soluble wheat protein is already in use.
Descriptors: soluble wheat protein; milk replacer; veal; digestibility; growth trial; skim-milk powder.

Introduction

For application in milk replacers, ingredients must have very specific characteristics. For instance, good quality milk replacers must have a good solubility, and have to be stable in emulsion. The milk also has to be readily drunk by the calves. Based on its physico-chemical characteristics (De Laporte & Demeersman, this book), soluble wheat protein can be an alternative to milk proteins in milk replacers for veal calves. When considering the amino acid composition, it is clear that skim-milk powder can be replaced on a theoretical base by soluble wheat protein and whey powder, if lysine and, to a lesser extent, threonine, are added.

Nutritionally, the digestibility of soluble wheat protein has to be determined for optimum formulation of diets. Finally, the real nutritional value of a product can only be determined in growth trials. To obtain unequivocal results, the experimental diet should differ by only one ingredient from the control; degree of incorporation has to be as high as possible.

To determine the applicability of soluble wheat protein in milk replacers for veal calves, the protein was evaluated in 3 phases:

- evaluation of milk replacer containing soluble wheat protein;
- digestibility of soluble wheat protein;
- comparative growth trial between calves fed with skim-milk powder or with soluble wheat protein based diets, respectively.

Table 1. Ingredients and composition of the diets E1 and E2.

	Diet E1	Diet E2
Ingredients (%)		
soluble wheat protein	10	20
skim-milk powder - fat concetrate	50	18
whey powder	15	10
fat	-	10.5
lactose/dextrose	19	31.3
starch	2	2
others	4	8.2
Composition		
crude protein (%)	22.5	24.0
crude fat (%)	18.2	18.0
metabolizable energy (kJ/kg)	18 560	18 552

Diets were balanced for amino acid composition by supplementing with synthetic amino acids.

Table 2. Ingredients and composition of experimental diets D1, D2 and D3.

	Diet D1 60% skim-milk powder	Diet D2 10% soluble wheat protein	Diet D3 20% soluble wheat protein
Ingredients (%)			
soluble wheat protein	-	10.0	20.0
skim-milk powder - fat concentrate	53.0	47.7	42.4
skim-milk powder	24.5	22.0	19.6
whey powder	10.0	9.0	8.0
lactose/dextrose	9.1	8.2	7.3
starch	1.0	1.8	1.6
others	1.4	1.3	1.1
Diet composition[1]			
dry matter (%)	96.4	96.0	95.5
ash (%)	6.3	6.0	5.8
crude protein (%)	22.5	28.0	33.8
fat (%)	16.8	15.8	14.8
iron (ppm)	66	54	50

[1] Values by analysis.

Material and methods

Experiment 1: Evaluation of milk replacer containing soluble wheat protein

Two experimental milk replacers containing 10 and 20% soluble wheat protein were evaluated using 2 groups of 3 male Dutch Friesian × Holstein Friesian calves. The calves had an age of about 8.5 weeks and a weight of about 80 kg. During the trial period of 5 days, the calves were bucket-fed twice a day with 6 l of milk replacer (12.5% d.m.). The diet was switched from a traditional milk replacer to the experimental diets in 2.5 days. The composition of both experimental diets is given in Table 1. The methods of evaluation are described by Tolman (1988).

Experiment 2: Digestibility of soluble wheat protein

The apparent faecal digestibility of dry matter, organic matter and crude protein of soluble wheat protein was determined with 3 groups of 5 male Dutch Friesian × Holstein Friesian calves, aged about 8 weeks and with a weight of about 70 kg. The calves were bucket-fed twice a day with diets prepared separately per calf at each feeding time. The apparent faecal digestibility was determined according to an indirect measurement of digestibility during 5 days (Tolman, 1989a).

In a pre-test period of 11 days, a commercial diet was fed to the calves. The calves were then divided into 3 groups with similar average weight and haemoglobin concentration of the blood. In the pre-test period 2 of 7 days, the calves were accustomed to the harnesses and the diets. In the test period of 8 days, the 3 experimental diets were fed. The ingredients and the composition of the diets are given in Table 2. The Reference Diet D1 contained 60% of skim-milk powder and 10% of whey powder. For formulation of the test diets, 10% and 20% of the Reference Diet was replaced by soluble wheat protein (Diet D2 and Diet D3, respectively).

Experiment 3: Comparative growth trial

In a growth trial, the performance of calves fed on a diet containing 20% of soluble wheat protein and 57% of whey powder (Diet G2) was compared with that of calves fed on a diet containing 66% of skim-milk powder and 12% of lactose (Diet G1). In the starting period at age 0–6 weeks, a commercial diet containing 55% skim-milk powder was fed to the calves (male Dutch Friesian × Holstein Friesian calves). The calves were bucket-fed twice a day. At an average age of six weeks, the calves were divided into 2 groups of 20 animals with similar average weight, average weight gain in the starter period and haemoglobin concentration of the blood.

During the adaptation period (2 × 2 weeks), the calves of the reference group were fed with Diet G1. The calves of the test group were changed gradually from Diet G1 with steps of 33% per 2 weeks to Diet G2. In the fattening period of 16 weeks, the animals received the Experimental Diets G1 or G2. The trial was described in detail by Tolman (1989b). The ingredients and the composition of the diets are given in Table 3.

Results and discussion

Experiment 1

Milk replacers containing soluble wheat protein and/or skim-milk powder and/or whey powder were prepared and fed to calves. A stable emulsion was formed without settling, even for milk replacers containing up to 20% of soluble wheat protein. The artificial milk containing this protein had no off-odours and had a typical floury smell. The calves had normal drinking behaviour with a normal drinking time of 2–3 minutes. There were no adaptation problems or feed refusals. The amount and consistency of faeces were normal in all cases; the faeces had a green to brown colour (Tolman, 1988).

Table 3. Ingredients and composition of diets G1 and G2.

	Diet G1 66% skim-milk powder		Diet G2 20% soluble wheat protein	
Ingredients (%)				
soluble wheat protein	-		20.0	
skim-milk powder - fat[1]	85.5		-	
whey powder - fat[1]	-		71.4	
fat	1.2		4.4	
lactose	12.3		-	
L-lysine-HCl	-		1.28	
L-threonine	-		0.16	
others	1.0		2.7	
Composition (%)[2]				
dry matter	96.4		97.0	
crude protein	22.9		25.2	
crude fat	17.1		19.6	
ash	5.6		6.5	
gross energy (kJ/kg)	20 682		21 070	
metabolizable energy (kJ/kg)[3]	18 405		18 418	
lysine	1.85	(1.86)[3]	1.68	(1.82)[3]
methionine + cystine	0.80	(0.94)	0.99	(1.06)
threonine	1.00	(1.02)	0.94	(1.02)
tryptophan	0.38	(0.22)	0.28	(0.25)

1 mixtures contain 20% fat
2 values by analysis
3 calculated values; for amino acids calculated valus are given in brackets

Table 4. Composition of faeces of calves fed diets D1, D2 and D3 (means ± standard deviation).

Faeces composition	Diet D1 60% skim-milk powder	Diet D2 10% soluble wheat protein	Diet D3 20% soluble wheat protein
Dry matter (%)	12.3 ± 2.0	14.4 ± 2.2	13.7 ± 2.0
Ash (%)	2.4 ± 0.4	2.6 ± 0.4	2.5 ± 0.4
Crude protein (%)	5.1 ± 1.2	6.3 ± 1.1	6.1 ± 1.2
Fat (%)	3.1 ± 1.1	2.6 ± 0.8	2.3 ± 0.8
Iron (ppm)	212 ± 41	210 ± 41	174 ± 30

Experiment 2

The digestibility test was performed without technical problems. Health, drinking behaviour and faeces were normal for all three groups. There were no feed refusals. Disturbances in the digestive process were not noticed; the composition of the faeces is given in Table 4.

The digestibility coefficients of the various components of the experimental diets and of soluble wheat protein are given in Table 5 (Tolman, 1989a). The digestibility of crude protein of soluble wheat protein was found to be 96%, which is comparable with the protein digestibility of skim-milk powder. An interaction between the level of soluble wheat protein in the diet (10 and 20%) and the digestibility of the components was not observed. The high digestibility of the soluble wheat protein might be explained, at least partially, by the flocculating properties (De Laporte & Demeersman, this book). Moreover, inclusion of soluble wheat protein in the diets did not influence the digestibility of fat, contrary to what is observed sometimes with other vegetable proteins (Troccon & Toullec, 1989). There were indications that the absorption of iron in the soluble wheat protein was remarkably low. This was in accordance with the comparable course of the haemoglobin concentration of the blood in calves fed with 60% skim-milk powder (Diet D1) and calves fed with diets containing soluble wheat protein (Diets D2 and D3).

Experiment 3

During the experiment, one calf of both groups died, and one calf was excluded because of poor health (Diet G2). The faeces of the preruminant animals were normal during the whole experiment, which demonstrates that no serious digestibility problems occurred. Some animals of both groups had drinking problems after 15 weeks. After improvement of the ventilation system, the problems with feed intake decreased gradually. Thus, these feed refusals were not related to any of the diets.

The results for growth performances, slaughter quality and meat quality are summarized in Table 6. The growth performance of calves fed on a non-skim-milk diet containing 20% of soluble wheat protein was about 4% worse than that of calves fed on the Reference Diet containing 66% skim-milk powder for the period 6–26 weeks. In total, 13 and 9 calves fed on Diet G1 and G2, respectively had feed refusals. The number of calves that had feed refusals of more than 2% was 5 and 4, respectively. When these calves were excluded, the growth performance could be determined for the 2 test groups with 14 calves each. In this way, final weight for the animals fed with Diets G1 and G2 was 237.2 ± 9.7 kg and 231.8 ± 12.0 kg, respectively. The cumulative weight gain over the period 6–26 weeks was 1240 ± 70 /d and 1204 ± 91g/d, respectively. The cumulative feed conversion for the same period was 1.78 ± 0.1 kg/kg and 1.84 ± 0.15 kg/kg, respectively.

Meatiness, meat colour and fat covering were comparable for both groups. The decrease and the final concentration of haemoglobin was also comparable for the two diets. The relatively red colour for both groups can be explained by the supplementary iron added from week 14 to week 21.

Conclusion

On the basis of its functionality and high digestibility and based on the growth performance of calves fed diets with 20% soluble wheat protein, this protein is a non-milk protein that appears to be able to compete with skim-milk powder or caseinate. Thus, depending on the market situation, high quality proteins can be replaced by soluble wheat protein. Today, partial replacement of milk protein by soluble wheat protein is already in use.

Table 5. Apparent faecal digestibility coefficients of various components.

Faeces composition	Diet D1 60% skim-milk powder	Diet D2 10% soluble wheat protein	Diet D3 20% soluble wheat protein
Diets[1]			
dry matter	96.6 ± 0.7	96.8 ± 0.9	96.0 ± 1.0
organic matter	97.1 ± 0.6	97.2 ± 0.8	95.5 ± 0.9
crude protein	94.1 ± 1.0	95.3 ± 1.1	94.9 ± 1.6
fat	94.9 ± 1.8	96.5 ± 1.5	95.6 ± 1.8
carbohydrates	99.1 ± 0.3	98.7 ± 0.4	98.1 ± 0.2
iron	13.7 ±26.3	20.3 ± 7.3	3.7 ±16.7
Soluble wheat protein[2]			
dry matter	-	98.7 ± 5.1	93.5 ± 2.7
organic matter	-	98.1 ± 4.4	94.1 ± 2.3
crude protein	-	98.2 ± 2.0	95.7 ± 1.6

1 mean ± standard deviation
2 mean ± standard error of mean

Table 6. Results of comparative growth trial.

		Diet G1 66% skim-milk powder		Diet G2 20% soluble wheat protein	
Growth results					
number of calves		19		18	
initial age (weeks)		6		6	
final age (weeks)		26		26	
initial weight (kg)		64.4	± 2.8[1]	64.2	± 2.8
final weight (kg)		236.2	± 9.9	227.5	± 13.5
average weight gain (g/d)		1 235	± 71	1 175	± 98
feeding schedule (kg)		321.5		321.5	
actual feed intake (kg)		303.6	± 4.0	304.3	± 5.0
feed conversion (kg/kg)		1.77	± 0.10	1.88	± 0.15
haemoglobin content (mM)		5.2	± 0.7	5.3	± 0.8
Quality results					
dressing percentage (%)		61.0	± 2.6	59.9	± 1.8
meatiness (%)	R	10		0	
	O	74		67	
	P	16		33	
meat colour (%)	3	11		11	
	4	79		72	
	5	10		17	
fat covering (%)	1	5		0	
	2	26		61	
	3	69		39	

1 mean ± standard deviation

References

Tolman, G.H., 1988. Toepasbaarheid van oplosbaar tarwe-eiwit in kunstmelk voor vleeskalveren. Fase 1. Beoordeling van oplosbaar tarwe-eiwit bevattende kunstmelk. ILOB rapport I 88 P-3626, Wageningen, NL.

Tolman, G.H., 1989a. Applicability of Kalpro S in a milk-replacer for veal calves. Phase 2. Digestibility of Kalpro S. ILOB report No. I 89-3631a, Wageningen, NL.

Tolman, G.H., 1989b. Nutritional value of Kalpro S in milk replacer for veal calves. Phase 3. Comparative growth trial. ILOB report No. I 89-3670a, Wageningen, NL.

Troccon, J.L. & R. Toullec, 1989. Aliments d'allaitement pour veaux d'élévage. INRA Production Animale 2(2):117–128.

9. Nutrition: antinutritional factors

Nutritional significance of dietary soya antigens in milk substitutes for calves

J.W. Sissons * & H.E. Pedersen **

* *AFRC Institute for Grassland and Animal Production, Shinfield, Reading RG2 9AQ GB*[1]

** *Aarhus Oliefabrik A/S, Aarhus, DK*

Summary

Suitable sources of protein alternative to skim milk must be found to meet the present demands for veal production. Amongst various vegetable proteins, soya has attracted most attention. So far, however, attempts to use soya protein as a complete replacement for milk protein have not proved successful. Fundamental studies of digestive function in calves have indicated that the inferior nutritional value of soya flour to casein is linked with a gut hypersensitive reaction to antigenic globulins of the soya bean. Commercial treatments which refine the soya protein and destroy its immunological activity have proved beneficial for calf feeding. The purpose of this paper is to present evidence that calves are intolerant to unrefined soya products because they suffer from a gastro-intestinal allergic reaction to the protein. The paper will also discuss the role of systemic anti-soya antibodies in mediating the disorder.
Descriptors: calves; soya antigens; systemic IgG antibodies.

Introduction

Although skim milk is the choice protein for feeding to young calves, there are times when it is in short supply. For this reason, there is much interest in replacing a large part of the milk protein with other protein sources. Most attention has been given to the use of soya products. Compared with other non-milk sources of protein, soya beans as a commodity are usually in constant supply. Also, in recent times, the soya-processing industry has developed technology for concentrating the protein as well as reducing the harmful effects of several anti-nutritional factors. However published digestibility studies indicate that, despite fat extraction and toasting, it is not possible to replace large amounts of the milk protein with soya products without severe depression in animal performance (Table 1). Clearly factors other than heat-labile components must influence the nutritive value of soya for the calf.

Intestinal hypersensitivity

Studies of digestive processes in preruminant calves reported by Sissons and co-workers (see review Sissons, 1989) have shown that heated and defatted soya flour is capable of inducing an adverse intestinal reaction which results in abnormal passage of digesta, impairment of nitrogen absorption and watery diarrhoea. Further detailed studies revealed that the digestive disorders were linked with mucosal tissue damage, crypt hyperplasia and exudation of fibrin casts.

1 Present address: Protein Technologies International, Checkerboard Square, St. Louis, MO, 63164, US.

Table 1. Antigenicity and apparent digestibility of nitrogen by calves of milk substitutes containing soya meal, flour or concentrate. Digestibility data from Sissons (1989).

Soya product	Typical antigenicity (ELISA log2 titre)	Proportion of milk protein replaced	Age of calves (days)	Digestibility of total nitrogen of the diet
Full-fat unheated meal (42% protein)	15	0.43	15 – 19	0.53
Defatted and heated flour (52% protein)	15	0.35	21 – 26	0.76
		0.50	21 – 26	0.68
Ethanol-extracted and heated concentrate (65% protein)	3	0.70	21 – 28	0.82

These changes together with a concomitant rise in circulatory IgG antibodies specifically against the major globulins, glycinin and B-conglycinin, led to the idea that calves fed certain soya products may suffer from hypersensitivity or allergy to the soya protein.

Indirect evidence in support of the allergy theory is drawn from studies of digestibility and growth performance of calves fed on products that were treated to destroy the putative allergens. Treating soya meal with hot aqueous ethanol or a combination of heat and pressure (extrusion) alters the protein structure so that it is not recognized by rabbit antibodies raised against native protein. It is supposed that such products are 'antigen-free' and interestingly they were found to be better utilized than toasted soya flour by young calves (Stobo et al., 1983). Moreover such products did not induce digestive disorders or cause villous atrophy when given as challenge feeds to calves which had been orally sensitized to heated soya flour (Pedersen, 1986; Sissons, unpublished data).

Systemic immune response

In immuno-physiological studies of soya fed calves (Kilshaw & Sissons, 1979), a rise in systemic anti-soya antibody titre was found to correspond with a marked increase in digesta flow through the small intestine which was indicative of digestive malfunction. Barratt et al. (1978) attributed the mucosal injury seen in calves fed on soya to complex formation between circulatory IgG antibodies and soya antigens. However it remains to be demonstrated whether a systemic immune response to soya protein is directly responsible for the digestive malfunction.

In a recent study (Heppell & Sissons, unpublished) soya naive calves were passively sensitized with serum containing IgG anti-soya antibodies raised in other animals given a soya-flour-based diet.

An ELISA revealed that high titres of antibody had been transferred into the blood of the recipient calves, but abnormalities in gut motor function (measured electromyographically) and diarrhoea only developed after repeated oral challenge with feeds containing heated soya flour, whilst the serum level of soya specific IgG and IgG1 subclass did not change. Thus, it appears that this class of antibody is not directly responsible for the adverse intestinal reaction

Table 2. Effect of passive transfer of anti-soya IgG antibodies and oral challenge with heated soya flour (HSF) on the occurrence of migratory myoelectric complexes (MMCs) along the small intestine. Migratory myoelectric complexes are the basis of muscular contractions which propagate down the small intestine. Mean values are given for 5 calves.

Order of giving challenge feeds	Soya-specific antibody titre (ELISA log2)				Number of MMCs between 0-6 h after feeding	
	IgG		IgG1			
	mean	s.e.	mean	s.e.	mean	s.e.
Casein (before transfer)	6.0	0.4	2.8	0.5	5.0	0.5
HSF 1st feed	11.9	0.6	8.8	0.3	5.0	0.4
HSF 2nd feed					4.4	0.4
HSF 5th or more feeds	11.0	0.4	8.4	0.5	10.0	0.7
Casein					5.2	0.4

to soya (Table 2). But it cannot be assumed that the transferred antibodies had penetrated the mucosal tissue which could be a prerequisite for an adverse immunological reaction.

Although the direct involvement of IgG antibodies in gut dysfunction is uncertain, they are indicative of 'immunological stress' which is thought to disturb intermediary metabolism. Recently Klasing (1989) observed in chicks that injections of bacterial immunogens reduced the methionine and lysine requirements. He proposed that stimulation of the immune system may alter the partitioning of dietary nutrients away from tissue accretion in favour of metabolic processes needed to support the immune response.

The concept of 'immunological stress' raises the question as to whether the level of the IgG antibody response relates directly to growth performance. As mentioned earlier, treatments which reduce the antigenic activity of the native soya protein have been linked with a low systemic immune response and improvements in the nutritional value of the product for the calf. But, such benefits from use of nitrogen may not have been entirely due to an alteration in immunological activity. For example, treatment with aqueous ethanol may impart some advantage for digestibility. Furthermore failure to detect systemic antibodies against undenatured soya protein after feeding a soya product cannot be taken to mean that an immune response was absent.

The sensitivity and specificity of the assay will affect the antibody measurements. Also new antigens could arise from the treatment or through partial digestion in the gut.

Besides IgG antibodies, others of the IgE class have also been detected in the sera of some calves fed on heated soya flour (Kilshaw & Sissons, 1979). IgE antibodies are known passively to sensitize mucosal mast cells. Indeed feeding on soya flour has been found to increase substantially the proliferation of these cells (Pedersen, 1986). It is thought that subsequent reactions with antigens at the mast cell surface can trigger the release of intracellular granules. These granules contain substances which act as mediators in inflammation of the gut tissue. Some success has been achieved in preventing diarrhoea in soya sensitive calves through the use of an anti-inflammatory drug given shortly before oral challenge with heated soya flour (Duvaux & Sissons, unpublished). Whether such treatment can be of benefit in improving the digestion of soya protein is unknown.

In conclusion, antigenic components of the soya bean appear to limit the amount of milk protein that can be replaced by soya in cow's milk substitutes for calves. The severity of the gut disorder may be dose-dependent since substitution of less than about 20% of the milk protein is tolerated by the calf, whilst higher levels greatly increased the risk of diarrhoea and depressed growth. Further understanding of the role of systemic antibodies in mediating digestive malfunction and of the survival of antigenic structures in the gut is needed to improve the commercial processing of soya and to increase the value of laboratory assays for predicting the suitability of new soya products for calf feeding.

References

Barratt, M.E.J., P.J. Strachen & P. Porter, 1978. Antibody mechanisms implicated in digestive disturbances following ingestion of soya protein in calves and piglets. Clinical Experimental Immunology 31:305–312.

Klasing, K.C., 1989. Nutritional implications of an immune response. Proceedings of the 10th Western Nutritional Conference, September 1989, Saskatoon, Canada. p. 193–202.

Pedersen, H.E., 1986. Studies of soyabean protein intolerance in the preruminant calf. PhD Thesis, University of Reading, England.

Kilshaw, P.J. & J.W. Sissons, 1979. Gastrointestinal allergy to soyabean protein in preruminant calves. Antibody production and digestive disturbances in calves fed heated soyabean flour. Res. Vet. Sci. 27:361–365.

Sissons, J.W., 1989. Aetiology of diarrhoea in pigs and preruminant calves. In: W. Haresign; D.J.A. Cole (Ed): Recent advances in animal nutrition. Butterworths Scientific Press, London. p. 261–282.

Stobo, I.J.F., P. Ganderton & H. Conners, 1983. Effect of level of inclusion of antigen-free soya protein concentrate in a whey based milk substitute for calves. Animal Prod 38:557.

Soya antigens and anti-soya-antibody formation in calves

G.H. Tolman
ILOB-TNO, Institute for Animal Nutrition and Physiology, P.O. Box 15, 6700 AA Wageningen, NL

Summary

Development of serum anti-soya-IgG was studied in young veal calves fed on diets containing soya. Calves receiving dietary soya flour developed within 1 to 4 weeks high serum levels of anti-soya-IgG. Development of serum anti-soya-IgG did not differ between calves fed every day a diet containing soya flour for several weeks or calves fed at intervals of 7 days for two days the diet containing soya flour. IgG antibodies were measured by an ELISA. Microtitre plates were coated with protein from soya flour. Calves receiving dietary 'antigen-free' soya concentrate developed moderate serum levels of anti-soya IgG. Development of serum antibodies specific to dietary antigens is not necessarily correlated with disappointing performance in calves.
Descriptors: calves; soya; antigens; antibodies.

Introduction

Substances that induce the formation of antibodies after oral administration are called feed antigens. A feed antigen or dietary antigen can be defined as a macromolecule of dietary origin that, after absorption, is recognized by the immune system as foreign, and that stimulates the immune system to produce antibodies to eliminate the antigen. In general, this reaction of the immune system is a normal beneficial phenomenon and the antigen is innocuous (Roitt et al., 1985). Proteins and glycoproteins are potent antigens, polysaccharides are weak antigens whilst pure lipids never stimulate the immune system to produce antibodies (Perlman, 1980). Although dietary antigens generally are innocuous, much work has been published about deleterious effects of soya products in calves arising from intestinal hypersensitivity reactions provoked by dietary soya antigens (Sissons, 1989; Toullec & Guilloteau, 1989). It is, therefore important to determine the antigenicity of feedstuffs *in vitro* and to assess which products cause antibody formation. To determine the antigenicity of feedstuffs *in vitro*, it is preferable to use specific antibodies raised against the antigens in the species under investigation. Such antibodies can be raised in two ways: by the parenteral route or the oral route.

Theoretically, the advantage of oral immunization is that serum antibodies are produced with the particular property of combining with those antigens that pass the mucosal barrier, whilst parenteral immunization induces the production of antibodies raised against all antigens present in the feedstuff. Therefore, antibodies produced after oral administration of antigens might be more selective and useful for determining the antigenicity of feedstuffs than antibodies produced after parenteral immunization.

The aim of the present study was to test whether feeding regime influences systemic formation of anti-soya-IgG. An experiment was conducted to produce anti-soya IgG by feeding calves on a diet containing lightly toasted soya flour. Two feeding regimes were applied:

1. feeding on the soya-flour-based diet for several weeks
2. feeding alternately for 2 days on the soya-flour-based diet and for 7 days on a milk-protein-based diet.

A second experiment was conducted to examine the systemic immune response to a commercial 'antigen-free' soya concentrate (hot-aqueous-ethanol-extracted soya flour).

Material and methods

Experiment 1

The experiment was carried out with male veal calves (Friesian Dutch × Holstein Friesian). Six newly born calves were fed on a milk-protein-based control diet (23% protein) for 3 weeks. Thereafter the animals were divided into 2 groups of 3 calves. All calves were gradually switched (25% every day) from the milk protein diet to a soya-flour-based test diet with 30% lightly toasted defatted soya flour (PDI = 70; crude protein (N × 6.25) = 52%; SBTI = 21 mg/g; HA = 40). This diet contained about 66% of the protein as soya flour.

After the switch, calves of the one group (Group 1) remained on the test diet, while calves of the other group (Group 2) were given the control diet. Seven days later, calves of Group 2 were again gradually switched in two days from the control diet to the test diet. Subsequently, these calves were returned to the control diet for 7 days. This feeding protocol was repeated 5 times. At the end of the feeding programme, the calves were about 10 weeks old. Jugular blood was sampled from the calves on arrival, 3 weeks later just before introducing the test diet and weekly after introducing the test diet. Serum anti-soya IgG level was measured in duplicate in each sample by an ELISA procedure. Briefly, microtitre plates (Costar) were coated with protein extract of the soya flour at 3 g/l. Plates were blocked with an excess of 3% BSA to prevent any subsequent non-specific binding of proteins and were then incubated with 5 doubling dilutions of test sample starting at 1/75, followed by incubation with peroxydase-conjugated rabbit anti-bovine IgG. Peroxydase-substrate (o-phenylenediamine in presence of 0.01% hydrogen peroxide) was added and, after incubation, absorbance was measured at 492 nm. Antibody level was expressed as absorbanceu at a dilution within the linear range of the titration curves. In this experiment, antibody level was determined at 1/1200. Weight gain, rectal temperature after the evening meal and quality of the faeces (consistency, smell, colour) were observed throughout the experiment.

Experiment 2

Four newly born male veal calves (Friesian Dutch × Holstein Friesian) were on fed a milk-protein-based control diet (23% protein) for 1 week. Thereafter, calves were gradually switched from the control diet to a soya-concentrate-based test diet. The test diet contained 15% 'antigen free' soya concentrate (crude protein (N × 6.25) = 66%). About 43% of the protein in the test diet was derived from the soya concentrate. Blood samples for the determination of serum anti-soya-IgG level were taken on arrival of the calves, 1 week later just before introducing the test diet, and weekly thereafter. Microtitre plates were coated either with a protein extract of lightly heated soya flour (as in Experiment 1) or with a protein extract of the soya concentrate. Other parameters measured were weight gain, feed conversion and quality of the faeces (consistency, smell, colour).

Results

Experiment 1

Figure 1 shows the soya-specific serum IgG response. Before introduction of the test diet, the response was, as expected, very low for all animals. One week after introduction, the response of Calf 1 (Group 1) and Calf 5 (Group 2) was high. The response of Calf 4 and Calf 6 (both Group 2) increased during a period of 3 to 4 weeks after introduction of soya flour.

Two weeks after switching to the test diet, the condition of Calf 3 was poor and this calf was transferred from Group 1 to Group 2. The same happened to Calf 1 about 4 weeks after

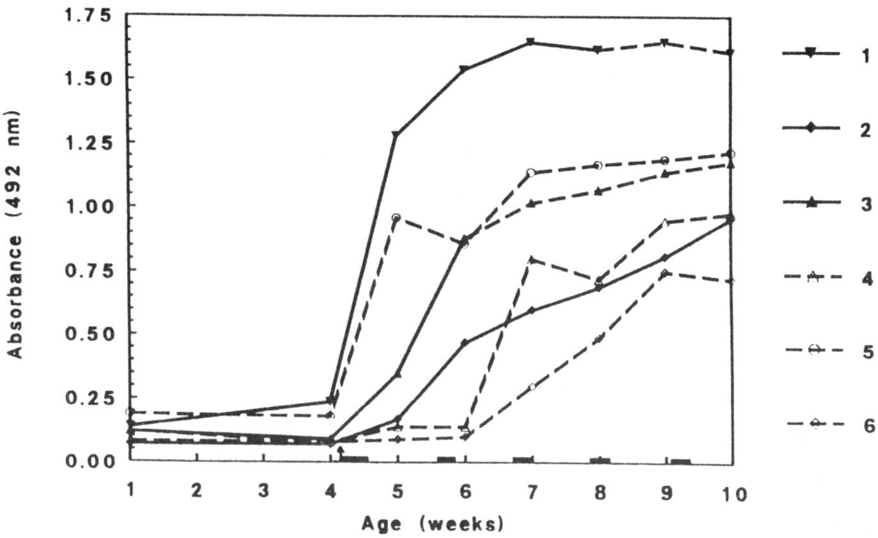

Figure 1. Development of anti-soya IgG in serum of preruminant calves fed on soya flour (serum dilution 1:1200). Arrow represents switch to soya flour. Blocks represent days of feeding calves of Group 2 (Calves 4, 5, 6) on the soya-flour diet. Calves of Group 1 (Calves 1, 2, 3) were fed on the soya-flour diet for all meals.

Table 1. Weight gain, rectal temperature and time with rectal temperature ≥ 39.5 °C or ≥ 40.0 °C, respectively), feed refusals (number) and time with diarrhoea.

Calf	Weight gain (g/d)	Rectal temperature (°C)		Time rectal temperature ≥ t (d)		Feed refusals	Diarrhoea (d)
		mean	sd	39.5 °C	40.0 °C		
1	293	39.3	0.3	12	1	-	-
2	372	39.2	0.3	8	0	13	2
3	419	39.5	0.5	18	8	17	2
4	744	39.3	0.6	14	6	3	-
5	781	39.0	0.2	0	0	7	1
6	870	39.1	0.2	2	0	-	-

switching to the test diet. Poor condition was seen as listlessness, almost zero weight gain and an elevated rectal temperature (only Calf 3). Nevertheless, both calves did not develop severe diarrhoea (Table 1). No distinct difference in anti-soya IgG response between calves of Group 1 and Group 2 was detected. Table 1 shows weight gain, rectal temperature and number of feed refusals of the calves as well as incidence of diarrhoea during the test period (4–10 weeks of age). Calves of Group 1 had a lower weight gain, somewhat higher rectal temperature and more feed refusals than calves of Group 2. The higher average rectal temperature of Calf 4

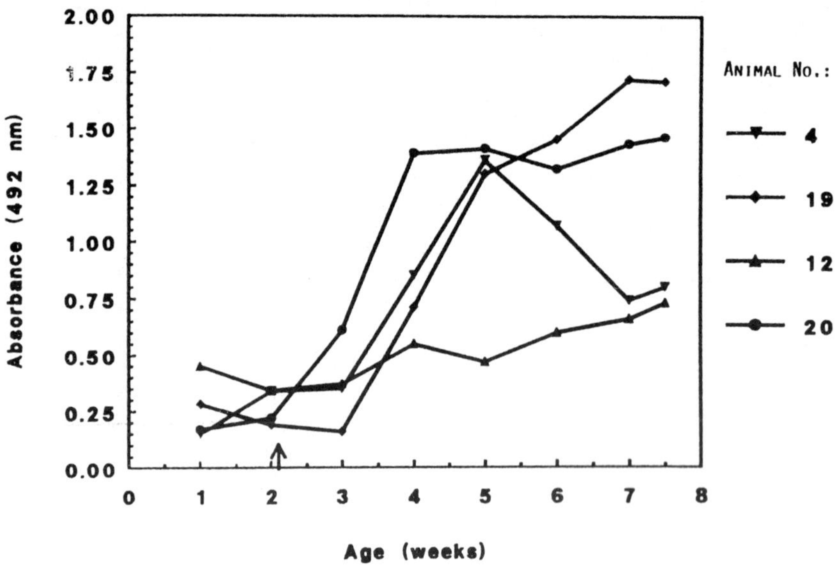

Figure 2. Development of anti-soya IgG in serum of preruminant calves fed on soya concentrate (serum dilution 1:300; microtitre plate coated with soya flour). Arrow represents switch to soya concentrate.

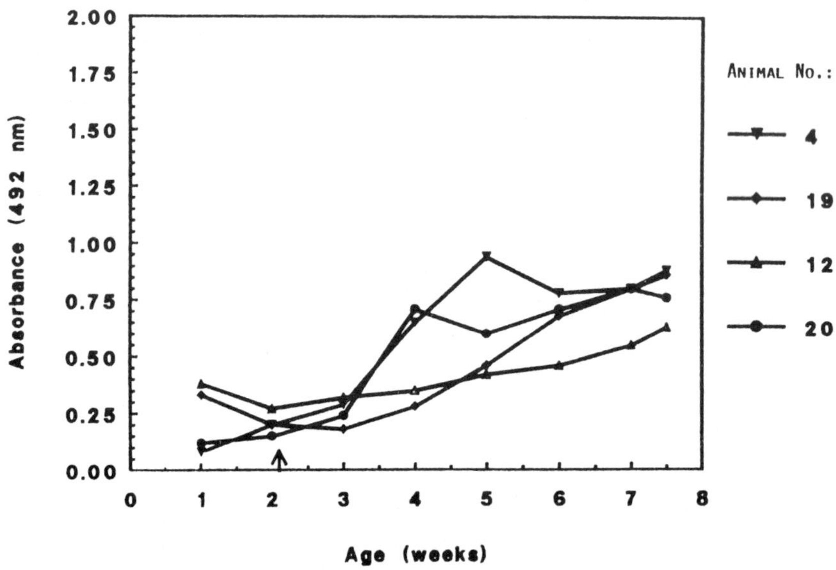

Figure 3. Development of anti-soya IgG in serum of preruminant calves fed on soya concentrate (serum dilution 1:150; microtitre plate coated with soya concentrate). Arrow represents switch to soya concentrate.

was mainly due to a period of 6 days in which rectal temperature was higher than 40.0 °C; in this period, the calf showed signs of pneumonia and was treated with an antibiotic. It was remarkable that Calf 1 never showed any feed refusal, in contrast to feed refusals of Calves 2 and 3, which were fed throughout on the test diet, and the feed refusals of Calves 4 en 5 when these calves were fed on the test diet. The calves fed on the test diet produced more and stiffer faeces than the other calves. The smell of the faeces of Calf 1 was unpleasant. The incidence of diarrhoea was very low and probably not associated with hypersensitivity.

Experiment 2

Figure 2 shows the soya-specific serum IgG response by ELISA where microtitre plates were coated with the protein extract of lightly heated soya flour. As in Experiment 1, the response was very low for all calves before introduction of the test diet.

Surprisingly, after introduction of the test diet, 3 out of 4 animals showed a clear increase in anti-soya IgG level. Only one (Calf 20) after feeding an 'antigen-free' concentrate, showed no clear increase. Calf 4 first showed an increase, but at an age of 5 weeks, the IgG response decreased. Although the average absorbance 4 weeks after introduction of the test diet was similar to that in Experiment 1, anti-soya IgG response was much lower because serum was diluted only 1:300 whereas in Experiment 1 serum was diluted 1:1200. Figure 3 shows the soya-specific serum IgG response according an ELISA where microtitre plates were coated with the protein extract of the 'antigen-free' soya concentrate. Hence, formation of anti-soya IgG is far less distinct, although serum was diluted only 1:150. The results of Experiment 2 indicated that feeding an 'antigen-free' soya concentrate can induce formation of anti-soya IgG and that these antibodies were detected more readily by coating microtitre plates with lightly heated soya flour than by coating with ethanol-treated soya concentrate.

Table 2 shows weight gain and feed conversion of the calves. Health of the animals was excellent and no diarrhoea occurred. Faeces seems to be a little stiffer than from milk-fed calves. A correlation between anti-soya IgG level and performance was apparently absent.

Table 2. Weight gain (g/d) and feed conversion (feed intake (kg) /weight gain (kg)) from 2 to 8 weeks of age.

Calf	Weight gain	Feed conversion
4	543	1.75
12	690	1.40
19	657	1.44
20	757	1.21

Discussion and conclusions

Experiment 1 shows that feeding on soya flour continuously for several weeks or alternately for 2 days followed by milk protein for 7 days, seems to have no effect on the development or level of anti-soya IgG response. A distinct effect of feeding regime on weight gain and to a lesser extent on rectal temperature and consistency and amount of faeces was observed. Calves fed continuously on the test diet showed a lower weight gain, produced more faeces than the other calves (indicating a lower digestibility of the test diet) and had slightly higher rectal temperatures, suggesting that the test diet may have induced some digestive malfunction possibly linked with inflammation of the gut mucosa (Sissons, 1989). Nevertheless, Experiment 1 showed that anti-soya IgG level was not necessarily correlated with health or performance. Experiment 2 showed that feeding an 'antigen-free' soya concentrate can induce formation anti-soya IgG. So the term 'antigen-free' is not appropriate, although the antigenicity of the product *in vitro* by ELISA was very low. The antibodies were detected more readily by coating microtitre plates with lightly heated soya flour (with a high antigenicity *in vitro*) than by coating with ethanol-treated soya concentrate, indicating that the sera contained immunoglobulins with greater avidity for soya flour than for proteins of the soya concentrate. This implies that only a few soya antigens are required to activate the immune system.

Ethanol denaturation of soya protein (i.e. unfolding of the protein molecule) seems insufficient to produce an antigen-free protein. Probably the only way to produce an 'antigen-free' soya protein is to hydrolyse the protein to low molecular peptides having no antigenic determinants.

Another possibility is to produce a protein that is highly digestible, so that in the digestive tract the macromolecules are hydrolysed quickly and antigenic determinants are broken down. However, it remains questionable how far a product needs to be 'antigen-free'. In agreement with Sissons & Pedersen (this book) and Toullec & Guilloteau (1989), the findings of the present experiments show that the development of systemic antibodies specific to dietary antigens is not necessarily correlated with performance in calves. It remains uncertain to what extent adverse reactions of the digestive system to soya protein are related with anti-soya IgG formation.

References

Perlman, F., 1980. Allergens. In: I.E. Liener (Ed.): Toxic constituents of plant foodstuffs. Academic Press, New York. p. 295–327.

Roitt, I.M., J. Brostoff & D.K. Male, 1987. Immunology. Churchill Livingstone, Edinburgh; Gower Medical Publising, London.

Sissons, J.W., 1989. Aetiology of diarrhoea in pigs and preruminant calves. In: W. Haresign & D.J.A. Cole (Ed.): Recent advances in animal nutrition, London, p. 261–282.

Toullec, R. & P. Guilloteau, 1989. Research into the digestive physiology of the milk-fed calf. In: E.J. van Weerden & J. Huisman (Ed.): Nutrition and Digestive Physiology in Monogastric Farm Animals. Pudoc, Wageningen, p. 37–55.

New developments in analysis of antinutritional factors

M.G. van Oort, R.J. Hamer & G.H. Tolman
TNO Cereals, Flour and Bread Institute, Wageningen, NL

Summary

Methods were developed for detection and quantification of antigenic soya proteins (by immunoblotting) and lectins (by FLIA). The antigenic soya proteins were selectively detected by antibodies obtained from calves orally immunized against soya proteins. In this way, antibodies were raised only against the antigenic proteins.

The functional lectin immunoassay (FLIA) selectively measured pathogenic lectins in feeds, since the lectins were detected by a combination of function and identity measurement.
Descriptors: antigenic proteins; allergenic proteins; soya proteins; lectins; lectin analysis; immunoblotting.

Introduction

Allergenic or antigenic proteins

Food hypersensitivity or allergy forms a serious problem in the feeding of young animals as calves and piglets. Allergy can be acquired by animals, although inheritance may also play a role. The mechanism of acquirement is not completely understood at this moment.

When young calves are fed with slightly toasted soya flour, they will form antibodies in a relatively short time against the antigenic soya proteins present in the soya flour.

A new method of analysis was developed to assay the antigenicity of soya products and to identify the antigenic proteins present in these products. In this test, use was made of the antisera from the orally immunized calves.

The identification of these antigenic soya proteins was possible through a combination of electrophoretic separation of the proteins followed by immunological identification.

Lectins

An important part in animal bioprocessing is the resorption of nutrients. For this purpose, the animal has its gut wall consiting of villous fingerlike structures. These structures provide a maximum resorptive surface. It is well known that lectins (which are sugar-binding proteins) can damage the gut wall resulting in less of these fingerlike structures and reducing resorptive capacity. Lectins are present in legumes and cereals. Detection of lectins is difficult with a traditional haemagglutination assay. In this assay, lectins must contain at least two binding sites for reliable detection. Only when two binding sites are present is agglutination of the red blood cells possible. Lectins with only one binding site will not agglutinate the red blood cells.

To overcome this problem, we developed a new lectin assay, the FLIA. This test detected lectins by functional carbohydrate binding followed by sensitive and specific immunological detection and quantification.

Material and methods

Electrophoresis

Electrophoresis was performed on a Biorad Mini Protean II Dual Slab Gel Electrophoresis apparatus, using 15% SDS polyacrylamide (PAA), 0.4% bisacrylamide (BAA) gels (10 cm × 0.8 cm, 0.8 mm thick) essentially according to Laemmli (1970). For separation of lower-mo-

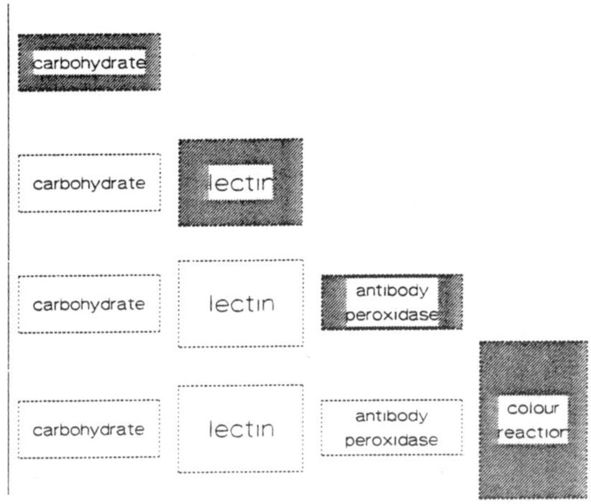

Figure 1. Schematic presentation of the FLIA principle.

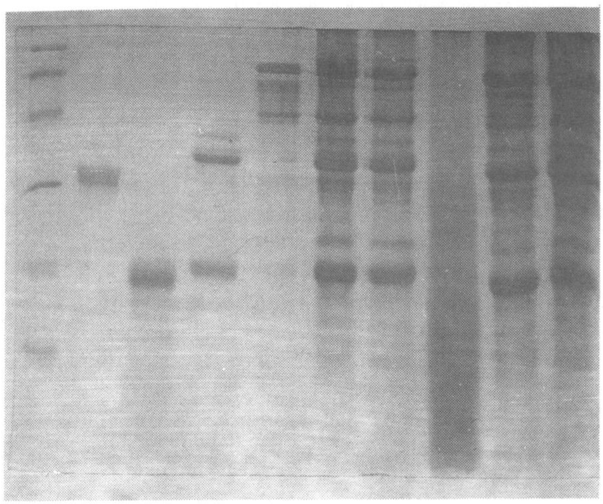

Figure 2. SDS-PAGE patterns of different soya proteins and soya products. From the left to the right: Lane 1, MW marker; Lane 2, SBA; Lane 3, KSTI; Lane 4, glycinin; Lane 5, β-conglycinin; Lane 6, soya concentrate; Lane 7, soya isolate; Lane 8, enzymatically treated soya flour; Lane 9, toasted soya flour; Lane 10, untreated soya flour.

lecular-weight protein bands (2000–20 000 D), similar 15% PAA and 0.4% BAA gels were used, with 15% urea added to the running gel solution before polymerization.

Blotting

The separated proteins were blotted onto nitrocellulose paper (Schleicher & Schull) by semi-dry electroblotting with a Pharmacia Multiphor II 2117-250 NOVA Blot system and a Pharmacia ECPS 2000/30 Power supply. The blotting was carried out at 0.8 mA/cm^2 for 1 h at room temperature.

Immunoblotting

The protein blots were washed and blocked with 5% casein solution for 2 h at 37 °C. The calf antiserum and the rabbit antiserum were diluted 50-fold and 2500-fold, respectively, in dilution buffer (1% bovine serum albumin (BSA), 0.1% Tween-20 in PBS). The blots were incubated in the antisera overnight at room temperature with constant agitation. After washing, the rabbit antiserum blots were incubated with a second antibody, i.e. goat-anti rabbit IgG antibodies conjugated to horse-radish peroxidase and the calf antiserum immunoblots were incubated with rabbit-anti-bovine IgG antibodies conjugated to peroxidase. Finally the blots were stained with 0.6 ml of diaminobenzene (DAB) and 4 ml of chloronaphtol in 44 ml of PBS containing 20 µl of hydrogenperoxide.

FLIA

The assay was carried out on 96-well microtitreplates essentially according to Hendriks et al. (1987). The principle of the assay is schematically shown in Figure 1. The first step was the binding of a carbohydrate matrix to a solid support. The second step was the incubation of samples containing lectin on this coating. Only the functionally active lectins can bind to the carbohydrates. The third step was the detection of the bound lectins by the antibodies (coupled to a peroxidase molecule) and finally the fourth step was the quantification of the bound antibodies by a colour-forming reaction.

In contrast to ELISA where the amount of total lectin is measured, the FLIA measured only the functional lectins, since the first step can only be accomplished by functionally intact lectins. As a carbohydrate matrix in this assay, a preparation of the small intestine of pigs was used as coating. With this gut-wall preparation, the so-called brush-border membranes (BBM), we could mimic the actual situation in the animal.

Results and discussion

By the electrophoretic technique, sodium dodecyl sulphate and polyacrylamide gelelectrophoreses (SDS-PAGE), the soya proteins were separated according to their molecular weight. The result of such a separation is shown in Figure 2.

This technique shows all proteins present in soya flour. The intensity of the bands gives an indication of the abundance of each protein, but no discrimination is made between the antigenicity or harmfulness of the proteins.

In such electrophoresi, the proteins are more or less locked in an acrylamide matrix. In this matrix, the proteins cannot be reached by antibodies. So for immunochemical detection, the proteins have to be removed from the inside of the gel to the surface of a nitrocellulose membrane. This transfer is called 'blotting'. After the blotting, the proteins on the surface of the membrane can be identified with antibodies. The total of transfer followed by immunochemical detection is called 'immunoblotting' or 'Western blotting'.

For immunoblotting detection, use was made of antibodies from the serum of the orally immunized calves. This results in the following pattern (Figure 3). A very important fact in this figure is that the intensity of the bands is an indication of the antigenicity and not of the abundance of the proteins. The bands that can be seen are proteins recognized by the antibodies in the calf serum and thus are proteins which are antigenic. It is remarkable that soya lectin, SBTI and the glycinin proteins were not or were hardly detected and were therefore not or were hardly antigenic. The β-conglycinin proteins, on the other hand, were strongly antigenic. This can be seen by the intense bands on the immunoblot.

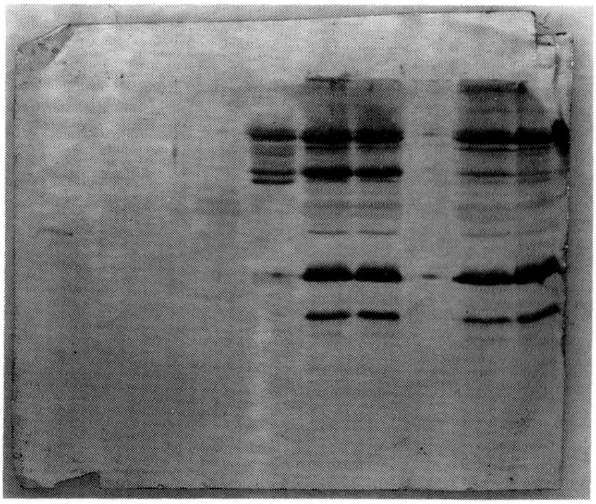

Figure 3. Immunoblot. Detection and staining with calf antibodies. Same soya proteins and soya products as shown in Figure 2.

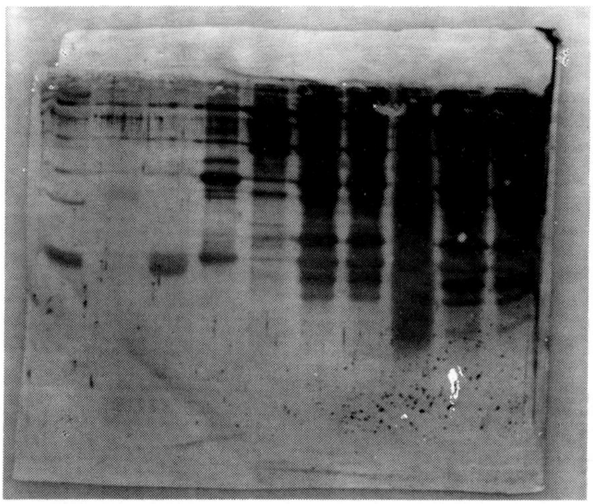

Figure 4. Immunoblot. Detection and staining with rabbit antibodies. Same soya proteins and soya products as shown in Figure 2.

The soya concentrate and the soya isolate both contained a number of antigenic proteins. These proteins were largely β-conglycinin proteins. Also some vague bands from the glycinin proteins can be seen. A second remarkable fact is the presence in the soya products of very intense protein bands in the range of MW of 15–20 000 D. Those bands were hardly visible on the SDS gel but were strongly fortified on the immunoblot. This means that these proteins are strongly antigenic. In the enzymically treated soya product, no antigenic bands were visible. So that treatment seems successful in decreasing the amount of antigenic proteins. The toasted and untreated soya flour look almost the same on this immunoblot. Difference in solubility of the proteins cannot be detected with this technique.

It is of course also possible to raise antibodies against soya proteins in rabbits. In fact, this is the usual way of obtaining antibodies. However in this way, antibodies will be raised against all proteins injected into the rabbits and not only against the antigenic proteins. Use of rabbit antibodies results in Figure 4.

With these antibodies, a very intense response is obtained. Although the rabbit antibodies were used in a 500 times more diluted form than the calf antibodies, the response was much better. This indicates that the oral immunization of calves is a less effective way for production of antibodies.

A second remarkable fact is that all proteins visible on the SDS gel are also visible on this immunoblot. The intensity of the protein bands was again an indication of abundance and not of the antigenicity. SBA, SBTI and the glycinin proteins were also detected. Furthermore, the very intensely antigenic proteins between 15 and 20 000 D seen on the calf immunoblot were hardly visible. The conclusion is that rabbit antibodies are very useful for the sensitive detection of soya proteins, but for detection of antigenic proteins antibodies from orally sensitized calves have to be used.

Quantitative analysis of antigenic soya proteins is also possible by a sandwich-type ELISA procedure (results not shown). In this procedure too, the calf antibodies can be used. The advantage of ELISA was that standardization was possible with a reference sample. A disadvantage was that only water-soluble proteins can be quantified. In combination with the immunoblotting method, ELISA proved very useful.

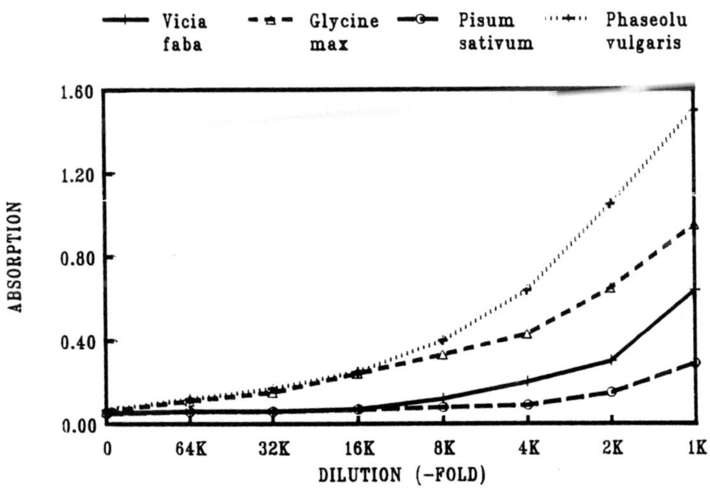

Figure 5. Response to FLIA-BBM of four different legume lectins.

Lectins

The result of FLIA for lectins from different legumes is shown schematically in Figure 5. It appeared that *P. vulgaris* lectins were sensitively detected. Also soya-bean lectin was sensitively detected. *Vicia faba* lectins were also detected, but sensitivity was much lower. The *Pisum sativum* lectin finally, hardly bound to these brush-border membranes. Obviously, these lectins had only very low affinity for the carbohydrate matrix of the gut wall.

Work is now in progress to develop these kind of FLIA with synthetic carbohydrate coatings. Although synthetic coatings are possibly less interesting than the BBM, they can very well be used for optimization of FLIA and for turning the principle of FLIA into a commercially available test kit.

References

Hendriks, H.G.C.J.M., J.F.J.G. Koninkx, M. Draaijer, J.E. van Dijk, J.A.M. Raaijmaker & J.M.V.M. Mouwen, 1987. Quantitative determination of the lectin binding capacity of small intestinal brush-border membrane. An enzyme linked lectin sorbent assay (ELLSA). Biochimica & Biophysica Acta 905:371–375.

Intolerance to pea protein in the preruminant calf

R.S. Bush[1], R. Toullec, I. Caugant & P. Guilloteau
Laboratoire du Jeune Ruminant, I.N.R.A., 65 rue de Saint-Brieuc, 35042 Rennes Cédex, FR

Summary

Five 2-month-old preruminant calves were fitted with a re-entrant ileo-caecal cannula. Two milk-substitute diets whose protein was provided either exclusively by skim-milk powder or partially (34%) by a dehulled raw pea flour, were given to the calves for 1 and 4 weeks, respectively. During the 1st week, the apparent ileal digestibility was lower with the pea diet than with the control diet (means ± SEM: 0.89 ± 0.01 instead of 0.93 ± 0.01 for total N; $P \leq 0.01$). Four of the calves exhibited intolerance to the pea diet, resulting in much lower digestibility during the 4th week (0.78 ± 0.05 for total N). The calves developed high titres of systemic antibodies against the main two globulins of pea (legumin and vicilin). Legumin was found to survive digestion in the small intestine, in amounts generally equivalent to 1–3% of intake.
Descriptors: veal calf; pea flour; intolerance; ileal digestibility; antibodies; legumin.

Introduction

Replacing milk protein by heated soya bean meal in the milk-substitute diet given to the preruminant calf has been shown to cause digestive disturbances including diarrhoea, decreased nutrient utilization and alteration of intestinal mucosa, resulting in increased absorption of macromolecules (review by Sissons, 1982). In sensitized calves, the high concentration of anti-soya antibodies in blood plasma and the rapid alteration of gut motility induced by a test meal of a diet containing heated soya bean meal suggest the disturbances to be of allergenic origin. Treatments denaturing the major globular storage proteins of soya bean meal (glycinin and β-conglycinin) suppressed the development of anti-soya antibodies and digestive disturbances. Calves given a diet containing raw pea flour were also found to develop antibodies against the main globulins of pea (legumin and vicilin) (Nunes do Prado et al., 1988). The present study was conducted to measure the changes with time in the ileal digestibility of a diet containing raw pea flour and the survival of legumin at the end of the ileum.

Material and methods

Five 2-month-old Holstein calves were fitted with a catheter in the abomasum and a re-entrant cannula which had the proximal part inserted in the distal ileum and the distal part in the caecum. Two experimental milk-substitute diets, whose protein was provided either exclusively by skim-milk powder (control diet) or partially (34%) by a dehulled raw pea flour of commercial cultivar Amino (pea diet), were compared. The content of skim-milk powder was 686 and 428 g/kg in the control and the pea diets, respectively, while the content in the powder of pea flour was 354 g/kg in the pea diet. The control diet contained in dry matter crude protein 255 and fat 213 and the pea diet crude protein 254 and fat 186 g/kg. The diets

1 Present address: Agriculture Canada, Research Station, P.O. Box 20280, Fredericton, N.B., Canada, E3B 4Z7.

were infused into the abomasum twice daily (at 08:30 and 16:30 hours), at a total DM rate of 58 g $kg^{-0.75}$ d^{-1}.

The calves were given the control diet for 1 week and the pea diet for 4 consecutive weeks. Three of them started with the control diet and 2 with the pea diet. The ileal digestibility was measured during the 1st week of distribution with both diets and also during the 4th week with the pea diet. There was a 1-week recovery period on the control diet between concurrent collections on the control and pea diets. The shifts from one diet to another were made during the 1st four meals of the week, starting on Friday evening. The re-entrant cannula was opened for the weekend and the ileal digesta were totally collected for 4 d beginning on Monday morning. The collection flasks contained 240 ml 1M NaOH (2 calves) or 60–100 ml 0.2 M $HgCl_2$ (3 calves) to limit microbial metabolism. The flasks were placed on a magnetic stirrer

Table 1. Apparent digestibility at the end of the ileum (means ± SEM); AAN as total nitrogen of assayed amino acids; values followed by unlike letters differed significantly ($P \leq 0.05$).

Component	Control diet	Pea diet	
	week 1	week 1	Week 4
Total nitrogen	0.93 ± 0.01^a	0.89 ± 0.01^b	0.78 ± 0.05
AAN	0.94 ± 0.01^a	0.91 ± 0.01^b	0.82 ± 0.05
Fat	0.96 ± 0.01^a	0.90 ± 0.01^b	0.70 ± 0.11
Nitrogen free extract	0.94 ± 0.01^a	0.76 ± 0.01^b	0.71 ± 0.02
Threonine	0.90 ± 0.01	0.86 ± 0.01	0.76 ± 0.04
Valine	0.95 ± 0.01^a	0.91 ± 0.01^b	0.82 ± 0.05
Cystine	0.83 ± 0.02^a	0.76 ± 0.03^b	0.58 ± 0.06
Methionine	0.97 ± 0.00^a	0.95 ± 0.00^b	0.89 ± 0.03
Isoleucine	0.95 ± 0.00^a	0.90 ± 0.01^b	0.81 ± 0.05
Leucine	0.96 ± 0.00^a	0.93 ± 0.01^b	0.86 ± 0.03
Tyrosine	0.96 ± 0.00^a	0.94 ± 0.01^b	0.90 ± 0.02
Phenylalanine	0.96 ± 0.01^a	0.92 ± 0.01^b	0.85 ± 0.03
Lysine	0.95 ± 0.01^a	0.93 ± 0.01^b	0.87 ± 0.03
Histidine	0.95 ± 0.01^a	0.92 ± 0.01^b	0.83 ± 0.04
Arginine	0.95 ± 0.01	0.94 ± 0.01	0.91 ± 0.01

Table 2. Amino acid composition of the diets and ileal digesta (g/100 g assayed amino acids); SAA as sum of assayed amino acids (g/16 g nitrogen); values followed by unlike letters differed significantly ($P \leq 0.05$).

Amino acid	Pea flour	Control diet	Pea diet	Digesta control	Digesta pea, week 1	Digesta pea, week 4
Threonine	3.5	4.1	4.0	7.6 ± 0.5^a	6.2 ± 0.3^b	5.9 ± 0.5^b
Serine	5.1	5.2	5.2	6.2 ± 0.4^a	5.6 ± 0.1	4.8 ± 0.5^b
Proline	4.5	9.5	7.7	5.8 ± 0.4	5.9 ± 0.1	6.8 ± 0.8
Glycine	4.3	1.9	2.7	4.7 ± 0.1^a	5.5 ± 0.2^b	6.3 ± 0.5
Arginine	8.8	3.3	5.2	3.2 ± 0.3	3.6 ± 0.2	2.9 ± 0.4
SAA	99.5	102.7	101.0	77.8 ± 1.2	85.0 ± 5.2	82.1 ± 4.7

and changed every morning. Four of the calves were used for an extra collection day after they had been given the pea diets for 3, 4, 1 and 1 week(s), respectively. Digesta were collected hourly for 10 h after the morning meal, in vessels containing 0.2 M $HgCl_2$ and 0.09 M phenyl-methyl-sulphonyl fluoride (protease inhibitor), in amounts corresponding to 2 and 1%, respectively, of the expected volumes of digesta. Plasma samples were obtained from a jugular vein on days 0, 7 and 28 of the pea-feeding period.

Legumin was extracted with borate buffer (0.15 M NaCl, 0.1 M $NaBO_3$, pH 8) from the pea flour and the digesta, except that collected in NaOH. Extracts of samples collected during the 4th and the 8th hour of the extra collection day were separated by gel filtration through a Sephadex G-200 column calibrated with pure legumin, vicilin, bovine serum albumin and ovalbumin (molecular weight 330, 130, 68 and 42.5 kDa, respectively). The concentration of legumin in the extracts and the separated fractions, as well as the antilegumin and anti-vicilin antibody titres in the plasmas were measured by the ELISA methods described by Nunes do Prado et al. (1989a). Samples of the experimental diets, pea flour and ileal digesta collected for 4 d were analysed for amino acids (AA) by ion-exchange-resin chromatography as previously reported (Guilloteau et al., 1986). Statistical significance was assessed by paired *t* test analysis since the values during the 4th week on the pea diet were directly related to the results of the 1st week. The AA compositions of digesta and diets were compared by the distance of χ^2 (Guilloteau et al., 1986).

Results and discussion

During the 1st week, the apparent digestibility was lower ($P \leq 0.05$) with the pea diet than with the control diet for total nitrogen, amino nitrogen of assayed AA, fat and nitrogen-free extract (Table 1). Also, the apparent digestibility was lower for all the essential AA, except threonine and arginine. Four of the calves exhibited intolerance to the pea diet, resulting in much lower digestibility during the 4th week. However because of large individual variations, changes were not significant. Assuming that the apparent digestibility of milk nitrogen was the same with the two diets (0.93), the value calculated for pea nitrogen was found to be 0.83 and 0.52 during the 1st and the 4th week, respectively. Nunes do Prado et al. (1989b) measured the ileal digestibility of a diet similar to the pea diet used here, except that the pea flour had been pregelatinized. Although digesta were collected during the 3rd week, digestibility was higher than during the 4th and even the 1st week of the present experiment (0.92 instead of 0.78 and 0.89 for total nitrogen; 0.80 instead of 0.71 and 0.76 for nitrogen-free extract). Pregelatinization probably facilitated starch digestion. It also denatured the protein since the concentration of immunoreactive legumin was only 1% of total protein in the treated pea instead of 21% in the raw pea.

With either diet, the AA composition of digesta protein was very different from that of dietary protein (Table 2), which resulted in large distances of χ^2: 188 between digesta and milk protein with the control diet, 178 and 254 between digesta and pea protein with the pea diet during the 1st and the 4th week, respectively. The results obtained with the control diet were in agreement with the observations of Guilloteau et al. (1986) showing that in the milk-fed calf, protein of ileal digesta were exclusively of endogenous and bacterial origin. With the pea diet during the 1st week, digesta protein contained less threonine and more glycine than with the control diet. However the differences were small since the distance of χ^2 was only 23. Similarly, the average differences between the 1st and the 4th week of pea feeding were slight ($\chi^2 = 19$). So whole pea protein did not appear to constitute a large proportion of the digesta protein in either period.

In the digesta collected over 4 d, immunoreactive legumin was found in amounts equivalent to 0.8 ± 0.3 and 1.2 ± 0.6% of intake during the 1st and 4th week, respectively. The corresponding values for the extra collection day were not very different for 3 of the calves

(0.9, 2.1 and 3.0%, respectively) but were much higher for the 4th calf (26 %). Legumin started to appear in digesta during the 3rd hour after feeding. It continued to be present until the end of the collection, but the flow rate was always very low during the 10th hour. A small part of this undigested legumin was apparently intact (330 kDa), but most of it was partially degraded (mainly fractions of about 160, 90 and 50 kDa). Therefore, some of the legumin escaped with only partial digestion in the small intestine, like glycinin in calves given a diet containing heated soya meal (Sissons & Thurston, 1984). Assuming that the proportion of total undigested pea protein was similar to that of the undigested legumin (about 2% of intake, according to the values recorded in the first 3 calves used for the extra collection day), pea protein can be calculated to constitute 7 and 4% of digesta protein during the 1st and the 4th weeks, respectively. These estimates are in agreement with the low variations observed in the AA composition. The lower and decreasing apparent digestibility of the pea diet appeared to be mainly due to an increase in losses of endogenous and bacterial protein; these losses could be a factor 1.6 and 3.1 as high in the 1st and 4th week with the pea diet as with the control diet.

The calves developed high systemic antibody titres against legumin and vicilin (0.9 ± 0.4 and 0.4 ± 0.1 initially, 5.4 ± 0.9 and 5.6 ± 1.5 on day 7 of pea feeding, 8.3 ± 0.4 and 9.4 ± 0.6 on day 28; $P \leq 0.5$ between the initial and the other values). A similar pattern was observed by Nunes do Prado et al. (1988) in calves given a diet containing raw pea flour. Also, these authors showed the presence of immunoreactive legumin in blood plasma. These phenomena could have been favoured by the long contact between immunoreactive proteins and the gut wall.

In conclusion, the development of intolerance to raw pea flour could have resulted from a gastro-intestinal allergic reaction. However, adverse effects of lectins or of micro-organisms present in the product cannot be excluded.

References

Guilloteau, P., R. Toullec, J.F. Grongnet, P. Patureau-Mirand, J. Prugnaud & D. Sauvant, 1986. Digestion of milk, fish and soya-bean protein in the preruminant calf: flow of digesta, apparent digestibility at the end of the ileum and amino acid composition of ileal digesta. British Journal of Nutrition 55:571–592.

Nunes do Prado, I., R. Toullec, J.P. Lalles, L. Hingand & J. Guéguen, 1988. Anticorps contre les protéines alimentaires et perméabilité intestinale aux macromolécules chez le veau préruminant recevant de la farine de pois. Reproduction, Nutrition et Développement 28 (supplément 1):157–158.

Nunes do Prado, I., R. Toullec, J.P. Lalles, J. Guéguen, J. Hingand & P. Guilloteau, 1989a. Digestion des protéines de pois et de soja chez le veau préruminant. I. Taux circulants de nutriments, formation d'anticorps et perméabilité intestinale aux macromolécules. Reproduction, Nutrition and Development 29:413–424.

Nunes do Prado, I., R. Toullec, P. Guilloteau & J. Guéguen, 1989b. Digestion des protéines de pois et de soja chez le veau préruminant. II. Digestibilité apparente à la fin de l'iléon et du tube digestif. Reproduction, Nutrition and Development 29:425–439.

Sissons, J.W., 1982. Effects of soyabean products on digestive processes in the gastrointestinal tract of preruminant calves. Proceedings of the Nutrition Society 41:53–61.

Sissons, J.W. & S.M. Thurston, 1984. Survival of dietary antigens in the digestive tract of calves intolerant to soyabean products. Research in Veterinary Science 37:242–246.

Screening test for trypsin inhibitor

J. Mol & J. Meijer
Denkavit Nederland B.V., P.O. Box 5, 3780 BA Voorthuizen, NL

Summary

Skim milk powder has been the main source of protein in milk replacer diets for veal calves during the past 35 years. Due to the increase in price and the decreased availability of skim milk powder, alternatively other sources of protein get more attention and application. This includes mainly proteins from vegetable origin.

Within the availability in the world market, soya products can be considered as a potential source of protein. By nature soya products contain anti-nutritional factors, in particular trypsin inhibitor, which activity can be reduced by a proper heat treatment.

The international method to determine the activity of trypsin inhibitor is known under the name Kakade method or adaptations of this method. This analytical procedure is time consuming and not adequate for process monitoring nor for a routine check for the milk-replacer industry. Moreover the applied pH is irrelevant.

The proposed method reduces the time for analysis to about 30 minutes and the applied conditions are more in harmony with the physiological parameters of the veal calf.
Descriptors: trypsin inhibitor; processed soya; veal calf.

Introduction

Milk replacers for fattening white veal calves were introduced in Holland in 1955. The product developed rapidly in the EC to about 1.8 million tons per annum. From the beginning skim-milk powder has been the main source of protein in milk-replacer diets for veal calves. Throughout the years there has been a regular increase in price for skim-milk powder, related to the Intervention system and the feed aid for skim-milk powder incorporated into milk replacers. During the past years the availability of skim-milk powder has changed dramatically from a surplus to a short position, mainly as a result of the EC policy.

Although researchers have been looking for replacement of skim-milk powder by other protein sources during the last 20 years, skim-milk powder retained its position in most of the member states of the EC. In the Netherlands the situation is different and due to its leading position in this particular market, the application of alternative proteins is of growing interest. This includes mainly proteins from vegetable origin.

Within the availability in the world market, soya products can be considered as a potential source of protein. By nature soya products contain anti-nutritional factors, in particular trypsin inhibitor, of which the activity can be reduced by a proper heat treatment.

The aim of this paper is to present a method for determining trypsin inhibitor activity to monitor process technology and to assure a proper quality, fitted for incorporation into milk-replacers for veal calves.

Soya bean trypsin inhibitor (SBTI)

Kunitz (1947) first described the isolation of SBTI. It is a protein of the globulin type and a part of the 2 S fraction with a molecular weight of 21 500 and showing two disulphides bridges. The Kunitz inhibitor acts on trypsin and to a lesser extent on chymotrypsin. Tan-Wilson et al. (1985) described at least 10 inhibitors occurring in raw soya beans. The Bowman-Birk inhibitor contains 7 disulphide bridges and as a result this polypeptide is more heat resistant. It acts on both trypsin and chymotrypsin. Rackis et al. (1985) reported levels of 3–4% trypsin inhibitors in raw soya beans.

Trypsin inhibitors loose most of their activity by denaturation under heat treatment, where the key factors are temperature, time and humidity. One should eliminate overheating in order to prevent a loss in the availability of certain amino acids.

In the past, urease activity was mainly applied to control the heat treatment of soya products. From our own investigations, it was found that urease activity is not a good guide to monitor process conditions.

Method

Kunitz (1947) first applied casein as a substrate to analyse the STBI activity. Kakade et al. (1969), introduced a synthetic substrate *N*α-benzoyl-DL-arginine-4-nitroanilide (BAPA) and achieved better results.

Collaborative studies have resulted into an international method (A.O.C.S. Ba 12-75). The disadvantages of this procedure are:

1. High pH (9.5) at extraction. Such a high pH does not occur in the digestive tract, so the obtained figure does not provide physiological implications.
2. Time consuming, which makes it unsuitable for practical application.

A method was developed to arrive at a more practical procedure.

Screening test for trypsin inhibitor activity in soya products

1. *Definition:* This method determines the residual trypsin inhibitor in soya products under the conditions of the test.

1.1. Scope: Applicable to processed soya products.

2. *Principle:* The substrate *N*α-Benzoyl-DL-arginine-4-nitroanilide is digested by a defined amount of trypsin. Under similar conditions an extract of soya is added. From the rate of change in δE. The inhibition is calculated and expressed as mg of trypsin inhibited by 1 kg of soya product.

3. *Equipment.*

3.1. Spectrophotometer, fitted with a cell holder, thermostatted at 25 °C, with 1 cm cuvettes.

3.2. pH meter.

3.3. Beakers 100 ml.

3.4. Graduated cylinder 25 ml.

3.5. Magnetic stirrer, stirring rod, glas.

3.6. Graduated pipette 10 ml.

3.7. Filters, Watman no 5 or 6.

3.8. Flasks, volumetric, 50, 100, 500, 1000 ml.

3.9. Water bath at 25 °C.

3.10. Pipettes 0.1, 0.2, 1.0 and 2.0 ml.

3.11. Capillator.

4. *Reagents.*

4.1. Hydrochloric Acid: HCl 0.5 mol/l.

4.2. Hydrochloric Acid: HCl 0.001 mol/l.

4.3. Sodium hydroxide: NaOH 2,0 mol/l.

4.4. Triethanolamine-HCl p.a. (TRA), Merck 8357.

TRA buffer solution: Dissolve 18.6 g Triethanolamine-HCl into 450 ml deionized water. Add 3.68 g $CaCl_2.2H_2O$ and 9.31 g EDTA. Dissolve by mixing. Adjust to pH 7.8 with

NaOH 2.0 mol/l. Dilute to 500 ml with deionized water and mix. (Store in the refrigerator at 4 °C.)

4.5. Calcium chloride. Merck 2382.

4.6. EDTA, Merck 8418.

4.7. Trypsin, 10 000 U/mg (substrate BAEE), Merck 24581.
Trypsin solution: Dissolve 7.5 mg trypsin into 100 ml hydrochloric acid 0.001 mol/l and mix.
This solution contains 15 μg trypsin per 0.2 ml. The trypsin solution should be kept in the refrigerator at 4 °C (will last 3–4 weeks).

4.8. *N*α-benzoyl-DL-arginine-4-nitroanilide-hydrochloride (BAPA), Fluka 12920.
BAPA solution: Weigh 100 mg BAPA into a flask of 100 ml. Add 80 ml hot (80 °C) deionized water. Heat in a water bath at 80 °C while mixing until complete dissolution. Temperature should not arise above 90 °C, otherwise the BAPA decomposes. After dissolution is completed, cool down and fill up with deionized water and mix. (Store at room temperature.)

5. *Sample preparation.*
Most of the commercial available products show a particle size of 75 μm. Products exceeding 150 μm must be ground under mild conditions.

6. *Procedure.*

6.1. Weigh 2.50–5.00 g of sample into a 100 ml beaker, add 25 ml (V_1) deionized water, suspend manually, adjust the pH to 6.5 (V_2) and agitate on a magnetic stirrer for 5 minutes at room temperature. Add dropwise 0.5 mol/l HCl (V_3) to pH 3.4 and so much 0.001 mol/l HCl (V_4) to arrive at a total volume ($V_1+V_2+V_3+V_4$) of 40 ml. Let it separate for 3 minutes and filter through Watman 5/6 to obtain a clear filtrate.

6.2. Condition a required volume of the extract, the TRA buffer, the BAPA solution and the trypsin solution in a waterbatch at 25 °C.
Condition the spectrophotometer, cell holder at 25 °C.

6.3. Add the conditioned cuvette 0.2 ml trypsin solution, 1.65 ml TRA buffer and 0.15 ml extract. Condition 7 minutes, add 1.0 ml BAPA solution, mix and read $\delta\, E/_{5\ \mathrm{min}}$ from minute 2 till minute 7.

6.4. Execute a blank $\delta\, E/_{5\ \mathrm{min}}$ using 0.15 ml deionized water. The blank should not drop below 0.225, otherwise a fresh trypsin solution must be prepared.

6.5. Calculation:

$$\frac{\delta\, E/5 \text{ min blank} - \delta\, E/5 \text{ min sample}}{\delta\, E/5 \text{ min blank}} \times \frac{F}{W} =$$

μg/g or mg/kg trypsin inhibited by 1 kg soya product.

$$F = \frac{40}{0.15} \times 15$$

40 = total volume of extract in ml
0.15 = ml in test
15 = g trypsin in test
W = sample in g

References

Kakade, M.L., N. Simons & I.E. Liener, 1969. An evaluation of natural versus synthetic substrates for measuring the antitryptic activity of soybean samples. Cereal Chemistry 46:518–526.

Kakade, M.L., J.J. Rackis, J.E. McGhee & G. Puski, 1974. Determination of trypsin inhibitor activity of soy products: a collaborative analysis of an improved procedure. Cereal Chemistry 51:376–382.

Kunitz, M., 1947. Crystalline soybean trypsin inhibitor. Journal of General Physiology 30:291–310.

Rackis, J.J., M.R. Gumbmann & I.E. Liener, 1985. The USDA trypsin inhibitor study. I. Background objectives and procedural details. Qual. Plant Foods Hum. Nutr. 35:213–242.

Tan-Wilson, A.L., S.E. Crosgriff, M.C. Duggan, R.S. Obach & K.A. Wilson, 1985. Bowman-Brik proteinase isoinhibitor complements of soybean strains. Journal of Agricultural and Food Chemistry 33:389–393.

Influence of conditions of processing soya beans on the content of antinutritional factors

N. Dubbeldam
Loders Croklaan B.V., P.O.Box 4, 1520 AA Wormerveer, NL[1]

Summary

Heat stability of trypsin inhibitors and antigenicity of soya bean proteins have been tested by using three types of processing equipment: a vertical conditioner, a twin screw extruder and a single screw extruder, under various conditions.
Descriptors: processing conditions; antinutritional factors; trypsin inhibitor activity (TIA); antigenic activity.

Introduction

Soya protein concentrates, if manufactured by an aqueous ethanol extraction, generally show a low content of antinutritional factors (ANF) content with selected processing conditions.

- Trypsin inhibitor activity (TIA) is reduced by heat treatment (toasting) after carbohydrate extraction;
- Antigenicity is inactivated by ethanol together with heat treatment (temperature maximum 80 °C), (Fukushima, 1969; Sissons, 1982).

Inactivation of the antigenicity of glycinin and betaconglycinin without the presence of ethanol, is reported to take place at high temperatures only, in the range of 145–170 °C. The positive effect of the presence of ethanol (Table 1) is due to the chemical structure of the protein molecule which shows hydrophillic properties on the surface, and hydrophobic properties inside the molecule. As glycinin and beta-conglycinin are found in the centre of the molecule, ethanol will be able to eliminate the activity.

Table 1. Antigenic activity of soya protein products.

Soya protein product	Antigen-titre[1] glycinin	betacon-glycinin
White flakes (untoasted soya flakes 50% protein)	13	13
Toasted defatted soya flour (protein 50%)	11	11
Water-extracted soya protein concentrate (protein 70%)	6	1
Water-ethanol-extracted soya protein concentrate (protein 70%)	< 1	< 1
Water-ethanol soya protein concentrate (low toasted; protein 70%)	3	1
Low-heat-treated soya protein isolate (protein 90%)	6	8

1 Values are expressed as the highest dilution (log2) which inhibits agglutination by antiglycinin and antibetaconglycinin.

1 Present address: Euroduna Rohstotte G.m.b.H., P.O. Box 1143, 2202 Barmstedt, DE.

Table 2. Raw material used in the tests.

Raw material	defatted untoasted soya flakes
Protein content (%)	50.7
Protein dispersibility index (PDI) (%)	68.0
Moisture (%)	9.0
TIA in product (mg/g)	21.5
Antigenicity	11.0

Table 3. Different test conditions in processing with a twin-screw extruder (type Berstorff).

	Temperature in tests (°C)		
	1	2	3
Stage 1 (entrance)	23	24	25
Stage 2	25	26	27
Stage 3	26	28	31
Stage 4	47	58	70
Stage 5	72	83	86
Stage 6	118	128	145
Stage 7	102	110	119
Die (outlet)	142	152	152–160
Addition of water (%)	20.5	18.4	14.4

Table 4. Moisture content, PDI, TIA and antigenicity of the soya product after processing with a vertical conditioner, processing temperature 70 °C (above) and 90 °C (below) in mixer; outlet temperature 20 °C.

Time (minutes	Moisture (%)	PDI (solubility)	TIA (mg/g product)	Antigenicity (Table 1)
0	12.0	62	21.0	11
15	12.0	60	21.0	11
30	11.8	62	21.0	11
45	11.9	59	21.0	11
0	13.1	51	18.0	11
15	12.8	40	14.0	11
30	12.3	34	11.0	11
45	11.6	30	10.0	11

The reduction of TIA is related to heat and moisture only (temperature maximum 120 °C). With ethanol followed by a heat treatment, most of the protein is completely denatured at the end of the production process (Figure 1).

As overheating of the protein will mostly damage the less heat-stable amino acids, such as lysine, this paper determines the influence of a moderate heat treatment of soya proteins on solubility, measured as protein dispersibility, antigenicity and TIA.

Material and methods

Defatted untoasted soya flakes were processed by three types of processing equipment:

- a vertical conditioner (mixer) with open steam injection.
- a twin-screw extruder
- a single-screw extruder

For all tests, the same raw material (Table 2) was used and protein content, moisture content, protein dispersibility index (PDI), TIA and antigenicity were determined.

The first test was done with the vertical conditioner at processing temperatures of 70 and 90 °C in the mixer and with an outlet temperature of 20 °C. The twin-screw extruder, consisted of 7 stages, all fitted with thermometers. Three tests were done under different conditions given in Table 3. The third test was done with a combination of a conditioner, as described above, and a single screw-extruder. It was decided to increase the moisture content in the conditioner, taken into account the disappointing results of the first test (Table 4).

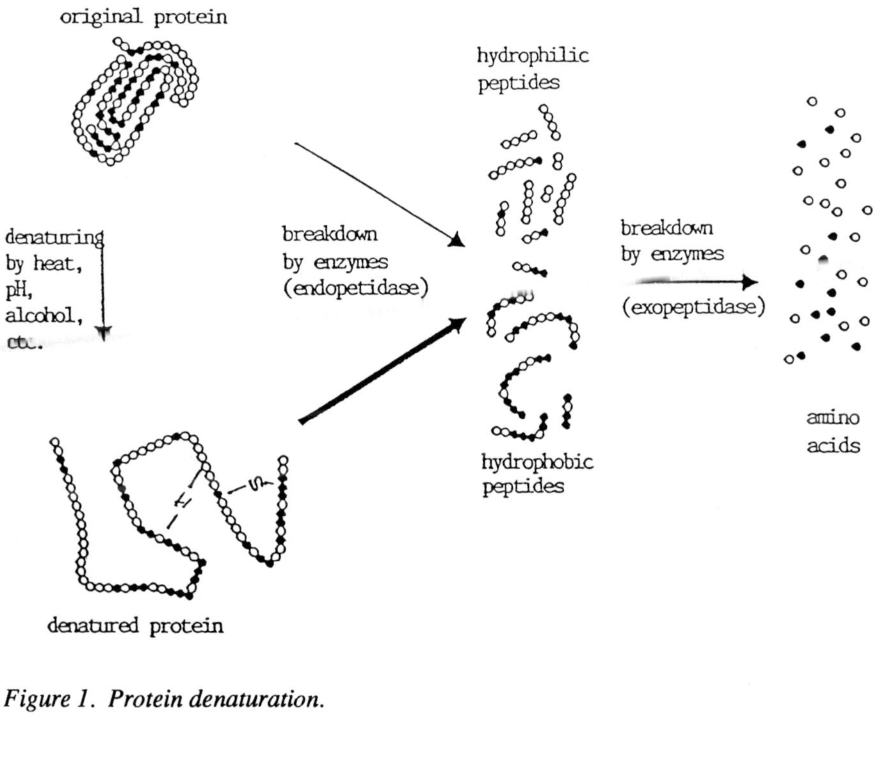

Figure 1. Protein denaturation.

Results and discussion

The results of the test with the vertical conditioner are described in Table 4. A temperature of 70 °C shows that even after a heating time of 45 minutes, the trypsin inhibitor activity and antigenicity are completely available. If the temperature is increased to 90 °C, the trypsin inhibitor activity starts to show some variability, whereas antigenicity is not decreased by the increase of the temperature.

The results of the test with the twin-screw extruder are described in Table 5. Because of the very high outlet temperatures of 142 °C and 152 °C, trypsin inhibitor activity and antigenicity had both very low levels. Unfortunately as a result of the high temperature, the quality of the protein, particularly the availability of lysin, may have suffered too.

The results of the test with the combination of the conditioner and the single-screw extruder are described in Table 6. If compared with the first test (Table 4) the increase of the moisture content in the conditioner, at an unchanged temperature has an influence on the stability of the trypsin inhibitor activity. Antigenicity remains very stable. After extrusion, both antinutritional factors show quite acceptable levels, even at a temperature of 120 °C.

Conclusions

- With selective processing conditions, the use of an extruder makes it possible to inactivate soya bean trypsin inhibitor and antigenicity.
- A temperature of 120 °C, not much higher than temperatures used in normal toasting processes, is sufficient to obtain low levels of both antinutritional factors.
- Feeding trials have to prove, whether the soya bean proteins, produced by an extrusion process, will perform better as a result of a better cell structure.

Table 5. Moisture content, PDI, TIA and antigenicity of the soya product after processing with a twin-screw extruder.

	Test 1	Test 2	Test 3
Moisture after extruder (%)	10.3	9.8	8.9
PDI solubility	6.7	6.8	8.9
TIA in product (mg/g)	< 1.0	1.2	< 1.0
Antigenicity (Table 1)	< 1.0	< 1.0	< 1.0

Table 6. Moisture content, PDI, TIA and antigenicity of the soya product after processing with a combination of a vertical conditioner and a single-screw extruder.

	Moisture (%)	PDI (solubility)	TIA (mg/g)	Antigenicity (Table 1)
Defatted flakes (untoasted before conditioner)	8.4	60	19.6	13
After conditioner (70 °C)	17.4	60	6.3	11
After extruder (120 °C)	11.7	10	2.0	< 1

References

Sissons, J.W., R.H. Smith & D. Hewitt, 1982. Prediction of the suitability of soya-bean products for feeding to preruminant calves by an in vitro immunochemical method. British Journal of Nutrition 47:311–318.

Fukushima, D., 1969. Denaturation of soybean proteins by organic solvents. Cereal. Chem. 46:156–163.

10. Nutrition: supplementation

Experimental data on the modes of action of probiotics

P. Raibaud * & J.P. Raynaud **

* *Laboratoire d'Ecologie et de Physiologie du Système Digestif (LEPSD), Centre de Recherches de Jouy-en-Josas 78350 Jouy-en-Josas, FR*

** *Fédération Européenne de la Santé Animale (FEDESA), 1, rue Defacqz, B 1050 Brussels, BE*

Summary

Many living micro-organisms referred to as probiotics are used today in animal feeds, either for short-term feeding as therapeutic agents to prevent or cure gastro-intestinal disorders before weaning or in stressed animals, or for long-term feeding as growth promoters or to prevent chronic disorders after weaning. Usual claims about the modes of action of probiotics are: colonization of the gastro-intestinal tract, prevention of pathogen overgrowth, neutralization of enterotoxins, modulation of the activity of some bacterial enzymes in the large intestine, enhancement of the digestive capacity of the small intestine, adjuvant effect on the immune system. Experimental data today do not allow us to conclude that probiotics can colonize the gastro-intestinal tract or prevent the growth of intestinal pathogens by production in situ of antimicrobial metabolites. Neutralization of enterotoxins can be demonstrated in vivo with very large numbers of probiotic cells. Controversial results were obtained for faecal β-glucuronidase activity in rats and in pigs fed on probiotics. In contrast, good evidence has been obtained that some probiotics can increase utilisation of lactose in the small intestine. Such results could account for a growth-promoting effect of probiotics. An increase in both phagocytic and lymphocytic activities has been demonstrated with some probiotics, and the clearance of injected enterobacteria can be increased by short-term probiotic feeding. This could account for the therapeutic effect of probiotics.

Descriptors: probiotics; mode of action; therapeutic agent; growth promoter.

Introduction

Human beings have always eaten viable bacteria in fermented foods. The idea that such alimentary bacteria, lactic acid bacteria, could be beneficial for health comes from Metchnikoff (1908). He believed that living lactic acid bacteria contained in the yoghurt could improve human health by preventing metabolism of pathogenic microbes in the intestine. The earliest cattle farmers also unwittingly used alimentary bacteria when they fed ill cattle on rumen contents from healthy cattle. Today silage is a fermented food that plays a large role in ruminant nutrition. In 1974, Parker defined probiotics as organisms that contribute to improved intestinal microbial balance. The probiotic industry is growing extremely rapidly and today many farmers give probiotics to monogastrics, birds and ruminants for two main purposes: as routine feed additives, which are believed to replace antibiotic supplements, or as drugs for a therapeutic purpose before weaning or to counteract stress. The first purpose is a long-term use and the main problem is to maintain the probiotics alive in the food during the whole feeding period. The number of viable micro-organisms needed in general reaches 10^6 to 10^7 colony-forming units per gram of food. For the second purpose, the number of viable microbes can be 10 times as high, but they are given for a short time. The legislation on probiotics is far to be adequate, since European law recognizes only two classes: additives and veterinary drugs. Thus, the same probiotic can fall into one or the other class according to its use. Probiotics can be used as basic dietary components or as dietetic supplements, and claims for a beneficial effect are forbidden in either case.

The aim of this contribution is to review some scientific data on possible modes of action of probiotics, in order to understand how they act and thus to gain scientific information so that we can choose and eventually create the best probiotic by genetic engineering.

Taxonomic position of probiotic used today

Probiotics belong to a variety of microbial species. According to Lloyd-Evans (1989), the following genera are involved: *Bacillus* (8 species), *Lactobacillus* (12 species), *Streptococcus* (9 species), *Propionibacterium* (2 species), *Pediococcus* (3 species), *Bifidobacterium* (7 species), *Clostridium* (1 species) and even *Bacteroides* (4 species), although *Bacteroides* is non-sporulating, gram-negative and strictly anaerobic. *Aspergillus* (2 species) and *Saccharomyces* (2 species) are also involved. However the common probiotics used in Europa belong to *Lactobacillus*, *Streptococcus* and *Bacillus*.

In general, the preparations contain a single strain, more rarely an association of two strains belonging to different genera. The main bacteriological problem of such probiotic preparations is the need for suitable tests for characterization and counting of the strain used. Phenotypic characterization is often unsatisfactory and must be sustained by genetic traits. Counts of viable cells of probiotics in commercial preparation are not a big problem, whereas those in faeces are hard to perform. Use of plasmid or chromosomal DNA probes seems to be the best method to ensure that counts are correct, mainly when probiotics belong to the same genera as indigenous bacteria, such as lactic acid bacteria. Other methods can be used, such as slide agglutination with specific antisera or immuno-enzymic techniques like ELISA. Very few works deal with this basic bacteriological problem.

Probiotics and the gastro-intestinal ecosystem

As probiotics are claimed to improve microbial intestinal balance, a brief survey on the gastro-intestinal (GI) ecosystem is relevant. Almost all the homeotherms and some poecilotherms harbour in their GI tract a huge number of viable microbial cells, mostly bacteria. The count in fresh faeces in the large intestine surpasses 5×10^{10} per g. More than 190 species have been recorded in the human faecal flora (Finegold et al., 1983). Most of the predominant bacteria are strictly anaerobic and some cannot be cultured today. In some farm animals, such as pigs or calves, the stomach also harbours 5×10^{8} to 10^{9} viable bacteria per g of contents, most of them facultative anaerobes and some of them adhering to the stomach wall. Crop and rumen also contain a large count of bacteria, 10^{9} and 10^{11} per g, respectively. By contrast, the small intestine is not a place where the bacteria can multiply in healthy state, because of the transit speed of the alimentary bolus. When bacteria proliferate in the small intestine, there are always great disturbances.

The microbial colonization of the GI tract is a complex process, which begins few hours after birth and includes several steps. It depends on multiple factors, namely the host and its dietary regimen. Scientific data concerning this sequential colonization are very poor (Raibaud & Ducluzeau, 1989).

The use of axenic (germ-free) and gnotobiotic animals (axenic animals inoculated with known bacterial strain(s)) provides us with valuable tools for studying the functions of the microbial flora. Comparison between axenic and conventional animals reveals the role of the whole flora. Comparison between gnotobiotic and both axenic and conventional animals evidences the specific role of associated bacterial strain(s). Use of gnotobiotic animals also allows us to study the activities of the bacterial enzymes under various environmental conditions. Such studies show that the expression of bacterial genes can be very different in vitro, in a culture tube, and in vivo, in the GI tract (Ducluzeau & Raibaud, 1980). The main functions of the GI flora can be summarized as follows: the microbes produce a huge number of metabolites, mostly in the large intestine, from endogenous materials and from indigestible fractions of the diet. Some of these metabolites can be absorbed from the large intestine. As a consequence, the anatomy of the GI tract is deeply modified, as well as many nutritional parameters, such as transit time and renewal of the enterocyte. Both local and systemic immune systems are strongly stimulated by the intestinal microflora. Bacterial antagonisms represent a strong barrier against exogenous bacteria (Ducluzeau et al., 1970). This function, also called resistance to colonization (van der Waaij et al., 1971) is probably the main one, because its maintains a rather stable, though fragile, equilibrium of the microbial flora in adults. The mechanisms involved are poorly understood. Freter et al. (1983) and Yurdusev et al. (1989) stated that a reversible bacteriostasis was induced by the inhibitory strains, which hinder exogenous target strains from utilizing available intestinal nutrients. Consequently, exogenous bacteria in general pass through the GI tract without any multiplication in situ, but they can remain viable, provided that they are not killed by GI secretions. Thus, the count of the population of transient bacteria cannot be high. It depends on the initial number of ingested bacteria, on the rate of death during transit and on dilution due to secretions. From several experiments in our laboratory with gnotobiotic animals, a bacterial strain cannot produce enough antimicrobial substances, such as organic acids or antibiotics, to prevent the proliferation of pathogenic strains in the GI tract if its count remains under 10^8 per g of contents.

Usual claims on the modes of action of probiotics

The usual claims on the modes of action of probiotics, as drawn from scientific or commercial articles, are as follows:

- probiotics colonize the digestive tract
- probiotics prevent proliferation of pathogens
- probiotics neutralize enterotoxins produced in situ
- probiotics modulate the activity of some bacterial enzymes
- probiotics enhance the digestive capacity of the small intestine
- probiotics exert adjuvant effects on the immune system.

Probiotics colonize the digestive tract

As previously noted, this claim is difficult to assess. Probiotics are exogenous bacteria and are thus submitted to the microbial barrier effects. A heterofermentative strain of *Lactobacillus*, isolated from the pars oesophagea of a piglet stomach was fed to piglets. No permanent establishment was observed in spite of this origin. However, the strain used did not improve the growth of piglets (Jonson, 1986). Thus, this claim has to be reinvestigated with adequate tools. Nevertheless, it is a poor commercial argument: if the probiotic were to become established, no permanent supply would be required; a single inoculum would be enough to improve growth or prevent GI disorders.

Probiotics prevent proliferation of pathogens

Several probiotics are active against pathogens when they are cultured in vitro. For instance, lactic acid produced by bacteria is a strong inhibitor for many pathogens. Some probiotics have been selected from their ability to produce antibiotics or bacteriocins. No experimental data are available to demonstrate that these antimicrobial substances can be produced in the GI tract. Even establishment of the active strains in the GI tract would remain a problem.

One can also imagine giving inhibitory strains soon after birth. Nurmi & Rentala (1973) showed that a mixture of more than 40 strains administered by mouth to newly hatched chicken protected them against *Salmonella enteritidis*, since this pathogen was no longer detected in the faeces of most of the treated chicken. Unfortunately, the active bacteria of the mixture were not identified. Dubos et al. (1984) succeeded in preventing diarrhoea and death of young hares by inoculating them with a complex mixture of bacterial strains immediately after birth. Both examples show that effective protection against pathogens can be provided by a single inoculation of inhibitory strains. Nevertheless, these complex mixtures cannot be considered as probiotics. Thus, this claim has also to be demonstrated by adequate experiments.

It has also been suggested that probiotics can compete for adherence receptors of some pathogens, such as virulent *Escherichia coli*, thus protecting the host against colibacillosis. This claim comes from the work of Underdahl et al. (1982). Using gnotobiotic piglets, these authors showed that the probiotic *Streptococcus faecium*, strain Cernelle, reduced the toxic effect of pathogenic *E.coli* and prevented generalized infection and death. However Underdahl et al. (1982) stated that competition for adherence sites was not responsible for the reduced diarrhoea.

Probiotics neutralize enterotoxin produced in situ

Mitchell & Kenworthy (1976) reported that cell-free extracts from a strain of *L. bulgaricus* showed anti-enterotoxin activity in ligated segments of intestine. When this strain was given to piglets challenged with an enterotoxic strain of *E. coli*, both weight gain and survival were improved. It was inferred, but not demonstrated, that these effects were likely to be caused by the anti-enterotoxin activity. More recently, Corthier (1988) showed that a heavy suspension (10^{10} per ml) of *Saccharomyces boulardii* given as drinking water afforded a high protection to gnotobiotic mice challenged with a highly pathogenic strain of *Clostridium difficile*. The enterotoxin was no longer detected, and the cytotoxin titre was 100-fold decreased in the faeces of surviving mice. However Corthier (unpublished data) did not found any protection against *C. difficile* in mice fed on yoghurt, containing 5×10^8 *S. thermophilus* and 5×10^8 *L. bulgaricus* per ml.

Probiotics modulate the activity of some bacterial enzymes

Goldin & Gorbach (1984) showed a significant decrease in the activity of bacterial β-glucuronidase and nitroreductase in faeces of meat-fed conventional rats, when they were fed on a strain of *L. acidophilus*. These results explain the decreased production of free amines from aromatic nitrocompounds and glucuronides in the faeces of treated rats. A reduction of amine production in situ is beneficial for the host. However, Lessard & Brisson (1987) did not find such a decrease of bacterial β-glucuronidase in weaned pigs fed on a *Lactobacillus* fermentation product, although this preparation increased weight gain and feed intake. Thus this claim remains controversial.

Probiotics enhance the digestive capacity of the small intestine

It is now well established that yoghurt enhances the lactose utilization in the small intestine of lactase-deficient human subjects (Kolars et al., 1984; Marteau et al., 1989). The mode of action of yoghurt has not been completely elucidated. Bacterial β-galactosidases could make up for the lack of endogenous lactase (Kolars et al., 1984), or lactase activity of the small intestine could be stimulated (Besnier et al., 1986). Heated yoghurt loses this effect. It would be of greatest interest to perform such experiments in calves, which are fed on a lactose diet for a long time. Presumably a better lactose utilization in the small intestine could account for a growth-promoting effect. Wong et al. (1983) showed an increase in the weight gain of rats fed on milk supplemented with *S. thermophilus* or with an extract of this strain. Unfortunately they did not study lactose utilization. Wolter et al. (1987) found a better protein efficiency coefficient in rats fed with some strains of *Lactobacillus*, whereas some strains of *Streptococcus* decreased it. They observed that strains enhancing this coefficient in rats could also enhance the weight gain of calves. There is no explanation of these findings, but one can speculate that enhancement of the digestive capacity of the small intestine could also be related to better protein utilization.

Probiotics exert adjuvant effects on the immune system

The adjuvant effect of ingested lactobacilli and streptococci on some immunological parameters has been studied in mice (Conge et al., 1980; Perdigon et al., 1988; de Simone et al., 1988), in pigs (Lessard et al., 1987), and in human beings (de Simone et al., 1989). Increase of seric immunoglobulins of the IgG type, increase of IgG antibodies, activation of phagocytic and lymphocytic activities have been reported. Perdigon et al. (1989) also showed an increase in secretory IgA antibodies against *Salmonella typhimurium* in the intestinal content of mice fed with *L. casei*. They also showed that the duration of *L. casei* feeding had a large influence on the clearance of injected *E. coli* and *S. typhimurium*, since a 2-day feeding led to complete elimination of those bacteria in liver and spleen 2 days after challenge, whereas a 5-day feeding led to a 10-fold increase in the number of injected bacteria in the same organs 2 days after challenge. This result could be a consequence of the intestinal inflammatory response observed with feeding for 5 days with *L. casei*.

Conclusion

Probiotic use is neither a voodoo nor a panacea. However some well documented field experiments are promising. The challenge for the future would be to understand how the probiotics act in various environmental conditions and in various farm animals. Basic research with valuable tools is lacking. Many claims about their modes of action are not supported by safe experimental data. However, some ideas may be drawn from the improvement in digestive capacity in small intestine and the adjuvant effects on the immune system.

References

Besnier, M.O., P. Bourlioux, J. Fourniat, R. Ducluzeau & A. Aumaitre, 1983. Influence de l'ingestion de yogourt sur l'activité lactasique intestinale chez des souris axéniques ou holoxéniques. Annales de Microbiologie (Institut Pasteur). 134 A:219–230.

Conge, G.A., P. Gouache, J.P. Desormeau-Bedot, F. Loisillier & D.Lemonnier, 1980. Effets comparés d'un régime enrichi en yoghourt vivant ou thermisé sur le système immunitaire de la souris. Reproduction Nutrition Développement 20 (4A):929–938.

Corthier, G., 1988. Ecologic means of protection against *C. difficile* infections in gnotobiotic mice. E. Bergogne-Berezin, Ed.: Microbial Ecology and Intestinal Infections. Springer Verlag France, Paris, 1989 p. 43–55.

de Simone, C., S. Tzantzoglou, L. Baldinelli, S. Di Fabio, B. Bianchi-Salvadori, E. Jirillo & R. Vesely, 1988. Enhancement of host resistance against *Salmonella typhimurium* infection by a diet supplemented with yogurt. Immunopharmacology and Immunotoxicology 10 (3):399–415.

de Simone, C., B. Bianchi Salvadori, E. Jirillo, L. Baldinelli, S. Di Fabio & R. Vesely, 1989. Yogurt and the immune response. In Syndifrais Ed.: Fermented milks, Current research, John Libbey Eurotext Ltd, London, Paris. p. 63–67.

Dubos, F., L. Martinet, J. Dabard & R. Ducluzeau, 1984. Immediate postnatal inoculation of a microbial barrier to prevent neonatal diarrhea induced by *Clostridium difficile* in young

conventional and gnotobiotic hares. American Journal of Veterinary Research 45(6):1242–1244.
Ducluzeau, R., M. Bellier & P. Raibaud, 1970. Transit digestif de divers inoculums bactériens introduits 'per os' chez des souris axéniques et 'holoxéniques' (conventionnelles): Effet antagoniste de la microflore du tractus gastro-intestinal. Zentralblatt für Bakteriologie, Parasitenkunde, Infektionskrankheiten und Hygiene, I Orig 213:533–548.
Ducluzeau, R. & P. Raibaud, 1980. Intérêt des systèmes gnotoxéniques pour l'étude des relations hôte-flore microbienne du tube digestif. Reproduction Nutrition Développement 20(5B):1667–1678.
Finegold, S.M., V.L. Sutter & G.E. Mathisen, 1983. Normal indigenous intestinal flora. In D.J. Hentges, ed.: Human intestinal microflora in health and disease. Academic Press, New York. p. 3–31.
Freter, R.E., E. Stauffer, D. Cleven, L.V. Holdeman & W.E.C. Moore, 1983. Continuous-flow cultures as in vitro models of the ecology of large intestinal flora. Infection and Immunity 39:666–675.
Goldin, B.R. & S.L. Gorbach, 1984. Alterations of the intestinal microflora by diet, oral antibiotics, and *Lactobacillus*: decreased production of free anmines from aromatic nitro compounds, azo dyes, and glucuronides. Journal of the National Cancer Institute 73 (3):689–695.
Jonson, E., 1986. Persistence of *Lactobacillus* strain in the gut of suckling piglets and its influence on performance and health. Swedish Journal of Agricultural Research 16:43–47.
Kolars, J.C., M.D. Levitt, M. Aouji & D.A. Savaiano, 1984. Yoghurt-An autodigesting source of lactose. The New England Journal of Medicine 310(1):1–3.
Lessard, M. & G.J. Brisson, 1987. Effect of *Lactobacillus* fermentation product on growth, immune response and fecal enzyme activity in weaned pigs. Canadian Journal of Animal Science 67:509–516.
Lloyd-Evans, L.P.M., 1989. Probiotic. PjB Publications Ltd, 18-20. Hill rise, Richmond, Surrey TW10 6UA, GB. p. 8
Marteau, P., B. Flourie, P. Pochart, P. Pellier, C. Franchisseur, J.F. Desjeux & J.C. Rambaud, 1989. Mesure directe de l'absorption du lactose du yaourt chez l'homme alactasique et étude du devenir de la lactase bactérienne du yaourt dans l'intestin. In Syndifrais Ed.: Fermented Milks, current research, John Libbey Eurotext Ltd, London, Paris. p. 179–182.
Metchnikoff, E., 1908. Prolongation of life. Putman & Sons, New york.
Mitchell, I. de G. & R. Kenworthy, 1976. Investigations on a metabolite from *Lactobacillus bulgaricus* which neutralizes the effect of enterotoxin from *Escherichia coli* pathogenic for pigs. Journal of Applied Bacteriology 41:163–174.
Nurmi, E. & M. Rantala, 1973. New aspects of *Salmonella* infections in broiler production. Nature (London) 241:210–211.
Parker, R.B., 1974. Probiotics, the other half of the antibiotic story. Animal Nutrition & Health 29:4–8.
Perdigon, G., M.E.N. De Macias, S. Alvarez, G. Oliver & A.P. De Ruiz Holgado, 1988. Systemic augmentation of the immune response in mice by feeding fermented milks with *Lactobacillus casei* and *Lactobacillus acidophilus*. Immunology 63:17–23.
Perdigon, G., S. Alvarez, M.E.N. De Macias & M. Medici, 1989. Effect of lactic acid bacteria orally administered and of yoghurt on the immune system. In Syndifrais Ed.: Fermented milks, Current research, John Libbey Eurotext Ltd, London, Paris. p. 77–84.
Raibaud, P. & R. Ducluzeau, 1989. Etude de la colonisation bactérienne du tractus gastro-intestinal à l'aide de modèles expérimentaux. Revue Scientifique et Technique de l'Office International des Epizooties 8(2):361–373.
Underdahl, N.R., A. Torres-Medina & A.R. Doster, 1982. Effect of *Streptococcus faecium* C-68 in control of *Escherichia coli*-induced diarrhea in gnotobiotic pigs. American Journal of Veterinary Research 43(12):2227–2232.

van der Waaij, D., J.M. Berghuis-de Vries & J.E.C. Lekkerkerk-van der Wees, 1971. Colonization resistance of the digestive tract in conventional and antibiotic treated mice. Journal of Hygiene 69:405–411.
Wolter, R., N. Henry, L. Jacquot, G. Briend, M. Blanchet, G. Delespaul & P. Dhoms, 1987. Probiotiques en alimentation animale. Etude expérimentale de leur efficacité chez le rat et chez le veau de boucherie. Recueil de Médecine Vétérinaire 163(12):1131–1138.
Wong, N.P., F.E. Mac Donough & A.D. Hitchins, 1983. Contribution of *Streptococcus thermophilus* to growth-stimulating effect of yogurt on rats. Journal of Dairy Science 66:444–449.
Yurdusev, N., M. Ladire, R. Ducluzeau & P. Raibaud, 1989. Antagonism exerted by an association of a *Bacteroides thetaiotaomicron* strain and a *Fusobacterium necrogenes* strain against *Clostridium perfringens* in gnotobiotic mice and in fecal suspensions incubated in vitro. Infection and Immunity 57(3):724–731.

Ferrokinetics in veal calves

G.A.J. Miltenburg *, Th. Wensing *, H.J. Breukink * & J.J.M. Marx **
* *Department of Large Animal Medicine and Nutrition, Veterinary Faculty, University of Utrecht, NL*
** *Department of Haematology, University Hospital, Utrecht, NL*

Summary

Transferrin-bound ^{59}Fe was injected intravenously into 2 normal veal calves and 1 five-week-old male veal calf with a subnormal concentration of haemoglobin in blood. The disappearance half-time of iron in plasma, the turnover of iron in plasma, the plasma volume and the uptake of iron by red cells were determined. The two normal calves showed a disappearance half-time of iron of 53 and 61 min and a turnover of iron (Fe) of 264 and 269 µmol l^{-1} d^{-1}. In the calf fed on an irondeficient diet, associated with a subnormal concentration of haemoglobin the disappearance half-time of iron and the turnover of iron were markedly lower, namely 24 min and 161 µmol l^{-1} d^{-1}. No difference was observed in the uptake of ^{59}Fe between the three calves. The maximum uptake was 92% (day 7) in the iron-deficient calf and 91% (day 7, day 9) in the other two.

Independent of the amount of iron administered in the milk replacer to young veal calves the uptake of iron was very high, suggesting a high requirement for erythropoiesis, even in 'normal' calves.

Descriptors: anaemia; ferrokinetics; iron-deficiency; ^{59}Fe.

Introduction

In human haematology, radionuclide methods have been used for many years to study various aspects of iron metabolism (Marx & Verzijlbergen, 1986). Kaneko & Mattheeuws (1966) introduced radionuclide studies in calves with haematologic disorders such as erythropoietic porphyria. In fattening veal calves, iron-deficiency anaemia is commonly observed due to the ration fed (Abdelrahim, 1983).

The purpose of the present study was to obtain more information about the internal cycle of iron metabolism in a calf fed on an iron-deficient diet. In addition to some general haematological data, transferrin-bound ^{59}Fe was used to investigate iron deficiency.

Material and methods

Three five-week-old male Friesian crossbred veal calves were used in this experiment. Two of them (Calves 1 and 2) were normal healthy calves (with haemoglobin concentrations of 7.2 and 6.6 mmol/l) and Calf 3 was a calf with a subnormal haemoglobin concentration of 5.6 mmol/l. The calves were housed individually in pens. They were fed twice daily with a commercial milk replacer[1], which contained an iron concentration of 8–10 ppm . 60 ppm iron (as $FeSO_4.7H_2O$) was added to the milk replacer for Calf 1 and 2. Calf 3 did not receive extra iron. The calves were weighed at the beginning and at the end of the experiment.

1 Tentofeed I, Tentego B.V., Animal feed specialities Import/Export, Mijdrecht, NL.

At the beginning of the experiment, sufficient blood was collected from each calf into heparin to yield about 10 ml of plasma. The blood was centrifuged and the plasma separated. About 367 kBq of ^{59}Fe, as ferric citrate, was added to the autologous plasma. After incubation of 30 min at room temperature, ^{59}Fe not bound to transferrin was removed by passing the mixture through an anion-exchange resin column (Cavill, 1971). After passing through a sterile 0.2 µm filter (Sartorius Minisart NML, 0.2 µm 600 kPa max) the eluate was drawn into a 20-ml syringe. A standard of this eluate was retained and a measured amount (about 10 ml) was injected by jugular-vein puncture. Immediately before injection of ^{59}Fe, 25 ml of blood was collected into heparin for measurement of background radioactivity and for determination of haemoglobin (Hb), haematocrit (Ht), red cell count, mean corpuscular volume (MCV), plasma iron, total iron-binding capacity (TIBC) and saturation of total iron-binding capacity.

Blood samples (15 ml) were collected from the opposite jugular vein into heparin tubes at 10, 20, 30, 45, 60, 90, 120, 240, 360, 480 and 600 min after the calves were given the transferrin-bound ^{59}Fe. Sampling was continued once daily for 16 days. A small part of these samples was used for determination of the haematological parameters and the other part was used for measurement of radioactivity in whole blood (4 ml) and plasma (4 ml). After the last day of the experiment, radioactivity of all the whole blood and plasma samples and the standard solutions was measured in a NaI scintillation gamma-counter (Automatic Gamma Counter PW 4800 Philips).The disappearance half-time ($T_{1/2}$) of iron, the iron turnover (PIT), the plasma volume (PV), the blood volume (BV), the red cell volume (RCV) and the iron uptake (RCIU) were determined and calculated as previously described (Davies 1971; Marx et al., 1986; Smith 1989).

Results

The haematological data of the three calves obtained on the first day of the trial are presented in Table 1. During the experiment the values of Hb, Ht, red cell count and MCV in the Calves 2 and 3 decreased; for Calf 1 these values remained constant. The TIBC increased in all three calves and iron concentration in Calf 3 decreased clearly.

The ferrokinetic data are presented in Table 2. A clear difference in the disappearance half-time and iron turnover was observed between the iron-deficient calf and the two normal calves. The RCIU was efficient. For all calves, more than 90% of the injected dose was incorporated into the erythrocytes. The maximum RCIU was observed on day 7 in the iron-deficient calf and on day 7 and 9 in the two normal calves. After day 7 and 9, the ^{59}Fe uptake decreased. Usually, the RCIU is determined on day 10 or 14 (Marx et al., 1986; Smith, 1989; Varela, 1971). In the present study the RCIU on day 14 was 79 and 69% in the two normal calves and 81% in the iron-deficient calf.

Discussion

In the present study, a clear difference for the plasma disappearance half-time and the iron turnover of ^{59}Fe bound transferrin was observed between the iron-deficient fed calf and the two normal calves. In accordance with the expectations, the $T_{1/2}$ in the iron-deficient fed calf was shorter than in the two normal calves (Marx et al., 1986; Smith, 1989). The low value of the PIT in the iron-deficient fed calf was probably due to the relatively high plasma iron concentration in the dietary iron-supplemented calves. This suggestion was supported by the fact that soon after the iron supplement had been stopped, all three calves presented a low plasma iron concentration at the end of the experiment.

It was remarkable, however, that $T_{1/2}$ was less than those published before for calves and human beings. In human beings, normal values for plasma $T_{1/2}$ ranges from 60 to 120 min

Table 1. Haematological data of the iron-deficient fed calf (Calf 3) and the two normal calves (Calves 1 and 2).

Item	Calf 1	Calf 2	Calf 3
Haemoglobin (mmol/l)	7.2	6.6	5.6
Haematocrit (l/l)	0.31	0.34	0.26
Red cell count (10^9/l)	9.84	10.28	7.70
Mean cell volume (10^{-15} l)	32	32	34
Plasma iron (µmol/l)	21	25	7
TIBC (µmol/l)	72	78	91
Saturation of the TIBC (%)	29	32	8

Table 2. Ferrokinetic findings in the iron-deficient fed calf (Calf 3) and the two normal calves (Calves 1 and 2).

Item	Calf 1	Calf 2	Calf 3
Plasma volume (ml)	3 731	3 127	3 485
Blood volume (ml)	5 364	4 727	4 709
Red cell volume (ml)	1 633	1 600	1 224
Plasma $T_{1/2}$ (min)	53	61	24
PIT (µmol/l plasma/day)	380	408	218
PIT (µmol/l blood/day)	264	269	161
RCIU maximum (%)	91	91	92

(Verwilghen & Punt, 1987). Kaneko & Mattheeuws (1966) mentioned values between 88 and 137 min for six-month-old normal Holstein-Friesian calves.

Another remarkable observation was the very high RCIU in the three calves in the present study. Verwilghen & Punt (1987) mentioned values between 70 and 90% for normal humans. Kaneko et al. (1966) found values between 66 and 82% for the above mentioned Holstein-Friesian calves. Possibly the two 'normal' calves, in spite of a ration with 60 ppm iron, were iron-deficient because of their fast weight gain or of malabsorption of the ingested iron. Within two weeks, each calf showed an average weight gain of about 18 kg. This weight gain also caused an increase in blood volumes during the experimental period. This increase could be an explanation for the reduction in RCIU after day 9 with dilution of radioactivity in a population of newly formed erythrocytes.

References

Abdelrahim, A.J., 1983. The role of iron supply in the fattening of veal calves. Doctoral thesis, Utrecht, NL.

Cavill, I., 1971. The preparation of ^{59}Fe-labeled transferrin for ferrokinetic studies. Journal of Clinical Pathology 24:472–474.

Davies, J.W.L., 1971. Blood volume studies. In: E.H. Belcher & H. Vetter (Ed.): Radioisotopes in medical diagnosis. Butterworths, London. p. 319–337.

Kaneko, J.J. & D.R.G. Mattheeuws, 1966. Iron metabolism in normal and porphyric calves. American Journal of Veterinary Research 27:923–929.
Marx, J.J.M. & J.F. Verzijlbergen, 1986. Radionuclide techniques in haematology. In: P.P. van Rijk (Ed.): Nuclear techniques in diagnostic medicine. Martinus Nijhoff Publishers, Dordrecht. p. 459–517.
Smith, J.E., 1989. Iron metabolism and its diseases. In: J.J. Kaneko (Ed.): Clinical biochemistry of domestic animals. Academic Press, Inc. p. 256–273.
Varela, J.E., 1971. Studies of iron absorption and metabolism. In: E.H. Belcher & H. Vetter (Ed.): Radioisotopes in medical diagnosis. Butterworths, London. p. 383–411.
Verwilghen, R.L. & K. Punt, 1987. Oorzaken, diagnostiek, gevolgen en behandeling van ijzergebrek. In: B.S. van Asbeck et al. (Ed.): IJzer, een onmisbaar element met schadelijke eigenschappen. Samson Stafleu, Alphen-aan-de-Rijn. p. 71–82.

Iron status of new-born calves and effects of supplementation with different amounts of iron in veal calf fattening

Th. Wensing, G.A.J. Miltenburg & H.J. Breukink
Department of Large Animal Medicine and Nutrition, Veterinary Faculty, State University of Utrecht, NL

Summary

The mean liver iron concentration in new-born calves was found to be 1746 μg/g DM (SD 1491, range 73–5592 μg/g DM). Between calf and dam, no correlation was found as far as liver iron concentration and plasma iron concentration were concerned.

When a supplement of 5 ppm extra iron was provided for veal calves during the whole fattening period, liver and spleen iron concentrations were clearly increased and the colour of the meat at slaughter was too red. Supplementation with more than 56 ppm of iron during the first six weeks of the fattening period can affect the colour of the meat.
Descriptors: veal calves; iron supply; liver; spleen; iron concentrations in newly born calves.

Introduction

It is well documented that in veal calf fattening, too much iron results in too red a colour of the meat, whereas too low a supplementation with iron induces anaemia, resulting in loss of appetite, decreased growth and poor food conversion (Welchman et al., 1988).

Our investigations dealt with (A) the iron status of the new born calf, (B) the effect of extra iron supply during the whole fattening period, (C) the effect of extra iron supply during the first six weeks of the fattening period only on the iron status of the calf and the colour of the meat at slaughter.

Material and methods

Iron status of the new born calf (A)

From 50 clinically normal Friesian-Holstein (FH) cows, blood and liver were sampled one week before parturition. From dam and calf, blood was sampled during parturition. After parturition, the liver of the calf was sampled. In plasma and liver, the iron concentration was measured.

Effect of extra iron supply (5 ppm) during the whole fattening period (B)

In two experiments I and II, 16 normal male FH calves were used, 8 experimental and 8 control calves, 5 ppm extra iron[1] was supplied. In Experiment I the extra iron was supplied until slaughter and in Experiment II until the 17th week of the fattening period. Blood was sampled bi-weekly during the fattening period. At slaughter, samples of liver and spleen were taken and the colour of the meat was assessed. Grading of the meat colour: 2–4, 'very good to acceptable'; 5, 'doubtful'; 6–8, 'dark to too dark'.

Effect of supplementation with different amounts of iron during the first six weeks of the fattening period (C)

In a number of experiments, different amounts of iron, 56[1] ppm; 100 ppm; 100 ppm + a subcutaneous injection of Imposil[2],150 ppm + a subcutaneous injection of Imposil[2]; were supplied to different groups of calves during the first 6 weeks of the fattening period.

Blood samples were taken fortnightly and liver and bone marrow samples were taken every four weeks.

Analytical procedures

In all experiments, blood haemoglobin, the plasma iron concentration and plasma iron saturation were estimated according to methods described earlier (Abdelrahim et al., 1983). In liver, spleen and muscle samples, the iron concentration was estimated by atomic absorption techniques (Abdelrahim et al., 1985). For the estimation of iron in the bone marrow, a histochemical technique was used (Franken et al., 1981).

In all experiments, the calves were fattened with the same commercial milk replacer containing 13 ppm of iron according to the same feeding schedule. The calves were housed individually in wooden pens; health care and management were as usual in veal calf fattening.

For every experiment, the milk replacer fed during the whole fattening period originated from one production batch and the calves used in a particular experiment were slaughtered on the same day.

Results and discussion

Iron status of new-born calves (A)

The mean iron concentration in the liver of new-born calves was 1746 µg/g DM (s.d. = 1491; range 73–5592 µg/g DM); in the liver of the cows one week before parturition, the mean value was 253 µg/g DM (s.d. = 112; range 47–533 µg/g DM).

No correlation was found between the iron concentration of plasma of the dam (mean 16 µmol/l, s.d. = 5.0) and that of the new-born calf (mean 23 µmol/l, s.d. = 8). The rather great variation in the liver iron concentration of new-born calves means that iron supply of veal calves cannot be based on the iron status of the very young calf for practical purposes.

A normal ration of commercial milk replacer, containing 50 ppm iron, during the first six weeks of the fattening period, and about 13 ppm during the rest of the fattening period clearly decreased mean iron concentration of liver between the 2nd week, 650 µmg/g DM and the 23rd week, 82 µmg/g DM and in the range of the liver iron concentration, 51–5000 µg/g DM in the 2nd week and 40–100 µg/g DM in the 23rd week. These data demonstrate that there is a clear decrease in that concentration in calves with a high liver iron concentration in the second week of the fattening period in spite of the presence of iron in the milk replacer (from 5000 µg/g DM in the 2nd to 100 µg/g DM in the 23rd week).

Apparently iron absorption in veal calves is at least partly regulated by the iron status of the calves, as is well known for other species too (Finch et al., 1982).

1 as $FeSO_4.7H_2O$ in the milk replacer.

2 Imposil, 10 ml, at the onset of the fattening period.

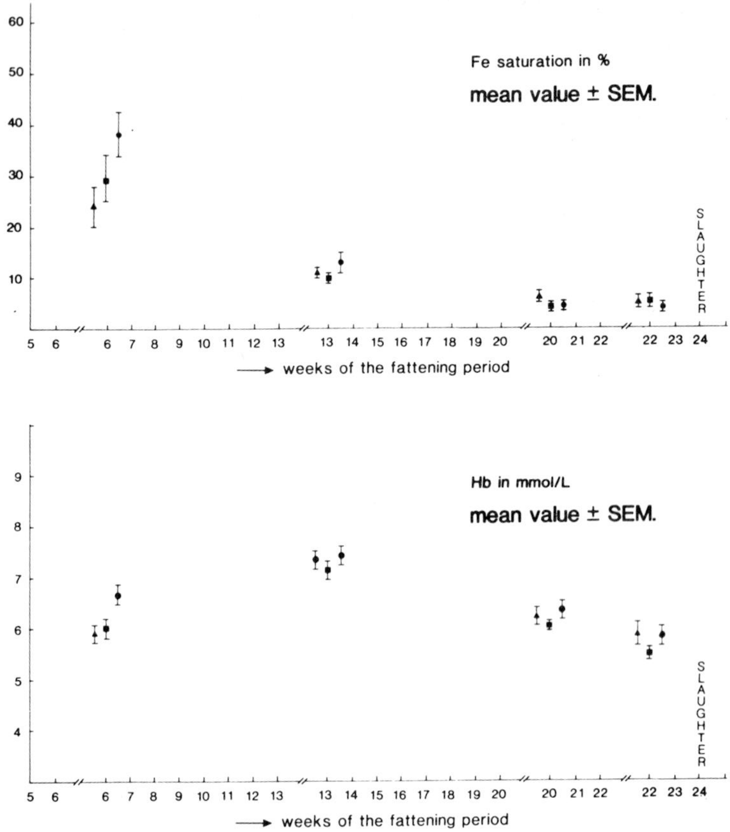

Figure 1a, 1b. Blood hemoglobin concentration in mmol/l plasma iron saturation in % during the fattening period in three groups of veal calves supplied with 56 ▲, 100 ■ or 150 ppm ● of iron during the first six weeks of the fattening period.

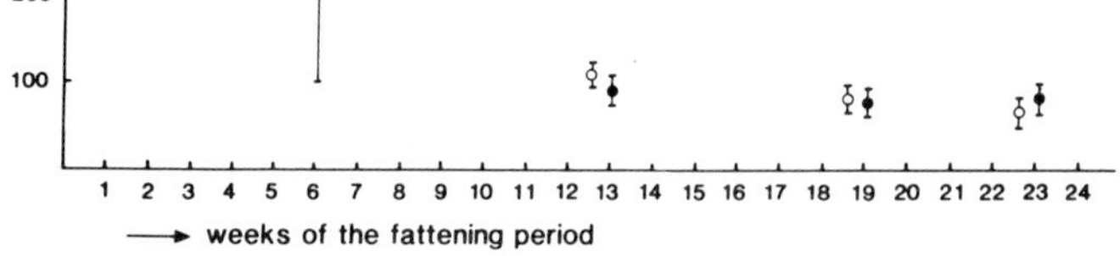

Figure 2. Course of the liver iron concentration in µg/g DM during the fattening period in two groups of veal calves supplied with 150 ppm iron ● and 150 ppm + 10 ml Imposil ○ during the first six weeks of the fattening period.

Effect of extra iron supply (5 ppm) during the whole fattening period (B)

Until the 17th week of the fattening period the course of the blood haemoglobin in Experiment I was the same in the group supplied with extra iron as in the control group. From the 17th week, haemoglobin was significantly higher in the group supplied with iron. However in Experiment II, where the extra iron was terminated in the 16th week of the fattening period, there was no significant difference between the course of haemoglobin concentration in blood of the group supplie with iron and the control group.

It is obvious from the results presented in Table 1 that supplying calves with 5 ppm extra iron until slaughter clearly increased concentration of iron in liver and spleen. This is in accordance with the observation that haemoglobin concentration was higher in the group that was supplied with extra iron until the end of the fattening period only (mean 5.5 mmol/l in the experimental group and 4.0 mmol/l in the control group). In addition, the colour of the meat of this group only was qualified as too red, mean 6.1 in the experimental group and 2.4 in the control group.

Table 1. Iron concentration in liver and spleen after supplementation of 5 ppm iron extra until the end of the fattening period in Experiment I and until the 17th week of the fattening period in Experiment II. I' and II", the experimental groups; I' and II", the control groups.

Iron (8 calves per group)	Group I' 5 ppm iron extra until slaughter	Group I" control	Group II' 5 ppm iron extra until the 17th week	Group II" control
Liver (μg/g DM)	117^a	83^b	63	62
Spleen (μg/g DM)	413^a	353^b	379	365
Total in liver (μg)	142 642^a	87 783^b	86 329	86 210
Total in spleen (μg)	60 957^a	43 566^b	52 829	45 416
Liver + spleen (μg)	203 599^a	131 349^b	139 158	131 626

^{a,b} significant, $P < 0.05$; Student's t test

Effect of supplementation with different amounts of iron during the first six weeks of the fattening period (C)

Supplementing veal calves with 56, 100 or 150 ppm iron during the first 6–7 weeks of the fattening period influenced neither the course of the blood haemoglobin nor that of the plasma iron saturation during the whole fattening period. This is in accordance with the findings of others (Welchman et al., 1988; Figure 1a, 1b). Supply of iron by means of scan injection of Imposil[2] in the first week of the fattening period was found to have no significant effect on the course of the blood haemoglobin, the plasma iron concentration and on the plasma iron saturation. On the course of the liver iron concentration (Figure 2), supplementation with extra iron by means of a subcutanous injection of Imposil[2] had no effect at all.

Supplementation with rather high amounts of iron in the first six weeks of the fattening period (150 ppm and 150 ppm + Imposil) caused a clear increase in the spleen iron concentrations whereas the liver iron concentration showed no clear changes. This observation suggests that the extra iron is stored predominantly in the spleen (Figure 3).

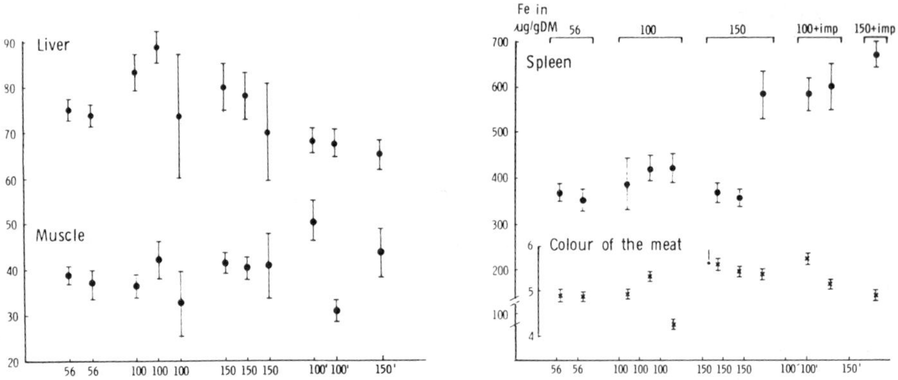

Figure 3. Colour of the meat and the iron concentration in liver, muscle and spleen of calves supplied with different amounts of iron (mean ± SEM).

Regarding the colour of the meat at slaughter, only supplementation of 56 ppm of iron in the first weeks of the fattening period always resulted in a mean qualification for the colour of the meat lower than 5 (acceptable) whereas higher dosages of iron, given in the first 6–7 weeks, gave variable results.

Irrespective of the amount of iron supplied in the aforementioned experiments, no iron could be detected in bone marrow of the veal calves.

References

Abdelrahim, A.I., Th. Wensing, P. Franken & A.J. Schotman, 1983. The influence of an extra supply of iron on some blood components and on meat colour at slaughter in fattening veal calves. Zbl. Med. Vet. A 30:325–331.

Abdelrahim, A.I., Th. Wensing & A.J. Schotman, 1985. Distribution of iron and copper in the liver and spleen of veal calves in relation to the concentration of iron in the diet. Research in Veterinary Science 40:209–211.

Finch, C.A. & H. Huebers, 1982. Perspectives in iron metabolism. New England Journal of Medidcine 306:1520–1528.

Franken, P., Th. Wensing & A.J. Schotman, 1981. The concentration of iron in the liver, spleen and plasma and the amount of iron in bone marrow of horses. Zbl. Vet. Med. A. 28:381–389.

Welchman, D. de B., O.P. Whelehan & A.J.F. Webster, 1988. Haematology of veal calves reared in different husbandry systems and the assesment of iron deficiency.Veterinary Record 123:505–510.

Dry feed supplements for fattening white veal calves

A.E. van de Braak & J. Mol
Denkavit Nederland B.V., P.O. Box 5, 3780 BA Voorthuizen, NL

Summary

Trials were carried out to see if small amounts of roughage could be used when rearing white veal calves.

In the first trial with 6 batches of calves (n = 238), feeding alkaline-treated straw pellets without starch (Group A) was compared with feeding similar pellets containing 35% starch (Group B). These straw pellets were fed in addition to the normal milk-replacer diet. The maximum daily allowance of straw pellets per calf was 300 g. No significant differences in weight gain were found. The meat was darker than that of veal calves raised on a milk-replacer diet; in Group B, the meat was slightly paler than in Group A. There tended to be more calves with health problems and insufficient weight gain in Group B than in Group A.

The second trial investigated the effect on weight gain of adding alkaline-treated straw pellets to the milk-replacer ration. In 14 different batches of calves (n = 858), half the calves received up to 200 g of straw pellets per day from age 6 weeks to slaughter. Their mean carcass weight was 1.9 kg higher and the feed conversion ratio tended to be lower than in the control group. The meat colour was excellent in both groups, and slightly better in the control group.

In a third trial, feeding a maximum amount of 300 g alkaline-treated straw pellets (Group A), was compared with feeding a maximum of 1000 g maize silage (28% dry matter) per day (Group B), in addition to the normal milkreplacer diet (n = 200 calves). Weight gain in both groups was identical. The feed conversion ratio tended to be lower in Group B. The meat colour was below standard for both groups, because of a high haemoglobin level at slaughter. It can be concluded that it may be possible to add small amounts of roughage to the ration of veal calves without negatively influencing the meat colour, provided that there is a very small amount of iron available from the roughage. The improved weight gain does not compensate for the cost of roughage and labour.
Descriptors: veal calves; roughage supply.

Introduction

Traditionally, veal calves are fed on milk replacer. With this type of diet it is easy to control the intake of iron and the meat quality. However it is contended that the welfare of calves fed only milk diets is impaired (Wiepkema et al., 1987). It would be physiologically and ethologically desirable to provide the calves with some roughage. This dry feed would develop the forestomachs and hence induce rumination. Van de Burgwal (1986) and Kooijman et al. (this book) showed that supplementing the milk replacer with roughage gave the calves something to do and therefore benefited their welfare. From other trials, van Putten et al. (1986) and van de Burgwal (1986) concluded that feeding veal calves small quantities of straw pellets may slightly increase weight gain.

This paper describes experiments to find out whether feeding alkaline-treated straw pellets to veal calves improves weight gain. Another trial studied addition of starch to straw pellets. The third experiment compared feeding straw pellets with feeding maize silage. In all these trials, the effect of roughage on performance and carcass quality was evaluated.

Material and methods

The experiments were done at the experimental station of our company, on Holstein Friesian male calves bought from markets about 1 week after birth. The initial weight of the calves was determined when the calves entered the fattening house. Until the age of 6 weeks the calves were fed a traditional milk replacer. At 6 weeks, the trials started. The calves were weighed and regrouped randomly into two groups according to weight, weight gain during the first 6 weeks and blood haemoglobin level. At slaughter, the carcasses were weighed and the meat colour was classified on a scale from 1 (red) to 10 (white). All calves received a commercial milk replacer twice daily. The amount was based on age, weight and length of the fattening period according to a feeding schedule used on commercial farms.

The first trial was done in 1985–1986. The calves ($n = 238$) were housed in groups of 5 per pen on slatted floors. The mean length of the fattening period was 23.5 weeks. From age 6 weeks until slaughter all the calves received a maximum amount of 300 g of alkaline-treated straw pellets once daily. The straw pellets for Group B were supplemented with wheat starch.

The second trial was done in 1987–1988. The calves were housed in individual boxes of 70 cm × 170 cm with slatted floors. The experiment was carried out with 14 batches of in total 858 calves randomized in two groups. Group A received a milk-replacer diet only and in Group B the milk-replacer was augmented with a maximum amount of 200 g/calf alkaline-treated straw pellets given once daily from age 6 weeks to slaughter. The mean duration of the fattening period was 25 weeks.

In the third trial in 1988–1989, calves were housed in groups of 5 per pen. The experiment included 200 calves from 4 different batches. The calves of Group A were fed a milk-replacer diet supplemented by a maximum amount of 300 g alkaline-treated straw pellets given once daily. For Group B, the straw pellets were replaced by a maximum amount of 1000 g maize silage with a dry matter content of 28%. The roughage was given from age 6 weeks until slaughter. Both the milk replacer and the roughage were put in a trough in front of each pen. The fattening period lasted 26 weeks.

The data were analysed with the Student's *t* test. Data from calves whose carcass weight deviated from the mean weight by more than twice the standard deviation were excluded.

Results and discussion

Trial I

Feed intake, the composition of the straw pellets and the results are detailed in Table 1. The feed conversion ratio was calculated from the 'net paid liveweight'. This live weight was calculated as follows: warm carcass weight minus 2% = cold carcass weight (liver included), multiplied by 100/65.

The trial started with 2 groups of 119 calves. Some calves became ill and died (A, 4; B, 5), some had a very poor growth rate and were sold off (A, 1; B, 3). These data provide some evidence that feeding straw pellets with starch may cause more zootechnical problems. Furthermore, calves in Group B had slight symptoms of bloat more frequently. Table 1 shows that adding starch to straw pellets was no advantage. The meat colour classification was generally below standard, because the blood haemoglobin levels of 3 of the 6 batches were too high at slaughter.

Table 1. Feed intake, composition of the straw pellets and results of Trial I.

	Group A	Group B	
Average intake (kg/d)			
– milk replacer	309	309	
– straw pellets	25	25	
Composition of straw pellets			
– starch content (%)	≤ 1	35	
– crude fibre (%)	35	18	
– iron (Fe) (mg kg^{-1})	75 – 110	90 – 135	
Weight (kg)			
– initial	41.5	41.4	
– at 6 weeks	69.7	69.7	(regrouped)
– carcass[1]	143.3 ± 10.5	144.1 ± 9.8	$P \geq 0.05$
Feed conversion ratio	1.769	1.759	$P \geq 0.05$
Meat colour	6.63	6.90	$P \geq 0.05$

[1] liver included

Table 2. Weight, feed conversion ratio and meat colour in Trial II.

	Group A	Group B	
Initial number of calves	445	413	
Initial weight (kg)	42.6	42.8	
Carcass weight (kg)	144.8 ± 11.3	146.7 ± 11.0	$P \leq 0.02$
Feed conversion ratio	1.845	1.812	
Meat colour	7.51	7.38	
Number of calves for stat. analysis	436 (–9)	404 (–9)	

Trial II

This trial was done on 14 batches of calves. The amount of milk replacer fed varied between the batches from 317–344 kg/calf, but was identical for the two groups. The intake of straw pellets varied slightly between the different batches: from 15 to 20 kg/calf. The results are summarized in Table 2.

Feeding with straw pellets improved carcass weight and feed conversion ratio. The meat colour of Group B tended to be inferior, but was excellent in both groups. Very few (less than 2% in both groups) calves were 'lost' during the trial, but the cost of the straw pellets (about 7 USD/calf) and labour far outweighed any such advantages.

Trial III

Trial III compared feeding small amounts of straw pellets with feeding maize silage. The straw pellets contained 100–140 mg kg^{-1} iron in the dry matter. The iron content of the maize silage was 120–190 mg kg^{-1} in the dry matter, with the exception of a batch which had an iron content of 450 mg kg^{-1}. This batch of maize silage was fed for 3 weeks to Group B in one trial. The results are summarized in Table 3.

Table 3. Results on weight, feed conversion, feed intake, meat colour and blood Hb levels in trial III.

	Group A	Group B	
Initial number of calves	100	100	
Number of calves for stat. analysis	94	97	
Feed intake (kg/d)			
– milk replacer	343	339	
– straw pellets	35	-	
– maize silage	-	93	
Initial weight (kg)	44.2	44.1	
Carcass weight (kg)	149.8 ± 16.7	150.1 ± 16.1	$P \geq 0.05$
Feed conversion ratio	1.844	1.816	$P \geq 0.05$
Meat colour	6.55	6.10	$P \leq 0.01$
Hb at slaughter (mmol/l)	6.1	6.4	

There were no significant differences in weight gain or feed conversion ratio between the groups. The standard deviation of the average carcass weight was very large, because the calves were fed from troughs, and their individual feed intake varied according to their appetite.

The score for meat colour was very low in both groups, but extremely low in Group B. This is related to the high blood haemoglobin content at slaughter. The iron analysis of maize silage indicated that iron content varied considerably, may be because of contamination with soil. These high iron levels clearly adversely affected meat colour.

At slaughter, it was observed that in the carcasses of the calves that ingested small amounts of roughage, the ribs were more 'rounded', because of the development of the rumen. This aberrant carcass conformation generally reduced the value of the carcass.

Conclusion

The trials showed, that it might be possible to add some roughage to the diet of veal calves without adversely affecting meat colour, provided that the amount of available iron in the roughage was small. An amount of 200–300 g of alkaline-treated straw pellets per calf per day may increase the weight gain. But the improved weight gain does not compensate for the cost of roughage and labour. Adding starch to straw pellets has no effect on performance, but tends to increase incidence of rumen bloat. Straw pellets can be replaced by maize silage, but the iron content should be taken into account.

References

van de Burgwal, J.A., 1986. Invloed strovorm op welzijn mestkalveren. Landbouwkundig Tijdschrift (98):30–32.

van Putten, G., W.J. Elshof & J.L.R.M. Overmeer, 1986. Het effect van het bijvoeren van stro of strobrok aan vleeskalveren. Instituut voor Veeteeltkundig Onderzoek Rapport B-268. Zeist, NL.

Wiepkema, P.R., K.K. van Hellemond, P. Roessingh & H. Romberg, 1987. Behaviour and abomasal damage in individual veal calves. App. Anim. Beh. Sci. (18):257–268.

Veal production with less milk replacer in the ration supplemented with concentrates and maize silage

P.L. Bergström & M. Dijkstra
Research Institute for Animal Production, P.O. Box 501, 3700 AM Zeist, NL

Summary

In a series of experiments calves have been fattened with reduced amount of milk replacer, supplemented with maize silage or a silage/concentrate mix. This resulted in a higher growth rate and lower feed costs but the meat colour was darker than in traditionally fed calves with only milk replacer. Attempts to influence the meat colour by return to a regime with exclusively milk, 8 weeks prior to slaughter had no effect. Compared with traditional veal calves and young bulls, the alternatively produced calves had an intermediate position for most of the meat quality traits. An analytical panel indicated only differences in taste/flavour between traditional veal and meat of one of the alternatively fed group of calves. No differences in tenderness or succulence were detected.
Descriptors: alternative veal production; meat colour; quality of veal.

Introduction

The financial returns from veal production in the Netherlands are rather poor because of the high prices of new-born calves and the high feeding costs in comparison to the price for the finished calves. Moreover there are many ethological objections to individual housing of calves and to feeding exclusively with milk replacer. So a series of experiments was started in which part of the milk was replaced by other products such as maize silage or a silage/concentrate mix. The experiments involved fattening to a final liveweight of 300 kg, which is somewhat higher than the usual final weight for veal calves in the Netherlands.

Material and methods

Several series of calves were fattened, for which part of the milk was replaced by maize silage or a silage/concentrate mix (mostly 50/50 on a dry matter basis). A group fed exclusively with milk (System A) served as reference on one side and groups fed only milk during the first 11 weeks and mix ad libitum to provide the least possible amount of milk as reference on the other side. Although the replacement of milk was achieved in different ways, we prefer in the present publication to refer only to two of the alternative systems, the system with the best growth and the system with the lowest feeding costs.

These are systems with a reduction of milk supply by 20% (from the beginning of the fattening period) and supplemented with a silage/concentrate mix ad libitum (System D) and the system, with only the first 11 weeks milk and silage/concentrate mix ad libitum (System E). Haemoglobin concentration of blood was determined and carcasses were appraised visiuallly. Colour was measured objectively by means of the Hunter apparatus. Haematin concentration, shear force resistance (Warner-Bratzler), and drip and cooking loss were determined in samples of the musc. longissimus thoracis between 6th and 8th thoracic vertebra.

In a one of the experiments, meat characteristics were compared between alternatively fed calves and traditional veal calves as well as young bulls. In another experiment, meat samples of traditional veal calves and alternatively fed calves were given to an analytical expert panel

(12 persons) in order to check for differences in taste/flavour, tenderness or succulence. All animals in the experiments mentioned here were Black and White.

Results and discussion

In Table 1, the results of a group of veal calves fed only with milk replacer are compared with the alternatively fed groups with the best growth (System D) and the group with the lowest feeding costs (System E). The meat colour was not different between the two alternatively fed groups, so if other products are fed in larger amounts alongside milk replacer, the amount of milk has no influence on the meat colour. In calves fed only with milk, the variation in Hb concentration was rather wide.

Table 2 refers to an experiment with alternatively fed calves of two different systems in which half of the calves returned to a regime exclusively with milk replacer. The aim was to see if a larger supply of milk in the period directly before slaughter would influence meat colour. In this experiment, the Fe reserve in the liver was considerably depleted but apparently this was not sufficient to decrease haemoglobin content of blood nor to affect the colour of meat over the 8-week period.

Because we were interested in the difference in meat quality characteristics between alternatively fed calves and traditional veal calves on one hand and young bulls on the other hand we determined several quality parameters of meat in alternatively fed calves from our own experiments and traditional veal calves and young bulls from which samples of the musc.

Table 1. Alternative system with the highest daily gain (D) compared with the systems with maximum (A) and minimum (E) supply of milk.

	System A only milk		System D 80% milk mix ad lib		System E 11 weeks milk mix ad lib	
	mean	s.d.	mean	s.d.	mean	s.d.
Number of calves	21		20		20	
Fattening period (days)	240		208		246	
Final liveweight (kg)	303	19	314	19	320	24
Daily liveweight gain (g)	1 088	77	1 309	85	1 130	92
Dressing %	58.9	2.5	57.6	1.2	55.0	1.4
Net carcass gain/day (g)	648	66	762	57	622	57
Haemoglobin (g/100 ml)	9.0	2.3	12.6	1.3	12.4	1.4
Visual appraisal of carcass						
conformation	2.7	0.5	2.7	0.4	2.6	0.4
fat covering	2.8	0.6	3.3	0.4	2.6	0.3
meat colour	1.8	0.6	3.0	0.5	2.9	0.4
Carcass yield (NLG)	1 758		1 640		1 578	
Feed + other costs (NLG)	1 474		1 069		711	
For calf + labour (NLG)	284		571		867	

Visual appraisal: for all traits 5 point scale 1 minimum development/intensity
5 maximum

Source: Dijkstra et al., 1988.

longissimus thoracis were bought from commercial slaughterhouses (Table 3). For most of the quality characteristics the alternatively fed calves were intermediate between the two other categories. The shear force resistance proved to be almost proportional to the weight differences between the categories.

The two groups of calves in Table 3 differed in weight and age, which could have had an effect on the meat colour and other meat quality traits irrespective of the feeding system. To study these possible effects, an experiment was done in which animals were slaughtered at about 24 and 30 weeks. Some data are given in Table 4. Significant differences were found between slaughters for colour characteristics of meat and drip loss. The differences in shear force were not significant between slaughters nor between the milk-fed and alternatively fed calves. For the majority of colour characteristics, the milk-fed and alternatively fed calves differed significantly. In Table 4, only significant differences between slaughters are indicated as well as slaughter × system interactions.

Samples of the musc. longissimus thoracis of the 2nd slaughter were given to a trained analytical panel. An analytical panel was chosen because the home consumption of veal is limited, so it would be difficult to interprete the results of a consumer panel. The analytical panel only indicates differences but not preferences for a certain product. In Table 5, only significant differences were indicated for taste/flavour between the calves fed only with milk and the alternative system with the largest amount of milk in the ration. No significant differences were found for tenderness and succulence between the milk-fed and alternatively fed calves.

Table 2. Results of an experiment with alternatively fed calves (Systems D and E), in which half of each of the treatment groups returned 8 weeks before slaughter to a regime exclusively with milk (System A).

	System D 80% milk mix ad lib.				System E milk mix for 11 weeks ad lib.			
	System D continued		return to System A		System D continued		return to System A	
	mean	s.d.	mean	s.d.	mean	s.d.	mean	s.d.
Number of calves	8		8		8		8	
Liveweight gain (g/d)	1 318	85	1 189	37	1 055	107	991	92
Dressing %	59.0	1.4	59.5	1.0	54.6	2.2	57.0	2.1
Net carcass gain (g/d)	785	50	716	26	575	61	570	58
Haemoglobin (g/100 ml)	12.4	0.9	12.4	1.6	12.1	1.0	11.9	0.8
Visual appraisal of carcass								
conformation	2.7	0.6	2.1	0.4	2.0	0.6	2.3	0.5
fat covering	3.2	0.3	3.0	0.3	2.5	0.4	2.5	0.6
meat colour	2.6	0.3	2.8	0.2	2.8	0.2	2.8	0.2
Meat colour CIELAB								
L* value (brightness)	43	2	42	2	44	2	45	4
a* value (redness)	17	2	17	1	17	2	17	2
Fe content in fresh liver (mg/kg)	32	8	19	8	32	6	19	10
Feed costs (NLG)	1 001	19	1 108	9	428	17	783	41

Source: Dijkstra & Bergström, 1989.

Table 3. Meat quality parameters of alternatively produced calves compared with traditional veal calves and young bulls.

	Traditional veal calves mean	s.d.	Alternatively fed calves mean	sd	Young bulls mean	sd
Number of calves	45		42		46	
Age (months)	5.5		8		16	
Carcass weight (kg)	148	16	172	11	302	29
Musc. logissimus thoracis samples						
meat colour visual (1–5)	2.0[a]	0.5	2.9[b]	0.3	3.3[c]	0.6
Meat colour CIELAB						
L* value (brightness)	51[a]	4	45[b]	2	41[c]	4
a* value (redness)	11[a]	2	15[b]	1	17[c]	2
Haematin muscle (mg/kg)	65[a]	14	99[b]	10	132[c]	24
Drip loss (%) (1 week 3 °C)	5.3[a]	1.3	5.7[a]	1.4	5.2[a]	1.3
Cooking loss (%) (1 hr 75 °C)	34.5[a]	1.6	34.6[a]	1.1	33.3[b]	1.5
Shear force resistance (daN)	3.0[a]	0.6	3.3[b]	0.7	4.6[c]	0.9

Groups with different superscripts are significantly different ($P < 0.05$).
Source: Dijkstra et al., 1990.

Table 4. Quality parameters of meat in traditional and alternatively fed calves slaughtered at two different final weights (Sl. I, Sl. II).

Trait/slaughtering	Sl.	System A milk only mean	s.d.	*n*	System D 80% milk mix ad lib mean	s.d.	*n*	System E 11 weeks milk mix ad lib mean	s.d.	*n*
Carcass weight (kg)	I	134	6	8	137	9	8	127	9	7
	II	166	12	11	180	7	7	167	12	8
Meat colour visual	I	1.8	0.4		3.1	0.2		2.9	0.2	
	II	2.2	0.6		2.9	0.3		3.0	0.3[b]	
Musc. longissimus thoracis										
Meat colour CIELAB										
L* value	I	56	1		46	1		48	1	
	II	54	3		45	1		46	3[a]	
a* value	I	11	1		13	1		13	1	
	II	12	2		15	1		14	1[a]	
Haematin content	I	47	6		89	10		80	10	
(mg/kg)	II	75	10		116	15		92	18[a]	
Shear force	I	2.9	0.7		3.4	0.5		3.0	0.5	
resistance (daN)	II	2.8	0.4		3.2	0.5		3.3	0.8[a]	
Drip loss (%)	I	6.2	0.8		5.1	1.3		5.0	0.5[a]	
	II	6.0	1.1		6.7	0.6		5.9	1.2[a]	

[a] Signif. differences between slaughters
[b] Signif. slaughter × system interactions. Source: Dijkstra et al., 1990.

Table 5. Results of analytical panel for comparison of samples of musc. longissimus thoracis. System A vs. D and System A vs. E; n.s., not significant.

	A vs. D	A vs. E
Taste/flavour	$P < 0.05$	n.s.
Tenderness	n.s.	n.s.
Succulence	n.s.	n.s.

Note that the panel only indicates difference, not preference.
Source: Dijkstra et al., 1990.

Conclusions

Veal production in which larger amounts of milk replacer in the ration are replaced by silage or a silage/concentrate mix proved to be possible with good growth and relatively low feeding costs. There were differences in meat characteristics between traditional exclusively milk-fed calves and alternatively fed calves. The most marked differences, however, were those in meat colour and colour parameters of meat. If larger amounts of milk in the ration were replaced by roughage or concentrates, there was no further difference in how far supply of milk reduced meat colour. In most of our experiments, the calves were slaughtered at a final weight of about 300 kg liveweight. At the moment, it is still uncertain what the prospects of these alternatively fed calves would be on market.

References

Dijkstra, M. & P.L. Bergström & K. Maatje, 1988. Alternatieve voersystemen voor de produktie van vleeskalveren. IVO publ. B-318 (Engl. summary).

Dijkstra, M. & P.L. Bergström, 1989. De invloed van enkele alternatieve voersystemen op de vleeskleur van vleeskalveren. IVO-publ. B-344 (Engl. summary).

Dijkstra, M. & P.L. Bergström, A.H. Bolink, B. Engel, B. Hulsegge, G.S.M. Merkus, A.W. de Vries & P.G. van der Wal, 1990. De vleeskwaliteit van met kracht- en ruwvoer gemeste vleeskalveren. IVO-publ. B-350 (Engl. summary).

Use of by-pass proteins and iso-acids in a heavy calf diet

L.A. Drevjany
Kemptville, Ontario, CA

Summary

Fourty Holstein bull calves weighing 76 kg were randomly assigned to four dietary treatments. Isonitrogenous diets, containing 40% protein equivalent in the form of urea (Treatment 1) were supplemented with blood meal as the source of bypass protein (Treatment 2), blood meal and a mixture of valeric and isovaleric acid (Treatment 3), or a combination of blood meal and distiller's grain (Treatment 4).

Performance of calves, receiving deficient amino acids in bypass protein form or in combination with iso-acids was significantly improved, while the blood urea levels were reduced.

Descriptors: bypass proteins; iso-acids; blood urea.

Introduction

The progress made by geneticists over the last few decades in dairy cattle breeding has led to conflict between the rapidly improving production capabilities of dairy animals and the ability of their digestive system to secure the necessary supply of nutrients. In the absence of similar genetic improvement in the food-processing capacity of ruminating animals, nutritionists' attention has been concentrated on enhancing the utilization of the existing digestive tract, in particular, on two distinct areas, the reticulo-rumen and the post-ruminal tract.

Among the techniques aimed at improving the synthesis of microbial proteins is the inclusion of deficient carbon (C) skeletons needed for synthesis of certain amino acids. In a study (Drevjany et al., 1984) with ruminating Holstein calves, the presence of 5-C valeric and isovaleric acids in the rumen fluid diminished when non-protein nitrogen was included in the diet, and the dependence of the animal's performance on the synthesis of microbial protein was increased. Deficiency of certain C-skeletons not only limits the total synthesis of microbial protein, but negatively influences the amino acid profile of microbial protein.

Another attractive approach is direct supplementation of the ration with bypass protein. The main role of the supplementary by-pass protein sources must be to fill the gap between the post-ruminal amino acid pool in the basic ration and the requirements for post-ruminal amino acids of the particular class of ruminating animals.

Material and methods

Fourty Holstein male calves, weighing on average 76.4 kg were housed in elevated stalls with wooden floors. At about 100 kg, they were moved to pens with double occupancy. For technical reasons, the experiment was terminated when the calves reached 150 kg. After randomization and formation of four treatment groups, calves were fed on ground ear corn and liquid protein supplement (LPS) (Treatment 1), supplemented with by-pass protein and a mixture of valeric and isovaleric acid (Treatment 3).

Table 1 offers the composition of experimental diets while Figure 1 indicates the expected fate of individual components of the ration during digestion.

Table 1. Composition of experimental diets[1].

Feed Component	Treatment 1 (kg)	Treatment 2 (kg)	Treatment 3 (kg)	Treatment 4 (kg)
Dry mix:				
ground ear corn	97.010	92.794	94.247	95.108
urea	2.850	1.286	2.479	2.274
sulphur	0.128	0.058	0.112	0.102
furazolidone	0.012	0.012	0.012	0.012
blood meal	-	5.850	3.150	2.118
distiller grain	-			0.386
Liquid mix:				
molasses	100.00	100.00	98.930	100.00
valeric acid	-	-	0.950	
iso-valeric acid	-	-	0.120	
Per 100 kg of mixed ration:				
dry mix	81.0	81.0	81.0	81.0
liquid mix	19.0	19.0	19.0	19.0

[1] Drevjany & Hooper, unpublished data, 1976.

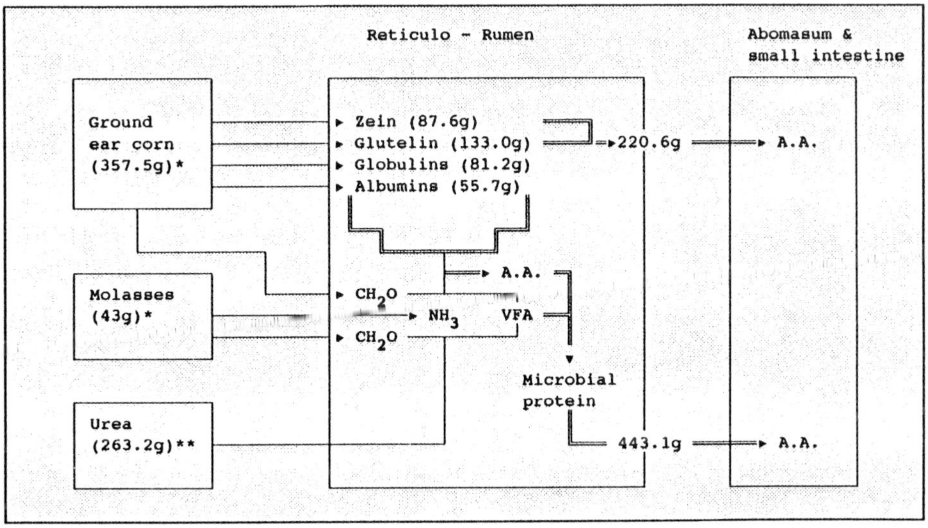

Figure 1. Predicted fate of dietary protein.

To estimate and then remove the deficiency of limiting amino acids in the post-ruminal area, the following procedure was used. The control ration (Table 1) was tested in a preliminary trial to establish its growth response and daily dry matter intake. After the differences in amino acid composition in the protein fraction of corn and differences in digestibilities of protein sources were taken into account, the estimated post-ruminal amino acid pool was compared

Table 2. Expected availability of selected amino acids (AA) (expected intake of ration of dry matter 4195 g/head/day)[1].

	Protein sources and their contribution to amino acid pool						
	Zein (52.6g)	Glutelin (79.8g)	Microbial protein (302.8)	Total AA pool available (g)	AA in 1270 g of exp. gain (=requirement)	Requirement × 1.67	Deficiency in g/head/day
Methionine (g)	0.68	1.92	6.05	8.65	5.61	9.37	−0.73
Lysine (g)	0.05	2.39	19.88	22.32	16.40	27.39	−5.07
Arginine (g)	1.63	3.27	12.07	16.97	15.40	25.72	−8.75
Tryptophan (g)	0.32	0.64	4.02	4.98	2.31	3.86	
Threonine (g)	1.37	2.71	12.73	16.81	9.22	15.40	
Isoleucine (g)	1.84	2.55	13.32	17.71	10.28	17.17	
Leucine (g)	9.31	9.58	18.22	37.11	17.89	29.88	
Phenylalanine (g)	3.16	3.91	13.34	20.41	8.24	13.76	
Histidine (g)	1.37	2.00	4.16	7.53	7.85	13.11	−5.58
Valine (g)	1.79	4.07	13.07	22.30	12.04	20.11	

[1] Adjusted from data by Lasztity, 1984.

Table 3. Effect of the diets on animal performance.

	Treatment 1 control	Treatment 2	Treatment 3	Treatment 4
Number of calves per group	10	10	10	10
Average gain (g/d)	680[a]	954[b]	936[b]	911[b]
Dry matter intake (g/head/day)	3 061[a]	4 020[a]	3 667[a]	3 962[a]
Convers. of diet. prot/gain (g/kg)	721[a]	668[a]	628[a]	623[a]
Blood urea (mg%)	3.26[a]	1.80[b]	2.58[c]	2.64[c]

[a,b,c] Values within the rows with different letters are significantly different at $P < 0.05$

with the estimated requirements. In the absence of established requirements for amino acids in growing calves, the content of amino acids in the expected gain multiplied by a factor of 1.67 to cover metabolic and other inefficiences was used instead (Table 2). Four amino acids were limited gain in the control group: arginine, histidine, lysine, and methionine (8.75, 5.58 and 73 g/head/day, respectively). By-passing properties of amino acids in selected protein sources were determined by nylon bag techniques.

Results and discussion

Performance and blood urea data are presented in Table 3. Average daily gain (ADG) was increased by over 40% in Treatment 2 (954 g/calf/day) where blood meal was the only supplementary source of by-pass protein. Calves on Treatments 3 and 4 increased their gain by 37.7% (936 g/calf/day) and 34% (911 g/calf/day), respectively ($P < 0.05$). Statistically, the ADG in Treatments 2, 3 and 4 were not different. In the last 14-day period prior to shipment, however, the calves reached the expected ADG of 2300 g. It seems that the use of individual elevated stalls and early shipment to market (150 kg liveweight) reduced the overall feed intake, and gains.

Compared to the control calves consuming 3061 g of dry matter (DM)/calf/day, the experimental calves increased the intake of DM by 31.3%, 19.8% and 29.4% for treatments 2, 3 and 4, respectively. The expected DM intake in the control group (4200 g/calf/day) was reached only in the last 14 days of the trial. Differences in daily DM intake as well as daily gains translated into differences in efficiency of DM and protein utilization. On both counts, calves on Treatment 3 converted the feed most effectively. Compared to the control group, they required 14.1% less DM and 12.9% less protein per kg of gain. Differences in feed conversion were not statistically significant ($P > 0.05$).

To obtain some indirect supporting information on the metabolic fate of amino acids from by-pass protein and iso-acids, blood urea values were measured at regular 28-day intervals. It was assumed that lower blood urea values would indicate that by-pass protein served as a source of the limiting amino acids and as such allowed improved utilization of all other amino acids. As a consequence, fewer amino acids were deaminated and less NH_3 was incorporated into the blood urea. Data in Table 3 supports this assumption. Blood urea values were lowered significantly ($P < 0.05$) in all experimental treatments. In addition, values in Treatment 2 were significantly lower ($P < 0.05$) than in Treatments 3 and 4.

Conclusion

Increasing the availability of post-ruminal amino acids seems to be the most challenging area facing ruminant nutritionists today. It may not only raise the plateau of animals' responses to new heights, but enable genetically superior animals to produce close to their potential without 'burning out' in the process. It appears that by-pass-protein research with heavy calves may offer a few important answers for other classes of ruminanting animals such as baby beef.

References

Drevjany, L.A., G.S. Hooper & W.V. Chandler, 1984. Processed poplar wood pellets as a source of energy for Holstein bull calves. Canadian Journal of Animal Science 64:1035–1044.

Klopfenstein, T., 1985. Animal protein products fed as bypass protein for ruminants. Feedstuffs 57:8:31–33.

Lasztity, R., 1984. The chemistry of cereal proteins. CRC Press, Inc. 2000 Corporate Blvd. N.W., Boca Raton, Florida, 33431.

Steel, R.G. & J.H. Torrie, 1960. Principles and procedures of statistics. McGraw-Hill Book Co. Inc. New York.

Storm, P. & E.R. Orskov, 1984. The nutritive value of rumen microorganisms in ruminants. British Journal of Nutrition 50:463–470.

Closing remarks: some observations

On behalf of the American veal farmer, I thank the chairman, the organizing committee, sponsors, and especially the Dutch people who have generously given their time and shared their culture. As the president of the American Veal Association, representing the veal producers of the United States, I appreciate the opportunity to learn from and interact with the European veal industry--from producers to allied industry to scientists.

We American veal producers are proud of the care and housing we provide our animals, the high liveability rate, and our quality end product. We have developed, many years ago, guidelines for the care and production of milk-fed veal calves. We have also developed, a year ago, a veal quality assurance program in response to concerns about food safety. Residue levels were reduced nearly three times, down to less than 1%, which is quite admirable for an industry to accomplish in a short time.

The position of all of American animal agriculture is that we cannot support any legislation which would impose regulations on animal production methods. The American Veal Association fully supports reliable research with reference to veal production with conditions specific to American farmers.

In the United States we have conducted several consumer attitude studies with reference to animal welfare concerns (Curtis, this book). All of American animal agriculture has begun consumer information programs to explain to consumers who are oftentimes two, three, or more generations removed from farm life, modern animal husbandry practices--what we do, how we do it, and why we do it.

For the future--our feed companies, packing companies, and producers are continuing to look optimistically towards the future. Producers are aware of the concern over animal agriculture and food safety and are responding in a positive, active manner.

It is obvious, even among the scientists gathered here at the symposium, that there are differing opinions on aspects of animal husbandry. Farmers and legislators are looking to the scientific community of researchers and veterinarians for guidelines. The International Veal Symposium was an excellent idea, occurring at the right time, allowing for presentations that unbiased, scientific minds can present are of great value. A forum for the discussion of veal issues has been needed and I would hope that this International Veal Symposium would continue.

In conclusion, any forced changes in animal husbandry systems, when not based on sound scientific research *and* the reality of day-to-day life on the farm can only result in the following:

1. a decrease in animal welfare.
2. an increase in food costs.
3. inefficient systems.
4. financial and/or labour difficulties for the farmer who, in the US, numbers only about 2% of the population.

It is in the best interest of everyone to provide efficient farm operations designed with the best interests of the animal in mind. I urge you to continue research in the area of the veal calf and to continue to have an open forum to share and discuss your findings. I would urge open communications with the veal producers in your countries to discuss our industry's research priorities and concerns. We would welcome this symposium to come to the US sometime in the future years and allow us to return your warm hospitality.

Barbara Huffman

President of the American Veal Association